MATHEMATICS
FOR HIGHER NATIONAL
CERTIFICATE

MATHEMATICS
FOR HIGHER NATIONAL CERTIFICATE

A TEXT BOOK FOR THE USE OF
FIRST YEAR (A1) STUDENTS

BY

S. W. BELL, B.Sc., F.I.M.A.
Formerly Head of Mathematics Department

AND

H. MATLEY, B.Sc., A.F.I.M.A.
Principal Lecturer in Mathematics
Norwich City College

VOLUME I
[METRIC]

CAMBRIDGE
AT THE UNIVERSITY PRESS
1971

Published by the Syndics of the Cambridge University Press
Bentley House, 200 Euston Road, London, NW1 2DB
American Branch: 32 East 57th Street. New York, N.Y.10022

ISBN: 0 521 04144 9

First edition 1955
Second edition 1959
Reprinted 1961 1963 1966
Reprinted with SI units 1971

Printed in Great Britain
at the University Printing House, Cambridge
(Brooke Crutchley, University Printer)

CONTENTS

PREFACE TO THE FIRST EDITION   *page* xi

PREFACE TO THE SECOND EDITION   xii

LIST OF ABBREVIATIONS   xiii

CHAPTER I
Revision and further differentiation. Applications

1·1. Function of a function   1

1·2. Differentiation of implicit functions   5

1·3. Parameters   7

1·4. Logarithmic differentiation   8

1·5. Formation of differential equations   10

1·6. Differentials. Errors and approximations   12

1·7. Continuity and differentiability   15

1·8. Maxima, minima and points of inflexion   18

1·9. The mean value theorem of differentiation   24

1·10. Newton's approximation to the root of an equation   25

     Miscellaneous exercises on Chapter 1   28

CHAPTER 2
Revision of integration and applications

2·1. Standard integration   32

CHAPTER 3
Hyperbolic functions

3·1. Definitions   37

3·2. Graphs of hyperbolic functions   37

3·3. Standard relationships between hyperbolic functions   38

3·4. Differentiation and integration of hyperbolic functions   40

CHAPTER 4
Inverse functions
4·1. Introduction *page* 44

4·2. The inverse trigonometric (or circular) functions 44

4·3. Inverse hyperbolic functions 50

4·4. Formation of differential equations 56

Miscellaneous exercises on Chapter 4 56

CHAPTER 5
Methods of integration
5·1. Use of $\int \dfrac{f'(x)}{f(x)} dx$ 59

5·2. Simple change of variables 60

5·3. Integration of rational algebraic fractions 62

5·4. Integration of irrational fractions of type $\dfrac{px+q}{\sqrt{(ax^2+bx+c)}}$ 71

5·5. Integration of trigonometric functions 73

5·6. Trigonometric and hyperbolic substitutions 77

5·7. Integration by parts 78

Miscellaneous exercises on integration 81

CHAPTER 6
Some applications of integration
6·1. More difficult areas and volumes 83

6·2. Lengths of curves 90

6·3. Surface areas of solids of revolution 93

6·4. Centroids and centres of gravity 95

6·5. Second moments: moments of inertia 100

6·6. Fluid thrusts and centres of pressure 109

6·7. Simple applications to electricity 118

CHAPTER 7

Curvature. Evolutes and Involutes

7·1. Definitions   *page* 122

7·2. Intrinsic equations. The cycloid   123

7·3. Radius of curvature in Cartesian co-ordinates   125

7·4. Newton's formula for radius of curvature   127

7·5. Evolutes   128

7·6. Involutes   130

CHAPTER 8

Infinite series. Convergence

8·1. Some types of infinite series. Definitions of
     convergence and divergence   135

8·2. Tests for convergence   136

CHAPTER 9

Expansion of functions as power series

9·1. General discussion   145

9·2. Importance of convergence   145

9·3. Maclaurin's expansion   146

9·4. Some expansions of standard functions   147

9·5. Odd and even functions   148

9·6. Other methods of expansion   150

9·7. Approximate values of definite integrals   151

9·8. Evaluation of limits using power series   152

9·9. Examples

CHAPTER 10

The catenary

10·1. Equations of a catenary   157

10·2. Approximations for a tightly stretched wire   160

10·3. Suspension bridge   160

CHAPTER 11
Complex numbers

11·1. Introduction *page* 163

11·2. Complex numbers as vectors 167

11·3. Exponential form of a complex number 175

11·4. De Moivre's theorem. Roots of a complex number 177

11·5. Application to certain integrals 181

CHAPTER 12
Partial differentiation

12·1. Introduction. Functions of more than one variable 183

12·2. Partial and total derivatives 183

12·3. Higher derivatives 188

12·4. Total increment 189

12·5. Total derivative. Total differential 192

CHAPTER 13
Ordinary differential equations

13·1. Introduction 195

13·2. First-order equations 197

13·3. Second-order equations 205

Miscellaneous exercises on Chapter 13 213

CHAPTER 14
Some applications of differential equations to Mechanical and Electrical Engineering

14·1. Motion in a straight line 216

14·2. Simple harmonic motion and damped harmonic motion 220

14·3. Bending of beams 226

14·4. Electrical applications 238

CHAPTER 15
Harmonic analysis

15·1. Introduction. Periodic functions *page* 244

15·2. Approximations by successive harmonics 256

15·3. Fourier series in engineering problems 257

15·4. Tabulatory method for a Fourier series 257

APPENDIX A
Curve tracing

A.1. Fundamental points in the case of Cartesian equations 262

A.2. Curves given by parametric co-ordinates 266

A.3. Polar equations 267

APPENDIX B
The straight line and the circle

B.1. The straight line 271

B.2. The circle 279

SPECIMEN A1 PAPERS 284

MISCELLANEOUS EXERCISES 291

INDEX 308

PREFACE TO VOLUME I

This book has been written for students reading mathematics as part of a course for the Higher National Certificate in Engineering. The authors realise that most of their readers will be part-time students at technical colleges, who have only a very limited time to devote to mathematics. Because of this the book is confined to what are considered to be the main essentials of such a course, with separate sections on applications to mechanical and electrical engineering.

As many worked examples and exercises as possible have been included, bearing in mind our effort to keep the book to a reasonable size.

Although a high standard of rigour is usually deemed unnecessary in a book of this type, we have tried to give no proofs which the student will have to unlearn if he ever proceeds to more advanced text-books.

The hyperbolic and inverse functions have been introduced early so as to be available for applications. Experience has shown the value of using integrals involving these functions immediately after their differentiation.

In the treatment of linear differential equations a method recently recommended by the Calculus Report of the Mathematical Association has been adopted.

When teaching, the authors build up a knowledge of curve tracing and the elements of co-ordinate geometry throughout the course, but to give a more connected treatment these topics have been dealt with in an appendix which can be referred to at any time. No treatment of conic sections has been given. One was planned, but the time factor really precludes any adequate treatment in an A 1 course.

Harmonic analysis has been introduced in this volume, although a rather advanced subject for first year work. Many students fail to choose mathematics as a subject of study for their A 2 examination, and it was felt undesirable for a course in engineering mathematics to be completed without some reference to this important topic. The authors have included harmonic analysis in their lectures to A 1 electrical students for some years.

It is realized that the syllabuses for A 1 mathematics vary at the different technical colleges throughout the country, but we hope this book will cover the majority of such syllabuses.

Many of the examples have been taken from the Intermediate B.Sc. and Part I examinations for a degree in engineering at London University; and also from the Joint Section A examination of the Institutions of Civil and Electrical Engineers. We would like to express our thanks to these authorities for permission to use their questions.

S. W. BELL

H. MATLEY

Norwich
November 1954

PREFACE TO THE SECOND EDITION

The opportunity has been taken to correct a number of errors and misprints. In the hope of making the book more useful for students following Sandwich courses, a set of Miscellaneous exercises, covering most of the topics dealt with in the text, has been included at the end of the book. In this connexion it should be noted that the Joint Section A examination of the Institutions of Civil and Electrical Engineers has, since 1955, been revised and renamed the Part I examination.

S. W. BELL

H. MATLEY

Norwich
May 1958

The opportunity has been taken to introduce SI units in the 1971 Reprint.

S. W. BELL

H. MATLEY

Norwich
1971

SOME ABBREVIATIONS USED IN THE TEXT

$\neq$	not equal to	$\equiv$	identically equal to
$\simeq$	approximately equal to	$+$ ve	positive
Σ	the summation of	$-$ ve	negative
lim	limit of	p.d.	potential difference
$\rightarrow$	approaches	e.m.f.	electromotive force
$<$	less than	e.s.u.	electrostatic units
$>$	greater than	M.V.	mean value
$\leqslant$	less than or equal to	r.m.s.	root-mean-square
$\geqslant$	greater than or equal to		value

SOME LETTERS OF THE GREEK ALPHABET IN COMMON USE IN MATHEMATICS

α	alpha	π	pi
β	beta	Π	capital pi
γ	gamma	ρ	rho
δ	delta	σ	sigma
Δ	capital delta	Σ	capital sigma
ϵ	epsilon	ϕ	phi
η	eta	ψ	psi
θ	theta	ω	omega
λ	lambda	Ω	capital omega
μ	mu		

REVISION & FURTHER DIFFERENTIATION. APPLICATIONS

Standard Derivatives and Rules

$$D_x(x^n) = nx^{n-1}, \quad D_x(e^x) = e^x, \quad D_x(\log_e x) = 1/x,$$
$$D_x(\sin x) = \cos x, \quad D_x(\cos x) = -\sin x, \quad D_x(\tan x) = \sec^2 x.$$
$$\frac{dy}{dx} = \frac{dy}{du}\frac{du}{dx}, \quad \frac{d}{dx}(uv) = u\frac{dv}{dx} + v\frac{du}{dx},$$
$$\frac{d}{dx}\left(\frac{u}{v}\right) = \frac{v\dfrac{du}{dx} - u\dfrac{dv}{dx}}{v^2}.$$

1·1. Function of a function

The use of the function of a function rule forms part of the Ordinary National Certificate Course. It is essential that the student should be capable of using this rule mentally, without any need for written substitution methods. The following example shows the three stages which are usually followed, the final one representing the standard that must be reached.

Example

Differentiate $\log_e \sin \sqrt{x}$ with respect to x.

(i) Let
$$y = \log_e \sin \sqrt{x} = \log_e \{\sin (x)^{\frac{1}{2}}\}.$$

Let
$$u = \sin (x)^{\frac{1}{2}} \quad \text{and} \quad v = x^{\frac{1}{2}}.$$

Then
$$y = \log_e u, \quad \therefore \frac{dy}{du} = \frac{1}{u} = \frac{1}{\sin \sqrt{x}}.$$

Also
$$u = \sin v, \quad \therefore \frac{du}{dv} = \cos v = \cos \sqrt{x}.$$

Finally
$$v = x^{\frac{1}{2}}, \quad \therefore \frac{dv}{dx} = \frac{1}{2}x^{-\frac{1}{2}} = \frac{1}{2\sqrt{x}}.$$

But
$$\frac{dy}{dx} = \frac{dy}{du}\frac{du}{dv}\frac{dv}{dx}$$

$$= \frac{1}{\sin \sqrt{x}} \cos \sqrt{x} \frac{1}{2\sqrt{x}} = \underline{\frac{1}{2\sqrt{x}} \cot \sqrt{x}}.$$

(ii) Here actual expressions, put in brackets, are used instead of substituting the separate letters u, v:

$$y = \log_e \{\sin (x^{\frac{1}{2}})\},$$

$$\frac{dy}{dx} = \frac{d(\log_e \sin x^{\frac{1}{2}})}{d(\sin x^{\frac{1}{2}})} \frac{d(\sin x^{\frac{1}{2}})}{d(x^{\frac{1}{2}})} \frac{d(x^{\frac{1}{2}})}{dx}$$

$$= \frac{1}{\sin x^{\frac{1}{2}}} \cos x^{\frac{1}{2}} \tfrac{1}{2} x^{-\frac{1}{2}}$$

$$= \frac{1}{2\sqrt{x}} \cot \sqrt{x}.$$

(iii) In this final stage the second line of (ii) above is omitted. The student holds in his mind the three 'types' of function occurring in this particular example. These may be thought of as 'a log to the base e', 'a sine function' and 'x to the power one-half'.

In this connexion it is helpful to replace all root signs by fractional indices and to make liberal use of brackets. If this is done the process becomes one of working from the outside through each bracket in turn. Thus

$$\frac{d}{dx} [\log \{\sin (x^{\frac{1}{2}})\}] = \frac{1}{\sin (x^{\frac{1}{2}})} \cos (x^{\frac{1}{2}}) \tfrac{1}{2} x^{-\frac{1}{2}}$$

$$= \frac{1}{2\sqrt{x}} \cot \sqrt{x}.$$

1·11. Some standard derivatives

Several standard differentiations are given below. The results should be memorized by those students who have not met them before.

$$\frac{d}{dx} (\tan x) = \frac{d}{dx} \left(\frac{\sin x}{\cos x}\right) = \frac{\cos x \cos x - \sin x(-\sin x)}{\cos^2 x}$$

$$= \frac{\cos^2 x + \sin^2 x}{\cos^2 x} = \frac{1}{\cos^2 x} = \sec^2 x.$$

$$\frac{d}{dx} (\sec x) = \frac{d}{dx} (\cos x)^{-1} = -1(\cos x)^{-2} (-\sin x)$$

$$= \frac{\sin x}{\cos^2 x} = \sec x \tan x.$$

$$\frac{d}{dx}(\cot x) = \frac{d}{dx}\left(\frac{\cos x}{\sin x}\right) = \frac{\sin x(-\sin x) - \cos x \cos x}{\sin^2 x}$$

$$= -\frac{\sin^2 x + \cos^2 x}{\sin^2 x} = -\frac{1}{\sin^2 x} = -\operatorname{cosec}^2 x.$$

$$\frac{d}{dx}(\operatorname{cosec} x) = \frac{d}{dx}(\sin x)^{-1} = -1(\sin x)^{-2}(\cos x)$$

$$= -\frac{\cos x}{\sin^2 x} = -\operatorname{cosec} x \cot x.$$

Examples

Differentiate with respect to x:

(1) $x^2 \log_e (2-x)$.   (2) $\dfrac{1+x}{\sqrt{(2x+x^2)}}$.   (3) $\sin^2 (x^2)$.

(1) $$f'(x) = 2x \log_e (2-x) + x^2 \frac{1}{2-x}(-1)$$

$$= 2x \log_e (2-x) - \frac{x^2}{2-x}.$$

(2) $$f'(x) = \frac{(2x+x^2)^{\frac{1}{2}} 1 - (1+x)\frac{1}{2}(2x+x^2)^{-\frac{1}{2}}(2+2x)}{2x+x^2}$$

$$= \frac{(2x+x^2) - (1+x)^2}{(2x+x^2)^{\frac{3}{2}}}$$

$$= -\frac{1}{(2x+x^2)^{\frac{3}{2}}}.$$

(3) $$y = f(x) = \{\sin (x^2)\}^2.$$

$$\therefore \ f'(x) = 2 \sin (x^2) \cos (x^2) 2x$$

$$= 2x \sin (2x^2).$$

Exercise 1

Differentiate the following with respect to x, simplifying the result where possible:

1. $x^3 \sin x$.

2. $e^x \cos x$.

3. $x^4 \log_e x$.

4. $\dfrac{x^2}{1+x}$.

5. $\dfrac{\tan x}{x}$.

6. $\dfrac{\sin x}{1+\cos x}$.

7. $(3-2x)^5$.

8. $\sin 2x \cos 3x$.

9. $x^2 e^{-2x}$.

10. $\log_e \cos x$.

11. $\dfrac{1}{\sqrt{(ax^2+bx+c)}}$.

12. e^{x^2}.

13. $x\sqrt{(1+x^2)}$.   **14.** $e^{3x}\sin 4x$.   **15.** $\log_e(\cot x)$.

16. $\sqrt{(\operatorname{cosec} 2x)}$.   **17.** $\sec^2 2x$.   **18.** $\dfrac{x^3}{\cos 3x}$.

19. $\dfrac{1}{(2x+1)^2}$.   **20.** $\sqrt{\dfrac{2x+3}{2x-3}}$.

21. $\log_e(\sec x + \tan x)$.   **22.** $\log_e\left(\dfrac{1+\sin x}{\cos x}\right)$.

23. $e^{2x}/\tan x$.   **24.** $e^{\sqrt{x}}\tan\sqrt{x}$.

25. $x\log_e(1+\cos x)$.   **26.** $\log_e\{x+\sqrt{(1+x^2)}\}$.

27. $\dfrac{x}{\sqrt{(x^2+1)}}$.   **28.** $x^2 e^x \sin 2x$.

29. $\cos(\sin 3x)$.   **30.** $\log_e\left(\dfrac{x^2+x+1}{x^2-x+1}\right)$.

Answers

1. $x^2(3\sin x + x\cos x)$.   **2.** $e^x(\cos x - \sin x)$.

3. $x^3(4\log_e x + 1)$.   **4.** $\dfrac{x(2+x)}{(1+x)^2}$.

5. $\dfrac{x\sec^2 x - \tan x}{x^2}$.   **6.** $\dfrac{1}{1+\cos x}$.

7. $-10(3-2x)^4$.

8. $2\cos 2x\cos 3x - 3\sin 2x\sin 3x$.

9. $2x(1-x)e^{-2x}$.   **10.** $-\tan x$.

11. $\dfrac{-(2ax+b)}{2(ax^2+bx+c)^{\frac{3}{2}}}$.   **12.** $2xe^{x^2}$.

13. $\dfrac{1+2x^2}{\sqrt{(1+x^2)}}$.   **14.** $e^{3x}(3\sin 4x + 4\cos 4x)$.

15. $-\dfrac{1}{\sin x\cos x}$.   **16.** $-\cot 2x\sqrt{(\operatorname{cosec} 2x)}$.

17. $4\sec^2 2x\tan 2x$.   **18.** $\dfrac{3x^2(\cos 3x + x\sin 3x)}{\cos^2 3x}$.

19. $-\dfrac{4}{(2x+1)^3}$.   **20.** $-\dfrac{6}{(2x-3)\sqrt{(4x^2-9)}}$.

21. $\sec x$.   **22.** $\sec x$.

23. $\dfrac{e^{2x}(2\tan x - \sec^2 x)}{\tan^2 x}$.

24. $\dfrac{1}{2\sqrt{x}}e^{\sqrt{x}}(\sec^2\sqrt{x}+\tan\sqrt{x})$.

25. $\log_e(1+\cos x)-\dfrac{x\sin x}{1+\cos x}$.

26. $\dfrac{1}{\sqrt{(1+x^2)}}$.

27. $(x^2+1)^{-\frac{3}{2}}$.

28. $xe^x(2\sin 2x + x\sin 2x + 2x\cos 2x)$.

29. $-3\cos 3x\sin(\sin 3x)$.

30. $\dfrac{2(1-x^2)}{x^4+x^2+1}$.

1·2. Differentiation of implicit functions

Explicit functions are those in which the dependent variable (y) is given explicitly, or clearly, in terms of the independent variable (x), e.g.

$$y=\sqrt{(a^2-x^2)}, \quad y=e^x\sin 3x, \quad y=x^2-\log x.$$

Implicit functions are functions in which the dependent variable is not given expressly in terms of the independent variable. In many cases it is not possible to solve for y in terms of standard expressions in x, e.g.

$$x^2+y^2=9, \quad \sin(y-x)+x=0, \quad \log x^2 y + 9y^2 = 3.$$

No new knowledge is required to differentiate implicit functions. The function of a function rule is used. Thus

$$\frac{d}{dx}(y^2)=(2y)\left(\frac{dy}{dx}\right), \quad \frac{d}{dx}(\sin y)=\cos y\left(\frac{dy}{dx}\right).$$

Expressions such as x^3y^2, x^4/y^3 are dealt with in the normal way as products and quotients.

EXAMPLES

Find dy/dx in terms of x and y for the following implicit functions:

(1) $x^3-3x^2y+y^2=5$.   (2) $\log_e(xy^3)-\sin xy=0$.

(1) Differentiating with respect to x:

$$3x^2-3\left(x^2\frac{dy}{dx}+2xy\right)+2y\frac{dy}{dx}=0.$$

$$\therefore \frac{dy}{dx}(2y-3x^2)=3x(2y-x).$$

$$\therefore \frac{dy}{dx}=\frac{3x(2y-x)}{(2y-3x^2)}.$$

(2) Simplifying the given expression:

$$\log_e x + 3\log_e y - \sin(xy) = 0.$$

Differentiating with respect to x:

$$\frac{1}{x} + 3\frac{1}{y}\frac{dy}{dx} - \cos(xy)\left(x\frac{dy}{dx} + y\right) = 0,$$

$$\frac{dy}{dx}\left(\frac{3}{y} - x\cos xy\right) = y\cos xy - \frac{1}{x},$$

$$\frac{dy}{dx} = \frac{y(xy\cos xy - 1)}{x(3 - xy\cos xy)}.$$

EXERCISE 2

Find dy/dx in the following cases:

1. $x^2 - 3xy + y^2 = 2$.   **2.** $x^3/y^2 = 5$ $(y \neq 0)$.

3. $\sin(x+y) = 2xy$.   **4.** $x = \tan y$.

5. $\log_e y = \sin x - \cos y$.   **6.** $e^{x+y} - 2x = 0$.

7. Find the slope of the tangent at any point on the circle $x^2 + y^2 = 25$. Hence find the equation of the tangent to this circle at the point $(3, 4)$.

8. Find the equation of the tangent to the ellipse

$$\tfrac{1}{9}x^2 + \tfrac{1}{25}y^2 = \tfrac{1}{2}$$

at the point $(\tfrac{3}{2}, \tfrac{5}{2})$.

9. If $y = x\log_e y$ prove $\dfrac{d^2y}{dx^2} = \dfrac{y^2(y-2x)}{x(y-x)^3}$.   (L.U.)

10. If $pv^\gamma = c$, show that

$$v^2\frac{d^2p}{dv^2} = \gamma(\gamma+1)p \quad \text{and} \quad p^2\frac{d^2v}{dp^2} = \frac{\gamma+1}{\gamma^2}v. \quad \text{[Sec. A]}$$

ANSWERS

1. $\dfrac{3y - 2x}{2y - 3x}$.   **2.** $\dfrac{3x^2}{10y}$.

3. $\dfrac{\cos(x+y) - 2y}{2x - \cos(x+y)}$.   **4.** $\cos^2 y$ or $\dfrac{1}{(1+x^2)}$.

5. $\dfrac{y\cos x}{(1 - y\sin y)}$.   **6.** $2e^{-(x+y)} - 1$.

7. $-x/y$; $3x+4y=25$.   **8.** $5x+3y=15$.

9. *Hint.* First solve for dy/dx in terms of x and y only; then differentiate again.

1·3. Parameters

Sometimes it is easier to deal with a function of x and y by expressing both x and y in terms of a third variable. This variable is called a parameter. In $x = a\cos^3 t$, $y = a\sin^3 t$, t is the parameter. In this case it is easy to eliminate t to get the (x, y), or Cartesian, equation:

$$\sin^2 t + \cos^2 t = 1 = (x/a)^{\frac{2}{3}} + (y/a)^{\frac{2}{3}}.$$

Thus

$$x^{\frac{2}{3}} + y^{\frac{2}{3}} = a^{\frac{2}{3}}.$$

In other cases the Cartesian equation is complicated and it is much easier to deal with the parametric equations.

The function of a function rule is used if dy/dx is needed. Thus, if

$$x = a\cos^3 t, \quad y = a\sin^3 t,$$

$$\frac{dx}{dt} = -3a\cos^2 t \sin t, \quad \frac{dy}{dt} = 3a\sin^2 t \cos t.$$

$$\therefore \ \frac{dy}{dx} = \frac{dy}{dt}\frac{dt}{dx} = \frac{dy}{dt} \Big/ \frac{dx}{dt} = -\frac{3a\sin^2 t \cos t}{3a\cos^2 t \sin t} = -\tan t.$$

Care must be taken if the second derivative d^2y/dx^2 is needed:

$$\frac{d^2y}{dx^2} = \frac{d}{dx}\left(\frac{dy}{dx}\right) = \frac{d}{dt}\left(\frac{dy}{dx}\right)\frac{dt}{dx}$$

$$= \frac{d}{dt}(-\tan t)\left(-\frac{1}{3a\cos^2 t \sin t}\right)$$

$$= +\frac{\sec^2 t}{3a\cos^2 t \sin t} = \underline{\frac{1}{3a\cos^4 t \sin t}}.$$

EXERCISE 3

1. If $x = \tan t$, $y = \tan pt$, find dy/dx.

2. If $x = \sin t$, $y = \sin pt$, find d^2y/dx^2.

3. If $x = a\cos t$, $y = b\sin t$, find the equation of the tangent at any point 't' on the curve.

4. Find the equation of the tangent at the point (at, at^n) on the curve $a^{n-1}y = x^n$.

5. If $\omega = f(\theta)$ and $y = \tan\theta$, show that $\dfrac{d\omega}{d\theta} = (1+y^2)\dfrac{d\omega}{dy}$ and find $\dfrac{d^2\omega}{d\theta^2}$ in terms of y, $\dfrac{d\omega}{dy}$ and $\dfrac{d^2\omega}{dy^2}$.   [L.U.]

6. If $x = a\sin 2\theta$, $y = a\cos\theta$, prove

$$\frac{d^2y}{dx^2} = -\frac{\cos\theta(1 + 2\sin^2\theta)}{4a\cos^3 2\theta}.$$

Sketch the curve, giving θ different values and finding the corresponding pairs of values of (x, y).

7. If $x = a(1 - \cos\theta)$, $y = a\sin\theta$, find dy/dx and d^2y/dx^2 in terms of θ.

8. If $x = \cos 2\theta + 2\theta\sin 2\theta$, $y = \sin 2\theta - 2\theta\cos 2\theta$, prove

$$\left(\frac{dx}{d\theta}\right)^2 + \left(\frac{dy}{d\theta}\right)^2 = 16\theta^2.$$ [Sec. A]

ANSWERS

1. $\dfrac{p\sec^2 pt}{\sec^2 t}$.

2. $\dfrac{p\cos pt\tan t - p^2\sin pt}{\cos^2 t}$.

3. $bx\cos t + ay\sin t = ab$.

4. $y - nt^{n-1}x = (1-n)at^n$.

5. Use $\dfrac{d\omega}{d\theta} = \dfrac{d\omega}{dy}\dfrac{dy}{d\theta}$; $\dfrac{d^2\omega}{d\theta^2} = (1+y^2)\left[(1+y^2)\dfrac{d^2\omega}{dy^2} + 2y\dfrac{d\omega}{dy}\right]$.

7. $\cot\theta$; $-\dfrac{1}{a\sin^3\theta}$.

1·4. Logarithmic differentiation

(a) If it is necessary to differentiate a continued product or quotient, it is easier to take logs *before* differentiating.

(b) If the function to be differentiated is already a log, as much simplification as possible should be done before differentiating, making use of the standard theorems on logarithms.

(c) To differentiate an expression with x in the index (an exponential expression), it is usually easier to take logs before differentiating.

EXAMPLES

(1) Differentiate $\dfrac{e^{2x}\sqrt{(x^2-1)}}{\sin 3x}$.

Let $y = \dfrac{e^{2x}\sqrt{(x^2-1)}}{\sin 3x}$.

Then $\qquad \log_e y = \log_e e^{2x} + \log_e \sqrt{(x^2 - 1)} - \log_e \sin 3x$

$$= 2x + \tfrac{1}{2}\log_e (x^2 - 1) - \log_e \sin 3x.$$

Differentiating both sides with respect to x:

$$\frac{1}{y}\frac{dy}{dx} = 2 + \frac{1}{2}\frac{2x}{(x^2 - 1)} - \frac{1}{\sin 3x} 3 \cos 3x$$

$$= \left(2 + \frac{x}{(x^2 - 1)} - 3 \cot 3x\right).$$

$$\therefore \frac{dy}{dx} = \frac{e^{2x} \sqrt{(x^2 - 1)}}{\sin 3x}\left\{2 + \frac{x}{(x^2 - 1)} - 3 \cot 3x\right\}.$$

(2) Differentiate $\quad y = \log_e\left\{e^x \sqrt{\dfrac{x - 1}{x + 1}}\right\}.$

Simplifying:

$$y = \log_e e^x + \tfrac{1}{2}\log_e\left(\frac{x - 1}{x + 1}\right)$$

$$= x + \tfrac{1}{2}\log_e (x - 1) - \tfrac{1}{2}\log_e (x + 1).$$

$$\therefore \frac{dy}{dx} = 1 + \frac{1}{2(x - 1)} - \frac{1}{2(x + 1)}$$

$$= \frac{2x^2 - 2 + x + 1 - x + 1}{2(x^2 - 1)}$$

$$= \frac{x^2}{(x^2 - 1)}.$$

(3) Differentiate $y = x^x$.

Taking logs: $\qquad\qquad \log_e y = x \log_e x.$

Differentiating: $\qquad \dfrac{1}{y}\dfrac{dy}{dx} = 1 \log_e x + x\dfrac{1}{x}.$

$$\therefore \frac{dy}{dx} = y(1 + \log_e x)$$

$$= x^x(1 + \log_e x).$$

EXERCISE 4

Differentiate with respect to x:

1. $\sqrt{\dfrac{2x + 3}{2x - 3}}.$

2. $\dfrac{\sqrt{(x^2 + 2)}}{\sqrt[3]{(x^3 - 3)}}.$

3. $\sqrt{\dfrac{\sin^3 x}{\cos 2x}}.$

4. $x^n \log 1/x.$

5. $(x - 1)(x - 2)(x - 3)^2.$

6. $e^{ax} \sin^3 x \cos^2 x.$

7. $a^x \tan x.$

ANSWERS

1. $-\dfrac{6}{(4x^2-9)}\sqrt{\dfrac{(2x+3)}{(2x-3)}}$.   **2.** $-\dfrac{\sqrt{(x^2+2)}}{\sqrt[3]{(x^3-3)}}\dfrac{x(3+2x)}{(x^2+2)(x^3-3)}$.

3. $\sqrt{\dfrac{\sin^3 x}{\cos 2x}}\,(\tfrac{3}{2}\cot x+\tan 2x)$.   **4.** $-x^{n-1}(1+n\log x)$.

5. $(x-1)(x-2)(x-3)^2\left[\dfrac{1}{(x-1)}+\dfrac{1}{(x-2)}+\dfrac{2}{(x-3)}\right]$.

6. $e^{ax}\sin^3 x\cos^2 x[a+3\cot x-2\tan x]$.

7. $a^x\tan x\left[\log a+\dfrac{\sec^2 x}{\tan x}\right]$.

1·5. Formation of differential equations

Any equation which involves the derivatives of a function is called a differential equation, e.g.

$$\left(\frac{dy}{dx}\right)^2 - x^2\sin y = 0, \quad 3\frac{d^2y}{dx^2}+5\frac{dy}{dx}-y=2\sin x.$$

They may be derived by the elimination of arbitrary constants from a function by repeated differentiation. Thus, if $x^2+y^2=a^2$ (a arbitrary), then $2x+2y\dfrac{dy}{dx}=0$, no matter what numerical value is given to a.

The differential equation would be $x+y\dfrac{dy}{dx}=0$.

In dealing with equations where a second differentiation is required it is often quicker to try and obtain a relationship between dy/dx and x and y before differentiating again. This, however, is not always possible if there are two arbitrary constants.

EXAMPLES

(1) If $x=\tan t$, $y=\tan pt$ show that $(1+x^2)\dfrac{d^2y}{dx^2}=2(py-x)\dfrac{dy}{dx}$.

Now $$\frac{dy}{dt}=p\sec^2 pt, \quad \frac{dx}{dt}=\sec^2 t.$$

$$\therefore \frac{dy}{dx}=\frac{p\sec^2 pt}{\sec^2 t}=\frac{p(1+\tan^2 pt)}{(1+\tan^2 t)}=\frac{p(1+y^2)}{(1+x^2)}.$$

Thus $$(1+x^2)\frac{dy}{dx}=p+py^2.$$

Differentiating again:

$$(1+x^2)\frac{d^2y}{dx^2} + 2x\frac{dy}{dx} = 2py\frac{dy}{dx}.$$

$$\therefore \ (1+x^2)\frac{d^2y}{dx^2} = 2(py-x)\frac{dy}{dx}.$$

(2) Show that $y = A/x + Bx^2 - \log_e x + \frac{1}{2}$ satisfies the equation $x^2\frac{d^2y}{dx^2} - 2y = 2\log x$. Find A and B given $y=0$, $\frac{dy}{dx}=1$ when $x=1$. [Sec. A]

Here we could find dy/dx, and d^2y/dx^2, not worrying whether A and B occur or not; put them in the given equation and test that it is satisfied. An alternative, in this case longer, is to eliminate A and B:

$$A = xy - Bx^3 + x\log_e x - \tfrac{1}{2}x.$$

Differentiating:

$$0 = \left(y + x\frac{dy}{dx}\right) - 3Bx^2 + (\log_e x + 1) - \tfrac{1}{2}.$$

Solving for $3B$:

$$3B = \frac{y}{x^2} + \frac{1}{x}\frac{dy}{dx} + \frac{1}{x^2}\log_e x + \frac{1}{2x^2}.$$

Differentiating:

$$0 = \left(\frac{1}{x^2}\frac{dy}{dx} - \frac{2}{x^3}y\right) + \left(\frac{1}{x}\frac{d^2y}{dx^2} - \frac{1}{x^2}\frac{dy}{dx}\right) + \left(\frac{1}{x^3} - \frac{2}{x^3}\log x\right) - \frac{1}{x^3}.$$

$$\therefore \ \frac{1}{x}\frac{d^2y}{dx^2} - \frac{2}{x^3}y = \frac{2}{x^3}\log_e x,$$

$$x^2\frac{d^2y}{dx^2} - 2y = 2\log_e x.$$

$$y = \frac{A}{x} + Bx^2 - \log_e x + \tfrac{1}{2}.$$

When $x=1$, $y=0$. $\therefore \ 0 = A + B + \tfrac{1}{2}$. $\hspace{2cm}$ (1)

$$\frac{dy}{dx} = -\frac{A}{x^2} + 2Bx - \frac{1}{x}.$$

When $x=1$, $dy/dx=1$.

$$\therefore \ 1 = -A + 2B - 1, \quad -A + 2B - 2 = 0. \hspace{1cm} (2)$$

From (1) and (2), $\underline{A = -1, \quad B = \tfrac{1}{2}.}$

EXERCISE 5

1. If $y = A \cos mx + B \sin mx$, prove $d^2y/dx^2 + m^2y = 0$.

2. If $y = A \log_e x + B$, prove $x\dfrac{d^2y}{dx^2} + \dfrac{dy}{dx} = 0$.

3. If $y = \dfrac{A e^{mx} + B e^{-mx}}{x}$, prove $\dfrac{d^2y}{dx^2} + \dfrac{2}{x}\dfrac{dy}{dx} - m^2y = 0$.

4. If $x = \sin t$, $y = \sin pt$, prove $(1 - x^2)\dfrac{d^2y}{dx^2} - x\dfrac{dy}{dx} + p^2y = 0$.

5. If $y = x e^{-x}$, prove $\dfrac{d^2y}{dx^2} + 2\dfrac{dy}{dx} + y = 0$.

6. If $y = x^n e^{ax}$, prove $\dfrac{dy}{dx} - ay = \dfrac{ny}{x}$.

1·6. Differentials. Errors and approximations

Students are often warned not to consider the first derivative dy/dx as a fraction. This is because dy/dx is defined as the limiting value of $\delta y/\delta x$ as δx approaches zero.

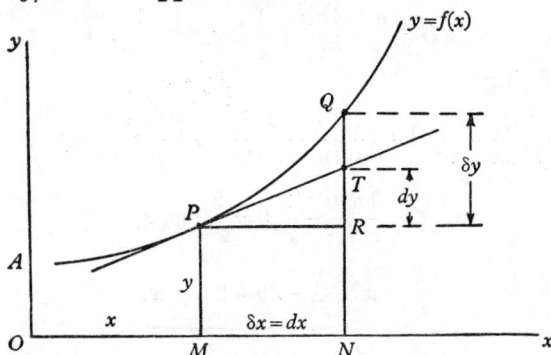

Fig. 1

The limit of $(\delta y/\delta x)$ is searched for treating it as a *single* expression, as obviously δy and δx taken separately each approach zero. dy/dx is thus defined as a single expression. If it is desired to treat dy and dx as separate expressions and $dy \div dx$ as being the first derivative, it is necessary to define quite separately what is meant by each of the symbols dy and dx. Care must be taken to see, also, that the definitions satisfy $dy \div dx = f'(x)$ (the first derivative).

Referring to fig. 1, let P be any point (x, y) on the curve of $y = f(x)$. Then $\qquad OM = x, \quad MP = y = f(x)$.

Let MN be any increase in x, *not necessarily small*. Then

$$MN = PR = \text{increase in } x = \delta x \text{ say.}$$

Let the tangent at P, PT cut NQ at T. Then the gradient of the curve at P is

$$f'(x) = \frac{RT}{PR}.$$

$$\therefore \ RT = PRf'(x) = f'(x)\,\delta x. \tag{1}$$

The increment of y along the curve $= RQ = \delta y$ say.

The increment of y along the tangent $= RT = f'(x)\,\delta x$.

The arbitrary change in x, δx is defined to be dx, and called the *differential of* x.

$RT = f'(x)\,\delta x = f'(x)\,dx$ is defined to be dy and is called the *differential of* y.

From (1) we have the defining relationship for dy:

$$dy = f'(x)\,dx. \tag{2}$$

By this definition $dy \div dx$ is $f'(x)$, the first derivative.

From (2) note that $f'(x)$ is the coefficient of the differential of $x(dx)$. Hence the name 'differential coefficient' is often given to the first derivative $f'(x)$.

Differentials may be of any size, but their ratio is always the first derivative. When convenient, therefore, the first derivative occurring in an equation is often replaced by the separate differentials. Thus $x\dfrac{dy}{dx} + y = 0$ may be written $x\,dy + y\,dx = 0$. If dx is small, then $RT \simeq RQ$. That is

$$\delta y \simeq dy.$$

This gives $\qquad \delta y \simeq f'(x)\,\delta x \quad$ or $\quad \delta y \simeq (dy/dx)\,\delta x,$

the well-known approximation for small changes.

EXAMPLES

(1) Give the differentials of the functions (a) $1/\sqrt{x}$, (b) $\sin^2 3x$.

(a) If $y = 1/\sqrt{x} = x^{-\frac{1}{2}}$,

$$dy = f'(x)\,dx = -\tfrac{1}{2}x^{-\frac{3}{2}}\,dx,$$

$$\underline{dy = -\frac{1}{2x\,\sqrt{x}}\,dx.}$$

(b) If $y = \sin^2 3x$,

$$dy = 2\sin 3x \ 3\cos 3x\,dx,$$

$$\underline{dy = 3\sin 6x\,dx.}$$

(2) The value of 'g' is calculated to be 9·81 from the formula $T = 2\pi \sqrt{(l/g)}$. If an error of 1 % is made in measuring T, and l is accurate, find (a) the actual error, (b) the percentage error in the value of g.

(a) $$T = 2\pi \sqrt{\frac{l}{g}}. \quad \therefore \quad g = 4\pi^2 l \frac{1}{T^2}. \tag{1}$$

$$\frac{dg}{dT} = -\frac{8\pi^2 l}{T^3} = -g\frac{2}{T} \quad \text{from (1)}.$$

Thus $$\delta g \simeq -\frac{2g}{T}\delta T. \tag{2}$$

But $\delta T/T =$ proportionate error in $T = 0{\cdot}01$.

$$\therefore \quad \delta g \simeq -2g(0{\cdot}01) = -0{\cdot}02 \times 9{\cdot}81 = \underline{-0{\cdot}1962}.$$

Note. It has been assumed that the error in T was positive (T measured too large); if T had been measured too small and the error therefore negative, it would simply mean a change in sign.

(b) From (2) $$\frac{\delta g}{g} = -2\frac{\delta T}{T}.$$

$$\therefore \quad \frac{\delta g}{g} \times 100 = -2\frac{\delta T}{T} \times 100.$$

That is, percentage error in $g = -2 \times$ percentage error in T

$$= -2 \times 1$$
$$= \underline{-2}.$$

Note. If percentage or proportionate errors only are required it is best to take logs before differentiating. In this way the proportionate error is found at once, e.g.

$$g = \frac{4\pi^2 l}{T^2},$$

$$\log_e g = \log_e (4\pi^2 l) - 2\log_e T.$$

Differentiating with respect to T:

$$\frac{1}{g}\frac{dg}{dT} = -\frac{2}{T}$$

or $$\underline{\frac{\delta g}{g} \simeq -2\frac{\delta T}{T}.}$$

Exercise 6

1. Find the differentials of the following functions:

   (i) $ax^n + b/x^n$, (ii) e^{3x}, (iii) $x^2 \sin 2x$, (iv) $\log \cos x$.

2. A formula for the variation of electrical resistance R of a platinum wire with temperature t is $R = R_0(1 + at + bt^2)$, where R_0, a, b are constants. Find the approximate change in resistance due to a small rise of temperature δt.

3. A ladder 15 m long rests with its upper end against a vertical wall and its lower end on the ground 4 m from the wall; if the lower end is pulled a distance 60 mm from the wall, how far will the upper end move?

4. The side c of a triangle is calculated from the formula $c = \dfrac{a \sin C}{\sin A}$, sides in metres. Find the percentage error in the value of c due to an error of 1 mm in the value of a. Find the percentage error if the formula $c^2 = a^2 + b^2 - 2ab \cos C$ had been used in the same circumstances.

Answers

1. (i) $n\left(ax^{n-1} - \dfrac{b}{x^{n+1}}\right)dx$,   (ii) $3e^{3x}dx$,

    (iii) $2x(x\cos 2x + \sin 2x)dx$,   (iv) $-\tan x\, dx$.

2. $R_0(a + 2bt)\delta t$.   **3.** 17 mm downwards.

4. $\dfrac{0\cdot 1}{a},\ \dfrac{0\cdot 1(a - b\cos C)}{c^2}$.

1·7. Continuity and differentiability

1·71. Continuous functions

A function is said to be continuous if small changes in x only produce small changes in y, the value of the function. In graphical terms this means there are no sudden breaks or jumps but that the curve is a continuous one. A function can be continuous for all but certain specific values of x. For such values the function is said to be discontinuous and the corresponding points on the graph *points of discontinuity*. At these points the y values jump from one to another with no perceptible change in x value. If the jump is finite it is called a point of finite discontinuity, and if immeasurably large, a point of infinite discontinuity. Simple examples are shown below in figs. 2, 3 and 4.

Fig. 2 represents the graph of $y = 1/x$. At $x = 0$ there is a point of infinite discontinuity. As $x \to 0$ from negative values $y \to -\infty$; as $x \to 0$ from positive values $y \to +\infty$. *At $x = 0$ no value can be stated for y.*

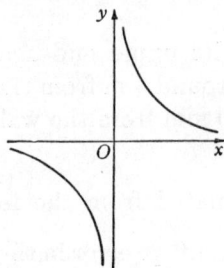

Fig. 2

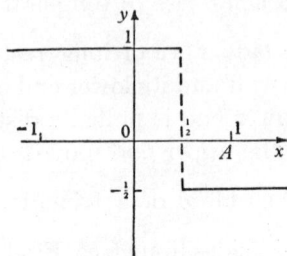

Fig. 3

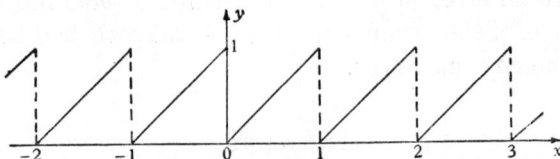

Fig. 4

Fig. 3 represents the graph of the function defined by

$$y = 1 \quad \text{when} \quad x \leqslant \tfrac{1}{2},$$
$$y = -\tfrac{1}{2} \quad \text{when} \quad x > \tfrac{1}{2}.$$

At $x = \tfrac{1}{2}$ there is a point of finite discontinuity.

Fig. 4 represents the graph of a function which is discontinuous for a set of periodic values of x, ..., -2, -1, 0, 1, 2, 3, ... Functions such as this latter will occur again in the chapter on Harmonic Analysis.

In general, a function, $y = f(x)$, is said to be continuous at a point $x = a$ if the limiting value of $f(x)$ as $x \to a$, from either lower or higher values, is equal to $f(a)$. Expressed in symbols:

$$\lim_{x \nearrow a} f(x) = \lim_{x \searrow a} f(x) = f(a).$$

If these limits are different the function is discontinuous for $x = a$. (Note the symbol $x \nearrow a$, meaning x approaches a from lower values, and $x \searrow a$, meaning x approaches a from higher values.)

An alternative way of expressing this is: $\delta y \to 0$ as $\delta x \to 0$ (from below or above) at the point $x = a$.

For example, for the function in fig. 3, at the point A where
$x = 1$: $\delta y \to 0$ as $\delta x \to 0$ (from below or above $x = 1$).

Alternatively, $\lim\limits_{x \nearrow 1} f(x) = \lim\limits_{x \searrow 1} f(x) = -\tfrac{1}{2}.$

Thus the function is continuous at $x = 1$.

1·72. Differentiability

Definitions of the first derivative with which the student should already be familiar are:

$$\frac{dy}{dx} = \lim_{\delta x \to 0} \left(\frac{\delta y}{\delta x} \right),$$

and, in functional notation,

$$f'(x) = \lim_{h \to 0} \frac{f(x+h) - f(x)}{h}.$$

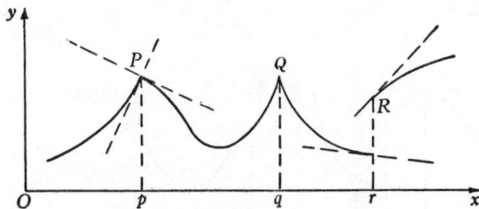

Fig. 5

For a completely general definition it should be stated that $\delta x \to 0$ from below or above, and h is either positive or negative.

If the above limit has a single value for any particular value of x, the function is said to be differentiable for that value; if the limit does not exist or is multivalued the function is *not* differentiable for that value of x.

Consider fig. 5 as representing the graph of some function of x, $y = f(x)$. At P there are two possible tangents and therefore no unique derivative. $f'(p)$ has a different value according as $x \to p$ from below or above. At Q both branches of the curve touch a line with infinite gradient. Here again there is no specific value for the derivative, as infinity is not a number. In this case it is usual to say $f'(q) \to \pm \infty$. At R the function has a finite discontinuity at $x = r$. Again there are two possible values for $f'(r)$, the slope of the tangent, depending on whether $x \to r$ from below or above.

In all three cases the function is not differentiable at the given points. Points P and Q are illustrations of the fact that a function may be continuous at a certain point but not differentiable at that point.

However, if a function is differentiable at a point it must be continuous at that point; for at the point $\delta y/\delta x$ must approach the same definite value from above or below, from which it follows that δy must approach zero as δx approaches zero from either above or below.

1·8. Maxima, minima and points of inflexion

The routine tests for maximum and minimum values of a function form part of the Ordinary National Certificate Course. A fuller consideration of the topic is given below.

Fig. 6, representing the graph of some function of x, will be used to bring out some of the more important points on the graph of a function.

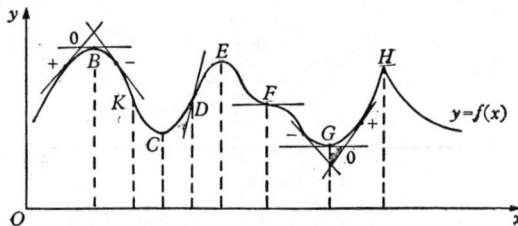

Fig. 6

1·81. Maxima and minima

Points B, E show typical maximum points.

Points C, G show typical minimum points.

(a) For a maximum value, y values on either side, however close, must be less than the y value at the maximum point. The reverse is the case for a minimum value.

(b) Tangent lines have been drawn around a maximum point B and a minimum point G. The signs of their gradients have been marked.

At both points the slope of the tangent is zero, i.e. $dy/dx = 0$.

Around a maximum point (B) the slope changes from $+$ to $-$ as x increases. The reverse is true around the minimum point G; the slope changes from $-$ to $+$ as x increases.

(c) Around a maximum point the value of dy/dx is decreasing as x increases ($+$ to 0 to $-$).

Thus its rate of change, $\dfrac{d}{dx}\left(\dfrac{dy}{dx}\right)$, is negative around the point. That is, d^2y/dx^2 is negative around a maximum point. Around a minimum point, d^2y/dx^2 is positive.

(d) Such a point as H satisfies the definition for a maximum point given in (a); but although dy/dx changes from $+$ to $-$ as x increases, at H there is no specific value for dy/dx. The function is not differentiable there. A point such as H is called a *cusp*.

In all subsequent discussion in this section it will be assumed $f'(x)$ exists for all relevant values of x.

The comments made in (b) and (c) are now collected to give tests for maximum and minimum values of a function.

Functional notation is used as it is more convenient.

The first essential for maximum *or* minimum is that $f'(x) = 0$ at the point. (Such points are often called stationary points.)

If $x = a$ makes $f'(x) = 0$, that is, $f'(a) = 0$, then there is a choice of tests for deciding whether the point is a maximum, minimum or neither.

(i) If $f(a+h), f(a-h)$ are both $\left(\begin{array}{c}\text{less than}\\ \text{greater than}\end{array}\right) f(a)$, where h is small and positive, then the point is a $\left(\begin{array}{c}\text{maximum}\\ \text{minimum}\end{array}\right)$.

(ii) If $f'(a-h)$ is $\left(\begin{array}{c}+\\ -\end{array}\right)$, $f'(a+h)$ is $\left(\begin{array}{c}-\\ +\end{array}\right)$, the point is a $\left(\begin{array}{c}\text{maximum}\\ \text{minimum}\end{array}\right)$.

(iii) If $f''(a-h)$ *and* $f''(a+h)$ are *both* $\left(\begin{array}{c}-\\ +\end{array}\right)$ the point is a $\left(\begin{array}{c}\text{maximum}\\ \text{minimum}\end{array}\right)$.

Note. With test (iii) the sign of $f''(a)$ itself will often suffice, but if it happens that $f''(a)$ is zero, then the test as given above must be applied. For example, if the student sketches the graph of $y = x^4$ there is an obvious minimum point at $x = 0$, but $f''(0)$ is zero. Any of the detailed tests in (i), (ii) or (iii) will, however, easily show that $x = 0$ *is* a minimum point.

1·82. Points of inflexion

The argument of § 1·81 (c) may be considered as showing that when a curve is concave downwards (often called convex) d^2y/dx^2 is negative, and when concave upwards d^2y/dx^2 is positive.

A point of inflexion is a point at which the curve changes its concavity. Points D and F in fig. 6 are examples. At each of these points the curve changes from concave upwards to concave downwards. Thus $f''(x)$ is changing from + to −. At the points, therefore, its value is zero (assuming $f''(x)$ is continuous).

A test for a point of inflexion is thus:

$f''(x)$ *is zero at the point and changes sign round it.*

K is another point of inflexion in fig. 6. Here the curve changes from concave downwards to concave upwards, and thus $f''(x)$ changes from − to +.

From the tangent lines drawn at points D and F it is seen that at a point of inflexion the tangent actually crosses the curve. At $D, f'(x)$, the slope of the tangent, is positive; at F it is zero. At a point of inflexion $f'(x)$ may have any value.

Points of inflexion are often called points of *contraflexure* because of their frequent occurrence in the shape of bent beams.

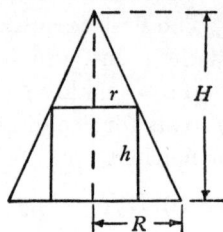

Fig. 7

Examples

(1) Show that the altitude of the cylinder of maximum volume that can be inscribed in a cone of altitude H and base radius R is $\frac{1}{3}H$.

Fig. 7 represents a plane section through the common axis of the cylinder and cone:

$$\text{Volume of cylinder} = V = \pi r^2 h. \tag{1}$$

By similar triangles

$$\frac{r}{H-h} = \frac{R}{H}. \quad \therefore \ r = \frac{R(H-h)}{H}. \tag{2}$$

Substituting for r^2 from eqn. (2) into eqn. (1)

$$V = \pi \frac{R^2}{H^2}(H-h)^2 h,$$

$$\frac{dV}{dh} = \frac{\pi R^2}{H^2}[(H-h)^2 - 2h(H-h)] = \frac{\pi R^2}{H^2}(H-h)(H-3h). \tag{3}$$

When $dV/dh = 0$, $h = \frac{1}{3}H$ or H.

When $h < \frac{1}{3}H$, dV/dh is $+$ $\left.\right\}$. $\therefore$ $\underline{h = \frac{1}{3}H}$ gives max. V.
When $h > \frac{1}{3}H$, dV/dh is $-$

Note. In a practical example of this type, once the values of the variable which make the first derivative zero have been found it is usual to argue from practical considerations which gives maximum or minimum. In the above example $h = H$ would mean the cylinder had zero radius and therefore zero volume; thus $h = \frac{1}{3}H$ gives maximum volume.

(2) The alternating current, i amp., in a circuit at t sec., is given by $i = 4 \sin pt + 1 \cdot 5 \sin 2pt$, where $p = 100\pi$. Find the maximum and minimum values of the current and the smallest positive values of t at which maximum and minimum current occur. [L.U.]

$$i = 4 \sin pt + 1 \cdot 5 \sin 2pt.$$

$$di/dt = 4p \cos pt + 3p \cos 2pt = 4p \cos pt + 3p(2 \cos^2 pt - 1)$$

$$= p(6 \cos^2 pt + 4 \cos pt - 3). \tag{1}$$

When $di/dt = 0$,

$$6 \cos^2 pt + 4 \cos pt - 3 = 0.$$

$$\therefore \cos pt = \frac{-2 \pm \sqrt{(4 + 18)}}{6}.$$

The negative value would make the numerical value of $\cos pt$ larger than unity; it is therefore disregarded.

Taking the positive value, $\cos pt = \dfrac{\sqrt{(22)}}{6} - \dfrac{1}{3} = 0 \cdot 4484$.

$\therefore pt = n \cdot 360° + 63° \, 22'$ or $n \cdot 360° - 63° \, 22'$ (n an integer),

$$pt = 2n\pi + 1 \cdot 106 \text{ or } 2n\pi - 1 \cdot 106 \quad \text{(in radians).} \tag{2}$$

Now $$\frac{d^2i}{dt^2} = -4p^2 \sin pt - 6p^2 \sin 2pt. \tag{3}$$

When $pt = 2n\pi + 1 \cdot 106$, both $\sin pt$ and $\sin 2pt$ are positive.

$$\therefore pt = 2n\pi + 1 \cdot 106 \text{ gives max.} \, i. \tag{4}$$

Thus $i_{\text{max.}} = 4 \sin 63° \, 22' + 1 \cdot 5 \sin 126° \, 44'$,

$$i_{\text{max.}} \simeq 4 \cdot 78 \text{ A.}$$

When $pt = 2n\pi - 1 \cdot 106$, $\sin pt$ and $\sin 2pt$ are both negative.

Thus, from (3), d^2i/dt^2 is $+$ and i is at a minimum.

$$i_{min.} = -4\sin 63° 22' - 1.5\sin 126° 44' \simeq -4.78 \text{ A.}$$

For $i_{max.}$, $pt = 2n\pi + 1.106$ (n an integer).

$$\therefore \; t = \frac{2n\pi}{100\pi} + \frac{1.106}{100\pi} \simeq 0.02n + 0.0035.$$

The lowest positive value of t for max. i is thus 0·0035 sec.

For $i_{min.}$, $pt = 2n\pi - 1.106$ (n an integer).

$$\therefore \; t \simeq 0.02n - 0.0035.$$

The lowest positive value of t for min. i is 0·02 − 0·0035 sec.

$$= 0.0165 \, sec.$$

Although this example is a little longer than most questions set for Higher National Certificate, it is worth careful study, as it shows the necessity for careful working when dealing with trigonometric functions.

EXERCISE 7

1. From a rectangular sheet of tin, the sides of which are of length a, b metres, equal squares are cut off at each corner and a box with an open top is formed. Find the length of the side of the square so that the box formed is of maximum volume.

2. If a beam, length L, is fixed at the ends and loaded uniformly by an amount w per unit length, the deflexion y, at distance x from one end, is given by

$$y = \frac{w}{12EI}\left(\tfrac{1}{2}x^4 - Lx^3 + \tfrac{1}{2}L^2x^2\right),$$

where E, I are constants. Find (i) the maximum deflexion of the beam, (ii) the distance of any points of contraflexure from the end.

3. Show that when current is taken from a battery or generator of constant e.m.f., the work done on the load (resistive) is greatest when the external resistance is equal to the internal resistance.

4. A figure consists of a semicircle with a rectangle constructed on its diameter. Given that the perimeter of the figure is 6 m, find its dimensions for maximum surface area.

5. Find the stationary points of $y = \dfrac{(4-x)^3}{(2-x)}$ and discriminate between them.

6. Prove that the function $2e^{0.5x} - 3x$ is a minimum when $x \simeq 2 \cdot 2$.

7. From the corner of the floor of a room 7 m long, 5 m broad, 4 m. high, an electric cable is to be led along a side wall and then an end wall to the opposite corner of the ceiling. To what height from the floor in the intervening angle of the room should the cable be led in order that it may be of minimum length?

8. The current sent through a resistance R by a battery consisting of a fixed number n of cells, each of voltage E and internal resistance r, arranged with x cells in series and n/x rows in parallel, is $\dfrac{nxE}{rx^2 + nR}$ amp. How many cells must be in series in order to give maximum current?

9. Find the stationary points of $\sin^3 x \cos x$ and discriminate between them.

10. The speed of signals along a certain cable is proportional to $x^2 \log_e 1/x$, where x is a positive ratio less than unity. Find the value of x for the greatest speed.

11. The power, P, of an engine can be expressed by $P = AnL^3 - Bn^3L^5$, where A and B are constants, $n = $ r.p.m. and $L = $ stroke length. Find (i) the value of n for max. H of engines of constant size, (ii) the value of L for max. P of engines of constant speed.

12. Prove that $\dfrac{3x^2 - 1}{x - \frac{2}{3}}$ has a minimum value of 6 and a maximum value of 2. Explain the apparent paradox by sketching a rough graph.

Answers

1. $\frac{1}{6}\{a + b - \sqrt{(a^2 - ab + b^2)}\}$ m.   **2.** (i) $\dfrac{wL^4}{384EI}$; (ii) $\frac{1}{6}L(3 \pm \sqrt{3})$.

4. Radius $\dfrac{6}{(4 + \pi)}$ m $=$ other side of rectangle.

5. Min. point $(1, 27)$; point of inflexion $(4, 0)$.

7. $2\frac{1}{3}$ m.   **8.** $x = \sqrt{\dfrac{nR}{r}}$.

9. Maximum value $\dfrac{3\sqrt{3}}{16}$ when $x = n\pi + \frac{1}{3}\pi$; minimum value

$-\dfrac{3\sqrt{3}}{16}$ when $x = n\pi - \frac{1}{3}\pi$; points of inflexion $(n\pi, 0)$.

10. $1/\sqrt{e}$ $(e = 2\cdot718)$.

11. (i) $n = \dfrac{1}{L}\sqrt{\dfrac{A}{3B}}$;   (ii) $L = \dfrac{1}{n}\sqrt{\dfrac{3A}{5B}}$.

1·9. The mean-value theorem of differentiation

Fig. 8 shows two neighbouring points A, B on the graph of $y = f(x)$ which is continuous and differentiable between A and B. As it is differentiable the curve of the function must possess a

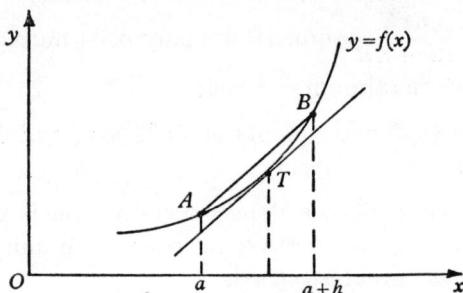

Fig. 8

tangent at every point between A and B. It appears evident from the figure that there is a point T, *between A and B*, at which the tangent is parallel to the chord AB. (A strict proof of this is beyond the scope of this book.)

Let the x values at A and B be $x = a$ and $x = (a+h)$. Then the slope of chord AB is $\dfrac{f(a+h) - f(a)}{h}$.

As T is between A and B its x value must be of the form $x = a + \theta h$, where $0 < \theta < 1$; that is, between a and $(a+h)$. The gradient of the tangent at T is therefore $f'(a + \theta h)$. Thus

$$\frac{f(a+h) - f(a)}{h} = f'(a + \theta h).$$

$$\therefore \ f(a+h) - f(a) = hf'(a + \theta h). \tag{1}$$

This result is called the (first) mean-value theorem for differentiation.

If A and B are very close and h consequently small, the approximation $f(a+h)-f(a) \simeq hf'(a)$ is obtained, i.e.

$$f(a+h) \simeq f(a) + hf'(a). \tag{2}$$

The mean-value theorem is important in the theory of the differential calculus. It will be used in a later chapter to obtain an important result in partial differentiation.

Equation (2) can be used for finding approximately the new value of a function when the variable is increased by a small amount.

EXAMPLE

Find an approximate value for $\cos 30° 6'$.

Here,
$$f(a) = \cos 30° = \tfrac{1}{2}\sqrt{3},$$
$$f'(a) = -\sin 30° = -\tfrac{1}{2},$$
$$h = 6' = \frac{\pi}{1800} \text{ radian.}$$

Using (2)
$$\cos 30° 6' \simeq \frac{\sqrt{3}}{2} - \frac{1}{2}\frac{\pi}{1800}$$
$$\simeq 0\cdot 8651.$$

EXERCISE 8

Without use of trigonometric tables calculate approximate values of the expressions:

1. $\sin 30° 6'$.　　　　2. $\cos 60° 10'$.　　　　3. $\tan 45° 8'$.

ANSWERS

1. $0\cdot 5015$.　　　　2. $0\cdot 4975$.　　　　3. $1\cdot 0047$.

1·10. Newton's approximation to the root of an equation

Let $x=a$ be an approximate solution of an equation $f(x)=0$.

Let $x=a+h$ be the accurate solution, i.e. $f(a+h)=0$.

If $x=a$ is a good approximation, h will be small.

Thus, from equation (2) in §1·9,

$$f(a+h) \simeq f(a) + hf'(a), \tag{1}$$

i.e.
$$0 \simeq f(a) + hf'(a),$$
$$h \simeq -\frac{f(a)}{f'(a)}.$$

If $x = a$ is a reasonably good approximate root of an equation $f(x) = 0$, then $x = a - \dfrac{f(a)}{f'(a)}$ is a closer one.

Having found this closer approximation, the process can be repeated indefinitely to give a solution correct to any required degree of accuracy.

Note that if $x = a$ is a bad approximation, then h will not be small and equation (1) is not at all accurate. Using the above process might then easily give a worse solution than the first approximation.

In practice the first approximation is usually found either from a table of values or from a graph of the function. In drawing up a table of values it should be noted that if, for example, $f(3)$ and $f(2)$ are of opposite signs, there is a solution to $f(x) = 0$ between $x = 2$ and $x = 3$. In this way the initial labour can be cut down.

EXAMPLE

Make a table of values for $y = 3x^3 - 5x^2 - 19x + 31$ between $x = 0$ and $x = 3$. From it deduce an approximate root of the equation $3x^3 - 5x^2 - 19x + 31 = 0$. Obtain the root correct to 4 significant figures.

x	0	1	2	3	1·7	1·6
y	+31	+10	−3	+10	−1·01	+0·088

$x = 1·6$ is taken as a first approximation.

$$f(x) = 3x^3 - 5x^2 - 19x + 31,$$
$$f'(x) = 9x^2 - 10x - 19.$$

For a second approximation:

$$f(1·6) = +0·088,$$
$$f'(1·6) = 9 \cdot (1·6)^2 - 10 \cdot (1·6) - 19 = -11·96,$$
$$\frac{f(1·6)}{f'(1·6)} = -\frac{0·088}{11·96} \simeq -0·007.$$

A closer root is thus
$$1·6 - (-0·007) = \underline{1·607}.$$

For a third approximation:

$$f(1·607) = +0·0022$$
$$f'(1·607) = -11·92$$
$$\frac{f(1·607)}{f'(1·607)} = -\frac{0·0022}{11·92} \simeq \underline{-0·0002}.$$

A closer root is thus $1 \cdot 607 - (-0 \cdot 0002) = 1 \cdot 607 + 0 \cdot 0002$.

As the correction, $0 \cdot 0002$, has its first non-zero figure in the fourth decimal place, then $x = 1 \cdot 607$ is correct to 4 significant figures.

Correct to four significant figures the root is $1 \cdot 607$.

The student should note that the working of the third approximation was necessary to ensure the validity of the figure 7 in the third decimal place.

It was essential to continue the process until the correction was less than $0 \cdot 0005$ in magnitude in this case.

Note that the table also shows another root, between $x = 2$ and $x = 3$.

EXERCISE 9

1. Sketch the curve $y = 3 \cdot 5x^2 - 10 \log_{10} x - 17 \cdot 5$ between $x = 1$ and $x = 4$, including the value $x = 2 \cdot 5$. From your figure deduce an approximate root of the equation $3 \cdot 5x^2 - 10 \log_{10} x - 17 \cdot 5 = 0$. Obtain the root correct to 4 significant figures.

2. Find a root of $x^3 + 5 \cdot 1x - 11 = 0$ correct to 4 significant figures.

3. Calculate the smaller positive root of the equation $\frac{1}{6}x^3 - 8x + 6 = 0$ correct to 2 decimal places.

4. Find the root of $x^3 + 3x^2 - 9x - 5 = 0$ which lies between 2 and 3 correct to 3 decimal places.

5. When a sphere, radius a, specific gravity $0 \cdot 75$, floats in water, the distance x of the centre below the surface is given by the real root of the equation $x^3 - 3a^2x + a^3 = 0$. Find x/a correct to 3 decimal places. [L.U.]

6. Show, graphically, that the equation $x = e^{-x}$ has only one root. Find it correct to 3 decimal places.

ANSWERS

1. $2 \cdot 475$. **2.** $1 \cdot 498$. **3.** $0 \cdot 76$.

4. $2 \cdot 180$. **5.** $0 \cdot 347$. **6.** $0 \cdot 567$.

Miscellaneous Exercises on Chapter 1

Note. Any examples on bookwork not covered in this chapter only entail knowledge of Ordinary National Certificate level.

Differentiate the functions in Exx. 1–18 with respect to x:

1. $\dfrac{1}{x^2}(2x^{3 \cdot 6} + 5x^{1 \cdot 4} + 3)$.

2. $x\sqrt{(4 - x^2)}$.

3. $\dfrac{\sin x}{\cos x + \sin x}$.

4. $x^2 \sin 3x$.

5. $(1 + e^{-x})^2$.

6. $x\sqrt{(x^2 + 1)}$.

7. $\sin 2x - \tfrac{1}{3}\sin^3 2x$.

8. $x^2 e^{-2x}$.

9. $\sqrt{(\cos 2x)}$.

10. $\dfrac{x^2 - 2x}{(x + 2)^2}$.

11. $x/\tan x$.

12. $(1 - x^2)^{\frac{1}{2}}\cos(1 - 2x^2)$.

13. $e^{-2x}\log_e(1 - 2x)$.

14. $\sin(\log_e x)$.

15. $x^x \log_e x$. **16.** $\log_e(1 + x + x^2) - \dfrac{x(2 + 3x)}{2(1 + x + x^2)}$. (L.U.)

17. $\log_e e^x \sqrt{\dfrac{x - 1}{x + 1}}$. **18.** $(2x + 3)^4 \sec^3(x - 2)\, e^{3x+4}$. (L.U.)

19. If $y = 4x\, e^{-0 \cdot 5x}$, solve (i) $dy/dx = 0$, (ii) $d^2y/dx^2 = 0$.

20. Given $3\sin\theta = 8\sin\phi$ and $x = 3\cos\theta + 8\cos\phi$, and θ increases at 4π radians/sec., find values for $d\phi/dt$ and dx/dt when $\theta = \tfrac{1}{3}\pi$. (Sec. A)

21. If $y = A\, e^{2x} + B\, e^{-2x}$, find A and B given $y = 0$ and $dy/dx = 6$ when $x = 0$. [Sec. A]

22. Find the maximum and minimum values of $\dfrac{x}{1 + x + x^2}$.

23. If $(1 + x)\, y = e^x$, prove $(1 + x)\dfrac{dy}{dx} = xy$. [Sec. A]

24. State the amplitude and period of $2 \cdot 5 \sin 10\pi t$.

If $x = 2 \cdot 5 \sin 10\pi t$, find dx/dt and d^2x/dt^2 and express each in terms of x. [Sec. A]

25. If $x = 4\tan\theta$ and $z^2 = 16 + x^2$, and x increases at 2 units/sec., find the rates of change of θ and z when $x = 3$ units and z is positive. [Sec. A]

26. Find the dimensions of the cylinder of maximum volume which can be inscribed in a sphere of radius R.

27. If $y = e^{x+3x^2}$, prove that $dy/dx = (1 + 6x)\, y$. [L.U.]

28. If $3x^2 - 7xy + 9y^2 = 2$, show that all points at which dy/dx is zero lie on the line $6x - 7y = 0$.

29. The total perimeter of a sector of a circle is given. Find the angle of the sector when its area is greatest.

30. If $y = xv$, where $v = f(x)$, express d^2y/dx^2 in terms of x and derivatives of v. [L.U.]

31. $x^3 - 3x^2 + 3{\cdot}9 = 0$ has two real roots close to 2. Find the larger one correct to 2 decimal places. [L.U.]

32. Find the positive root of $1 - 4x = e^{-10x}$ correct to 3 decimal places. [L.U.]

33. Find, correct to 3 significant figures, the smallest positive root of the equation $\tan\theta = 2\theta$ (θ in radians). [L.U.]

34. If $y = a\, e^{-2x} \sin 3x$, prove $\dfrac{d^2y}{dx^2} + 4\dfrac{dy}{dx} + 13y = 0$.

35. If $e^y = a \cos px$, prove $\dfrac{d^2y}{dx^2} + \left(\dfrac{dy}{dx}\right)^2 + p^2 = 0$.

36. Find the equation of the tangent at any point on the curve $x = 3at^2$, $y = at^3$.

37. Find dy/dx and d^2y/dx^2 in terms of t, when
$$x = a(\cos t + t \sin t), \quad y = a(\sin t - t \cos t).$$

38. Calculate $\sin 61°$ correct to 3 decimal places, without use of trigonometric tables.

39. If a small object is placed distant u in front of a convex lens of focal length f, the image is distant v behind, where $1/v + 1/u = 1/f$. Put $(u + v)$ in terms of v and f. Hence show the minimum distance between image and object is $4f$.

40. The curve $y = ax^3 + bx^2 + cx + d$ passes through the points $(1, 3)$ and $(0, 1)$ and has a tangent parallel to the x-axis at the point $(2, 1)$. Prove the maximum and minimum values of y occur at $x = \frac{2}{3}$, $x = 2$ respectively. Sketch the curve. Are there any points of inflexion? If so, find them. [L.U.]

Answers

1. $3 \cdot 2x^{0 \cdot 6} - 3x^{-1 \cdot 6} - 6x^{-3}$.

2. $\dfrac{2(2-x^2)}{\sqrt{(4-x^2)}}$.

3. $\dfrac{1}{(\cos x + \sin x)^2}$.

4. $3x^2 \cos 3x + 2x \sin 3x$.

5. $-2e^{-x}(1+e^{-x})$.

6. $\dfrac{2x^2+1}{\sqrt{(x^2+1)}}$.

7. $2\cos^3 2x$.

8. $2x(1-x)e^{-2x}$.

9. $-\dfrac{\sin 2x}{\sqrt{(\cos 2x)}}$.

10. $\dfrac{2(3x-2)}{(x+2)^3}$.

11. $\cot x - x \operatorname{cosec}^2 x$.

12. $4x\sqrt{(1-x^2)}\sin(1-2x^2) - \dfrac{x}{\sqrt{(1-x^2)}}\cos(1-2x^2)$.

13. $-2e^{-2x}\left[\log(1-2x) + \dfrac{1}{(1-2x)}\right]$.

14. $\dfrac{1}{x}\cos(\log_e x)$.

15. $x^x \log_e x \left(1 + \log_e x + \dfrac{1}{x\log_e x}\right)$.

16. $\dfrac{x^2(4x+5)}{2(1+x+x^2)^{\frac{3}{2}}}$.

17. $\dfrac{x^2}{(x^2-1)}$.

18. $(2x+3)^4 \sec^3(x-2) e^{3x+4}\left[\dfrac{8}{(2x+3)} + 3\tan(x-2) + 3\right]$.

19. (i) $x=2$; (ii) $x=4$.

20. $\dfrac{d\phi}{dt} = \dfrac{12\pi}{\sqrt{(229)}} \simeq 0 \cdot 793\pi$;

$\dfrac{dx}{dt} = -6\sqrt{3}.\pi\left(1 + \dfrac{3}{\sqrt{(229)}}\right) \simeq -12 \cdot 45\pi$.

21. $A = \frac{3}{2}$, $B = -\frac{3}{2}$.

22. Maximum $\frac{1}{3}$, minimum -1.

24. Amp. $2 \cdot 5$; period $\frac{1}{5}$; $dx/dt = \pm 10\pi\sqrt{(6 \cdot 25 - x^2)}$; $d^2x/dt^2 = -100\pi^2 x$.

25. $\dfrac{d\theta}{dt} = \dfrac{8}{25}$; $\dfrac{dz}{dt} = \dfrac{6}{5}$.

26. $r = \sqrt{\frac{2}{3}}.R$; $h = \dfrac{2}{\sqrt{3}}R$.

29. $\theta = 2$ radians.

30. $2\dfrac{dv}{dx} + x\dfrac{d^2v}{dx^2}$.

31. 2·18.   **32.** 0·223.   **33.** 1·17.

36. $2y - tx + at^3 = 0$.

37. $\dfrac{dy}{dx} = \tan t;\ \dfrac{d^2y}{dx^2} = \dfrac{1}{at}\sec^3 t$.

38. 0·875.   **39.** $(u + v) = \dfrac{v^2}{v - f}$.

40. Curve is $y = 2x^3 - 8x^2 + 8x + 1$. Point of inflexion at $(\tfrac{4}{3}, \tfrac{59}{27})$.

Note that as the curve passes through points $(1, 3)$, $(0, 1)$ and $(2, 1)$ and $dy/dx = 0$ when $x = 2$, there are four equations to find the four unknowns a, b, c, d.

CHAPTER 2

REVISION OF INTEGRATION
AND APPLICATIONS

Standard Integrals and Methods

$$\int x^n \, dx = \frac{x^{n+1}}{n+1} + c \text{ unless } n = -1.$$

$$\int \frac{dx}{x} = \log_e x + c. \quad \int e^x \, dx = e^x + c.$$

$$\int \sin x \, dx = -\cos x + c. \quad \int \cos x \, dx = \sin x + c.$$

If $\int f(x) \, dx = F(x) + c$ then $\int f(ax+b) \, dx = \frac{1}{a} F(ax+b) + c.$

Also $\int_{x=a}^{x=b} f(x) \, dx = [F(x)]_a^b = F(b) - F(a).$

The area between a curve $y = f(x)$, the x-axis and the ordinates at $x = a$ and $x = b$ is $\int_a^b y \, dx$.

The volume of the solid of revolution formed by revolving the above area about the x-axis is

$$\int_a^b \pi y^2 \, dx.$$

The mean value of $f(x)$ from $x = a$ to $x = b$ is

$$\frac{1}{b-a} \int_a^b f(x) \, dx.$$

The root-mean-square value of a function is the square root of the mean value of the square of the function.

2·1. Standard integration

The integrals and applications given above are covered in most Ordinary National Courses. To these are added a few others which may not be familiar to some students.

(*a*) From § 1·11

$$\frac{d}{dx} (\tan x) = \sec^2 x \quad \text{and} \quad \frac{d}{dx} (\cot x) = -\operatorname{cosec}^2 x.$$

Thus $\quad \int \sec^2 x \, dx = \tan x + c, \quad \int \operatorname{cosec}^2 x \, dx = -\cot x + c.$

Reversing two more differentiations proved in § 1·11 gives

$$\int \sec x \tan x \, dx = \sec x + c,$$

$$\int \operatorname{cosec} x \cot x \, dx = -\operatorname{cosec} x + c.$$

(b) The method for integrating the squares of sines or cosines makes use of the double-angle formulae of trigonometry, viz.

$$\cos^2 A = \tfrac{1}{2}(1 + \cos 2A) \quad \text{and} \quad \sin^2 A = \tfrac{1}{2}(1 - \cos 2A),$$

e.g. $\qquad \int 3 \cos^2 2x \, dx = \int \tfrac{3}{2}(1 + \cos 4x) \, dx$

$$= \tfrac{3}{2}x + \tfrac{3}{8} \sin 4x + c.$$

This type of integral occurs frequently when finding root-mean-square values in electrical work.

EXAMPLES

(1) Evaluate: (i) $\int (3x - 2)^3 \, dx$; (ii) $\int_1^2 \dfrac{dx}{2x + 1}$;

(iii) $\int \tan^2 x \, dx$; (iv) $\int_0^{\frac{1}{3}\pi} \sin^2 \theta \, d\theta.$

(i) $\int (3x - 2)^3 \, dx = \tfrac{1}{3} \cdot \tfrac{1}{4}(3x - 2)^4 + c$

$$= \tfrac{1}{12}(3x - 2)^4 + c.$$

(ii) $\int_1^2 \dfrac{dx}{2x + 1} = [\tfrac{1}{2} \log_e (2x + 1)]_1^2$

$$= \tfrac{1}{2} \log_e 5 - \tfrac{1}{2} \log_e 3 = \tfrac{1}{2} \log_e \tfrac{5}{3}.$$

(iii) $\int \tan^2 x \, dx = \int (\sec^2 x - 1) \, dx = \tan x - x + c.$

(iv) $\int_0^{\frac{1}{3}\pi} \sin^2 \theta \, d\theta = \dfrac{1}{2} \int_0^{\frac{1}{3}\pi} (1 - \cos 2\theta) \, d\theta$

$$= \tfrac{1}{2}[\theta - \tfrac{1}{2} \sin 2\theta]_0^{\frac{1}{3}\pi}$$

$$= \dfrac{1}{2}\left[\left(\dfrac{\pi}{3} - \dfrac{\sqrt{3}}{4}\right) - (0 - 0)\right] = \dfrac{1}{2}\left(\dfrac{\pi}{3} - \dfrac{\sqrt{3}}{4}\right).$$

(2) Find the area between the curve $y = x(x - 1)(x - 3)$ and the x-axis from $x = 0$ to $x = 2$.

$$y = x(x - 1)(x - 3) = x^3 - 4x^2 + 3x.$$

Note the expression for y is a cubic expression in x, and will have the typical cubic shape.

Also $y = 0$ when $x = 0$ or 1 or 3; when $x = \frac{1}{2}$, y is positive, when $x = 2$, y is negative.

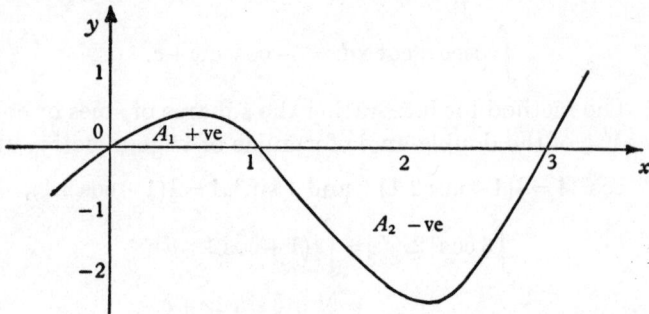

Fig. 9

Calculating the area in two parts since areas above the x-axis are positive, and below the x-axis are negative (fig. 9), we have

$$A_1 = \int_0^1 (x^3 - 4x^2 + 3x)\,dx = [\tfrac{1}{4}x^4 - \tfrac{4}{3}x^3 + \tfrac{3}{2}x^2]_0^1$$

$$= (\tfrac{1}{4} - \tfrac{4}{3} + \tfrac{3}{2}) - 0 = \tfrac{5}{12},$$

$$A_2 = \int_1^2 (x^3 - 4x^2 + 3x)\,dx = [\tfrac{1}{4}x^4 - \tfrac{4}{3}x^3 + \tfrac{3}{2}x^2]_1^2$$

$$= (4 - \tfrac{32}{3} + 6) - \tfrac{5}{12} = -\tfrac{13}{12}.$$

$\therefore$ Area required is $\frac{13}{12} + \frac{5}{12} = 1\frac{1}{2}$ sq. units.

(3) Using integration methods, prove that the volume of a sphere, radius r, is $\frac{4}{3}\pi r^3$.

The sphere is formed by revolving a semicircle of radius r (fig. 10) about a diameter. Thus its volume is $\int \pi y^2\,dx$.

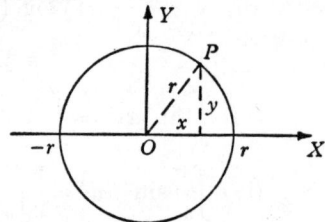

Fig. 10

The limits of integration are from $x = -r$ to $x = +r$.

$$\therefore V = \int_{-r}^{r} \pi y^2\,dx.$$

But $x^2 + y^2 = r^2$ for any point on the circle.

$$\therefore y^2 = r^2 - x^2.$$

Thus
$$V = \int_{-r}^{r} \pi(r^2 - x^2)\, dx$$
$$= \pi[r^2 x - \tfrac{1}{3}x^3]_{-r}^{r}$$
$$= \pi[(r^3 - \tfrac{1}{3}r^3) - (-r^3 + \tfrac{1}{3}r^3)]$$
$$= \tfrac{4}{3}\pi r^3.$$

EXERCISE 10

Evaluate the integrals nos. 1–21:

1. $\displaystyle\int x\sqrt{x}\, dx.$ **2.** $\displaystyle\int \frac{dx}{x^4}.$ **3.** $\displaystyle\int \left(\sqrt[3]{x} - \frac{1}{\sqrt[3]{x}}\right) dx.$

4. $\displaystyle\int \frac{dx}{x+2}.$ **5.** $\displaystyle\int (3\sin\theta + 4\cos\theta)\, d\theta.$

6. $\displaystyle\int e^{-2x}\, dx.$ **7.** $\displaystyle\int \frac{dx}{1-x}.$ **8.** $\displaystyle\int \frac{dx}{\sqrt{(2x-3)}}.$

9. $\displaystyle\int \sec^2(\theta + \tfrac{1}{4}\pi)\, d\theta.$ **10.** $\displaystyle\int 2\cos(\omega t - \alpha)\, dt.$

11. $\displaystyle\int (\cos^2\theta - \sin^2\theta)\, d\theta.$ **12.** $\displaystyle\int_1^2 \frac{x^3 - 1}{x^2}\, dx.$

13. $\displaystyle\int_0^1 (\sqrt{x} - 1)^2\, dx.$ **14.** $\displaystyle\int_0^{\frac{1}{12}\pi} \cos^2 x\, dx.$

15. $\displaystyle\int_1^2 \frac{dx}{e^x}.$ **16.** $\displaystyle\int_1^2 \frac{dx}{3x-1}.$ **17.** $\displaystyle\int_0^{\frac{1}{4}\pi} 3\sin^2 2x\, dx.$

18. $\displaystyle\int \sec 2x \tan 2x\, dx.$ **19.** $\displaystyle\int \frac{e^x - 1}{e^{2x}}\, dx.$

20. $\displaystyle\int_0^{\frac{1}{6}\pi} (\cos 3\theta + 2\sin 3\theta)\, d\theta.$

21. $\displaystyle\int_0^{2\pi/\omega} (3\cos^2\omega t - 2\sin\omega t)\, dt.$

22. Find the area cut from the curve $y = 4x - x^2$ by the x-axis. Find also the volume formed when this area is rotated about the x-axis.

23. Sketch the curve $y^2 = 1 - \cos 2x$ from $x = 0$ to $x = \pi$. Find the volume formed when this part of the curve is rotated about the x-axis.

24. If $d^2y/dx^2 = -10/x^2$, find y in terms of x, given that $dy/dx = 3$ when $x = 5$, and $y = 1$ when $x = 1$. ⠀⠀⠀⠀⠀[Sec. A]

25. Find the ordinate which bisects the area between $y = 2\sqrt{x}$, the x-axis and $x = 4$. ⠀⠀⠀⠀⠀[Sec. A]

26. Find the volume formed by the revolution about the x-axis of the area between the curve $y = 1 + 2/x$, the x-axis and the lines $x = 1$, $x = 4$. ⠀⠀⠀⠀⠀[Sec. A]

27. If $EI\dfrac{d^2y}{dx^2} = -\tfrac{1}{2}wx$ and $y = 0$ when $x = 0$, $\dfrac{dy}{dx} = 0$ when $x = \tfrac{1}{2}l$, find y in terms of x and also the maximum value of y.

28. If $L\,di/dt = 30\sin 10\pi t$, find a formula for i in terms of t, given $L = 2$ and $i = 0$ when $t = 0$.

29. Find the mean value of $i = 1 + \cos \omega t$ over a period.

30. Find the root-mean-square value of the function in Ex. 29.

ANSWERS

Arbitrary constants have been omitted.

1. $\tfrac{2}{5}x^{\frac{5}{2}}$. ⠀⠀⠀⠀⠀2. $\dfrac{-1}{3x^3}$. ⠀⠀⠀⠀⠀3. $\tfrac{3}{4}x^{\frac{4}{3}} - \tfrac{3}{2}x^{\frac{2}{3}}$.

4. $\log_e(x+2)$. ⠀⠀⠀⠀⠀5. $4\sin\theta - 3\cos\theta$. ⠀⠀6. $-\tfrac{1}{2}e^{-2x}$.

7. $\log_e\dfrac{1}{(1-x)}$. ⠀⠀⠀⠀⠀8. $\sqrt{(2x-3)}$. ⠀⠀⠀⠀⠀9. $\tan(\theta + \tfrac{1}{4}\pi)$.

10. $\dfrac{2}{\omega}\sin(\omega t - \alpha)$. ⠀⠀11. $\tfrac{1}{2}\sin 2\theta$. ⠀⠀⠀⠀⠀12. 1.

13. $\tfrac{1}{6}$. ⠀⠀⠀⠀⠀⠀⠀⠀⠀⠀⠀14. $\tfrac{1}{8}(\tfrac{1}{3}\pi + 1)$.

15. $1/e - 1/e^2 \simeq 0\cdot 233$. ⠀⠀16. $\tfrac{1}{3}\log_e 2\cdot5 \simeq 0\cdot3054$.

17. $\tfrac{3}{4}\pi$. ⠀⠀⠀⠀⠀18. $\tfrac{1}{2}\sec 2x$. ⠀⠀⠀⠀⠀19. $\tfrac{1}{2}e^{-2x} - e^{-x}$.

20. 1. ⠀⠀⠀⠀⠀⠀21. $3\pi/\omega$. ⠀⠀⠀⠀⠀22. $\tfrac{32}{3}$; $\tfrac{512}{15}\pi$.

23. π^2. ⠀⠀⠀⠀⠀24. $y = 10\log_e x + x$.

25. $x = 2\sqrt[3]{2}$. ⠀⠀⠀26. $\pi(6 + 8\log_e 2)$.

27. $EIy = -\dfrac{wx^3}{12} + \dfrac{wl^2x}{16}$; $\dfrac{wl^3}{48EI}$.

28. $i = \dfrac{3}{2\pi}(1 - \cos 10\pi t) = \dfrac{3}{\pi}\sin^2 5\pi t$.

29. 1. ⠀⠀⠀⠀⠀30. $\sqrt{\tfrac{3}{2}}$.

CHAPTER 3

HYPERBOLIC FUNCTIONS

These are new types of functions which frequently occur in problems in both mechanics and electricity.

3·1. Definitions

The hyperbolic sine of x, written $\sinh x$, is defined as $\frac{1}{2}(e^x - e^{-x})$.
The hyperbolic cosine of x, written $\cosh x$, is defined as
$$\tfrac{1}{2}(e^x + e^{-x}).$$
The hyperbolic tangent of x, $\tanh x$, is then defined as
$$\frac{\sinh x}{\cosh x} = \frac{e^x - e^{-x}}{e^x + e^{-x}} = \frac{e^{2x} - 1}{e^{2x} + 1}.$$

The names are usually pronounced as 'shine', 'cosh' and 'tansh' or 'than'.

These functions behave in many ways similarly to the trigonometric functions, hence the occurrence of the words sine, cosine, etc., in their names. They are also intimately connected with the standard equation of a hyperbola; hence the 'hyperbolic'.

Hyperbolic secants, cosecants and cotangents are defined as:

$$\operatorname{sech} x = \frac{1}{\cosh x} = \frac{2}{e^x + e^{-x}},$$

$$\operatorname{cosech} x = \frac{1}{\sinh x} = \frac{2}{e^x - e^{-x}},$$

$$\coth x = \frac{1}{\tanh x} = \frac{e^x + e^{-x}}{e^x - e^{-x}} = \frac{e^{2x} + 1}{e^{2x} - 1}.$$

3·2. Graphs of hyperbolic functions

It will benefit the reader if he himself draws the graphs. Tables of $\sinh x$, $\cosh x$ and $\tanh x$ are to be found in most mathematical tables.

Fig. 11 (a) shows sketches of e^x, e^{-x}, $\sinh x$ and $\cosh x$. Fig. 11 (b) shows a sketch of $\tanh x$.

The main points to note are:

(i) None of the hyperbolic functions is periodic.

(ii) $\cosh x$ is never less than 1, and is symmetrical about the y-axis. It is actually the curve taken up by a freely hanging chain.

(iii) $\sinh x$ passes through the origin and can have all values from $-\infty$ to $+\infty$.

(iv) $\tanh x$ lies completely between the values ± 1, and approaches ± 1 as x approaches $\pm \infty$.

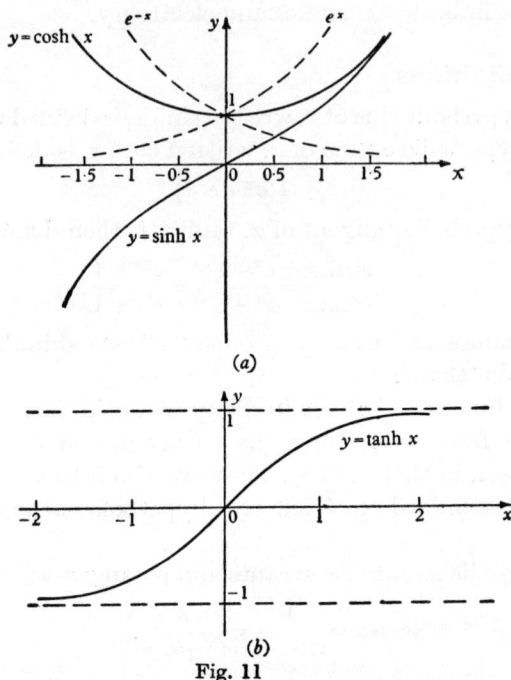

(a)

(b)

Fig. 11

3·3. Standard relationships between hyperbolic functions

To every relationship between the trigonometric functions there corresponds one for the hyperbolic functions.

Thus to $\sin^2\theta + \cos^2\theta = 1$ there corresponds

$$\cosh^2 x - \sinh^2 x = 1.$$

Although the formulae are very similar great care must be taken to avoid mistakes. Many 'hyperbolic' formulae are the same as those for 'trigonometric' formulae; but not all.

Shortly, a rule will be given from which 'hyperbolic' formulae may be derived from the 'trigonometric' ones. This is convenient for memory work, but the reader should always be prepared to prove them.

Several standard formulae are proved below:

(i) $\cosh^2 x - \sinh^2 x = 1$

$$\cosh^2 x - \sinh^2 x = (\cosh x + \sinh x)(\cosh x - \sinh x)$$
$$= e^x e^{-x} = \underline{1}.$$

(See the definitions of $\cosh x$ and $\sinh x$.)

(ii) $\cosh^2 x + \sinh^2 x = \cosh 2x$

$$\cosh^2 x + \sinh^2 x = \frac{(e^x + e^{-x})^2 + (e^x - e^{-x})^2}{4}$$
$$= \frac{(e^{2x} + 2 + e^{-2x}) + (e^{2x} - 2 + e^{-2x})}{4}$$
$$= \frac{2(e^{2x} + e^{-2x})}{4} = \frac{e^{2x} + e^{-2x}}{2}$$
$$= \underline{\cosh 2x}.$$

(iii) From (i) and (ii):

$$\cosh 2x = \cosh^2 x + \sinh^2 x = \cosh^2 x + (\cosh^2 x - 1) = \underline{2\cosh^2 x - 1}$$
$$= (1 + \sinh^2 x) + \sinh^2 x = \underline{1 + 2\sinh^2 x}.$$

(iv) $\sinh 2x = 2 \sinh x \cosh x$

$$2 \sinh x \cosh x = 2 \frac{(e^x - e^{-x})(e^x + e^{-x})}{4} = \frac{e^{2x} - e^{-2x}}{2}$$
$$= \underline{\sinh 2x}.$$

(v) $\sinh(x + y) = \sinh x \cosh y + \cosh x \sinh y$

$\sinh x \cosh y + \cosh x \sinh y$
$$= \frac{(e^x - e^{-x})(e^y + e^{-y}) + (e^x + e^{-x})(e^y - e^{-y})}{4}$$
$$= \frac{2\{e^{x+y} - e^{-(x+y)}\}}{4}$$
$$= \underline{\sinh(x + y)}.$$

Rule. To transform a relationship between trigonometric functions into the corresponding hyperbolic forms, change the sign ($+$ to $-$, or $-$ to $+$) of any term in the trigonometric formula which contains the product of two sines, actual or implied.

As an example of an 'implied' product of two sines, note that $\tan^2 x = \sin^2 x / \cos^2 x$, and therefore really contains $\sin x \times \sin x$.

This rule will be proved in the work done on complex numbers in vol. II. Examples are:

$$\cos(A+B) = \cos A \cos B - \sin A \sin B,$$

$$\cosh(x+y) = \cosh x \cosh y + \sinh x \sinh y.$$

$$\sin(A-B) = \sin A \cos B - \cos A \sin B,$$

$$\sinh(x-y) = \sinh x \cosh y - \cosh x \sinh y.$$

$$\tan(A+B) = \frac{\tan A + \tan B}{1 - \tan A \tan B},$$

$$\tanh(x+y) = \frac{\tanh x + \tanh y}{1 + \tanh x \tanh y}.$$

Exercise 11

1. Prove that $\dfrac{1+\tanh x}{1-\tanh x} = e^{2x}$.

2. Prove, from definitions, that
$$\cosh(x-y) = \cosh x \cosh y - \sinh x \sinh y.$$

3. Simplify $\qquad \cosh x - 2\sinh^2 \tfrac{1}{2}x$.

4. Express $\cosh^2 \tfrac{1}{2}x + 2\sinh^2 \tfrac{1}{2}x$ in terms of $\cosh x$.

5. Prove, from definitions, that
$$\sinh x + \sinh y = 2\sinh \tfrac{1}{2}(x+y)\cosh \tfrac{1}{2}(x-y).$$

Answers

3. 1.   4. $\tfrac{1}{2}(3\cosh x - 1)$.

3·4. Differentiation and integration of hyperbolic functions

There is *no* rule for finding the derivatives and integrals of hyperbolic functions from those of the corresponding trigonometric functions. Standard results must be proved and remembered.

3·41. Differentiation

$$D_x(\sinh x) = \frac{d}{dx}\frac{(e^x - e^{-x})}{2} = \frac{e^x + e^{-x}}{2} = \underline{\cosh x}.$$

$$D_x(\cosh x) = \frac{d}{dx}\frac{(e^x + e^{-x})}{2} = \frac{e^x - e^{-x}}{2} = \underline{\sinh x}.$$

$$D_x(\tanh x) = \frac{d}{dx}\left(\frac{\sinh x}{\cosh x}\right) = \frac{\cosh^2 x - \sinh^2 x}{\cosh^2 x} = \frac{1}{\cosh^2 x} = \underline{\operatorname{sech}^2 x}.$$

More complicated functions can be differentiated using these standard results, together with standard methods: function of a function rule, product and quotient rules, etc.

3·42. Integration

From 3·41 above it immediately follows:

$$\int \cosh x \, dx = \sinh x + C, \quad \int \sinh x \, dx = \cosh x + C,$$

$$\int \text{sech}^2 x \, dx = \tanh x + C.$$

$\cosh^2 x$ and $\sinh^2 x$ can be integrated using the facts that:

$$\cosh^2 x = \tfrac{1}{2}(1 + \cosh 2x), \quad \sinh^2 x = \tfrac{1}{2}(\cosh 2x - 1).$$

(See 3·3 (iii).)

EXAMPLES

(1) Differentiate:

(i) $\sinh 2x \cos 3x$; (ii) $\log_e (\sinh x)$; (iii) $e^{\cosh x}$.

(i) $f'(x) = \sinh 2x(-3 \sin 3x) + 2 \cosh 2x \cos 3x$

$\qquad = 2 \cosh 2x \cos 3x - 3 \sinh 2x \sin 3x.$

(ii) $f'(x) = \dfrac{1}{\sinh x} \cosh x = \coth x.$

(iii) $f'(x) = e^{\cosh x} \sinh x.$

(2) Integrate: (i) $\sinh (3x - 2)$; (ii) $3 \sinh^2 2x$.

(i) $\displaystyle\int \sinh (3x - 2) \, dx = \tfrac{1}{3} \cosh (3x - 2) + C.$

(ii) $\displaystyle\int 3 \sinh^2 2x \, dx = 3 \int \tfrac{1}{2}(\cosh 4x - 1) \, dx$

$$= \frac{3}{2}\left(\frac{\sinh 4x}{4} - x\right) + C.$$

(3) Find the minimum value of $y = 5 \cosh x + 3 \sinh x.$

$$y = 5 \cosh x + 3 \sinh x,$$

$$dy/dx = 5 \sinh x + 3 \cosh x.$$

When $dy/dx = 0$: $\qquad 5 \sinh x = -3 \cosh x,$ \hfill (1)

$$d^2y/dx^2 = 5 \cosh x + 3 \sinh x = y. \hfill (2)$$

From (1), as $\cosh x$ cannot be negative, $\sinh x$ must be negative.

Squaring: $\qquad 25 \sinh^2 x = 9 \cosh^2 x,$

$$25 \sinh^2 x = 9(1 + \sinh^2 x),$$

$$16 \sinh^2 x = 9,$$

$$\sinh x = \pm \tfrac{3}{4}.$$

But $\sinh x$ is negative, therefore $\underline{\sinh x = -\tfrac{3}{4}}$ when $dy/dx = 0$.
Thus, from (1), $\underline{\cosh x = +\tfrac{5}{4}}$.
From (2) for these values, $d^2y/dx^2 = \tfrac{25}{4} - \tfrac{9}{4} = +\tfrac{16}{4} = +4.$
Therefore for these values, y is a minimum.

$$\therefore \quad \underline{\text{Minimum}\, y = +4.}$$

EXERCISE 12

1. If $y = c \cosh x/c$, $s = c \sinh x/c$, show that $y^2 = c^2 + s^2$.

2. Express $3 \cosh^2 x - 5 \sinh^2 x$ in terms of $\cosh 2x$.

3. Define $\tanh x$ in terms of e^x. Hence obtain $\tanh 2x$ in terms of $\tanh x$. If $\tanh x = 0.5$, evaluate $\sinh 2x$ without the use of tables. [L.U.]

4. Differentiate with respect to x:

(i) $x^2 \sinh 2x$;   (ii) $\dfrac{\sinh x}{1 + \cosh x}$;   (iii) $\coth x$;

(iv) $\operatorname{sech} x$;   (v) $\sinh^2 x$;   (vi) $e^{3x} \sinh^2 x$;

(vii) $\dfrac{e^{3x}}{\sinh^2 x}$;   (viii) $\log_e \sqrt{(\tanh 2x)}$;   (ix) $\cos 2x \cosh 5x$;

(x) $\dfrac{\sin 3x}{\sinh 2x}$.

5. Find the values of:

(i) $\displaystyle\int \sinh 3x\, dx$;   (ii) $\displaystyle\int \sinh^2 x\, dx$;   (iii) $\displaystyle\int_{-1}^{1} 2 \cosh \tfrac{1}{2}x\, dx$.

6. Express $\coth^2 x$ in terms of $\operatorname{cosech}^2 x$. Hence find $\displaystyle\int \coth^2 x\, dx$.

7. Find the area between $y = a \cosh x/a$, the x-axis and $x = 0$, $x = 2a$.

Also find $\displaystyle\int_{0}^{2a} \sqrt{\left\{ 1 + \left(\dfrac{dy}{dx}\right)^2 \right\}}\, dx$.

(This is the length of the curve from $x = 0$ to $x = 2a$.)

8. If $y = \tanh x$ show that $\dfrac{d^3y}{dx^3} + 2\left(\dfrac{dy}{dx}\right)^2 + 2y\dfrac{d^2y}{dx^2} = 0$.   [L.U.]

ANSWERS

2. $4 - \cosh 2x$.

3. $\dfrac{2\tanh x}{1 + \tanh^2 x}$; $\sinh 2x = \frac{4}{3}$.

4. (i) $2x(\sinh 2x + x\cosh 2x)$;   (ii) $\dfrac{1}{1 + \cosh x}$;

    (iii) $-\operatorname{cosech}^2 x$;   (iv) $-\operatorname{sech} x \tanh x$;

    (v) $\sinh 2x$;   (vi) $e^{3x}(3\sinh^2 x + \sinh 2x)$;

   (vii) $\dfrac{e^{3x}}{\sinh^3 x}(3\sinh x - 2\cosh x)$;

  (viii) $\dfrac{1}{\sinh 2x \cosh 2x}$ or $2\operatorname{cosech} 4x$;

   (ix) $5\cos 2x \sinh 5x - 2\sin 2x \cosh 5x$;

    (x) $\dfrac{3\sinh 2x \cos 3x - 2\sin 3x \cosh 2x}{\sinh^2 2x}$.

5. (i) $\frac{1}{3}\cosh 3x + C$;   (ii) $\dfrac{1}{2}\left(\dfrac{\sinh 2x}{2} - x\right) + C$;

   (iii) $8\sinh(0\cdot5) \simeq 4\cdot169$.

6. $1 + \operatorname{cosech}^2 x$; $x - \coth x + C$.

7. $a^2 \sinh 2 \simeq 3\cdot63a^2$; $a\sinh 2 \simeq 3\cdot63a$.

Chapter 4

INVERSE FUNCTIONS

4·1. Introduction

If y is given as a function of x and x is solved in terms of y, the new function is called the *inverse* of the original one.

Thus the inverse of $y = e^x$ is $x = \log_e y$; the inverse of $y = 4x^2$ is $x = \pm \frac{1}{2} \sqrt{y}$.

It is seen that although the original function may be single-valued (i.e. to each value of x there corresponds only one value of y), the inverse function may be double-valued, or even multi-valued in some cases (one value of y giving more than one value of x).

The trigonometric functions are cases of single-valued functions which have multi-valued inverse functions. For example, if $y = \sin x$, there is only one value of y corresponding to each value of x. The inverse function is given by 'x is the angle whose sine is y'. For a given value of y (the sine) there are multiple values of x. If α is one value of x, $r\pi + (-1)^r \alpha$ (r any integer) are also values for x.

4·2. Inverse trigonometric (or circular) functions

4·21. Symbols

If $y = \sin x$, the inverse relation is:

x is the angle whose sine is y.

This is cumbersome and a symbol has therefore been invented for it.

'The angle whose sine is y' is written as $\sin^{-1} y$.

Thus $\cos^{-1} y$ stands for 'the angle whose cosine is y', and so on.

The symbol '-1' must *not* be confused with a power. Although $\sin^2 x$ is a short way of writing $(\sin x)^2$, if $\sin x$ to the *power* -1 is needed it must be written $(\sin x)^{-1}$. This avoids ambiguity.

For example:

$\tan^{-1} x$ stands for 'the angle whose tangent is x'.

$(\tan x)^{-1}$ stands for '$(\tan x)$ raised to the power -1'.

The three main inverse trigonometric functions, $\sin^{-1} x$, $\cos^{-1} x$, $\tan^{-1} x$, will now be considered in turn.

4·22. Sin⁻¹ x

(a) *Graph.* The graph of $y = \sin^{-1} x$ is shown in fig. 12. The reader is advised to draw up a table of values and repeat the sketch for himself.

The sine of an angle cannot numerically exceed unity; thus the graph lies within the range $-1 \leqslant x \leqslant 1$. Note that if the axes Ox, Oy be relabelled Oy, Ox, the curve is simply that of $y = \sin x$. This gives a quick method of sketching the inverse trigonometric functions.

(b) *Derivative.* If

$$y = \sin^{-1} x \quad \text{then} \quad x = \sin y,$$

$$dx/dy = \cos y = \pm \sqrt{(1 - \sin^2 y)}$$

$$= \pm \sqrt{(1 - x^2)},$$

$$\therefore \quad \frac{dy}{dx} = \pm \frac{1}{\sqrt{(1 - x^2)}}. \qquad (1)$$

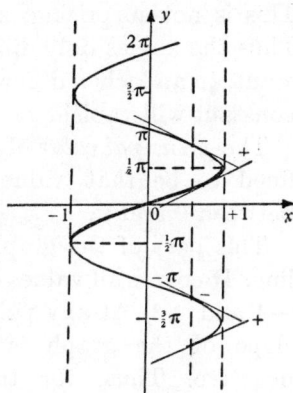

Fig. 12

It is no surprise that for a given value of x $(-1 < x < 1)$, the slope of the tangent has two values, equal but opposite in sign. This is easily seen from the graph.

To avoid ambiguity what is called the *principal value* of $\sin^{-1} x$ is defined.

From fig. 12 it is seen that within the range $-\frac{1}{2}\pi$ to $+\frac{1}{2}\pi$ all the slopes of the tangents are positive. This range also covers all possible values of x from -1 to $+1$, which give real values for $\sin^{-1} x$.

The principal value of $\sin^{-1} x$ is defined as that value which lies between $-\frac{1}{2}\pi$ and $+\frac{1}{2}\pi$. This part of the graph is shown in thick line.

For this principal value,

$$\frac{d}{dx}(\sin^{-1} x) = + \frac{1}{\sqrt{(1 - x^2)}}. \qquad (2)$$

Henceforth, unless otherwise stated, $\sin^{-1} x$ will be taken as meaning the principal value of $\sin^{-1} x$. Thus $\sin^{-1}(-\frac{1}{2})$ would be taken as $-\frac{1}{6}\pi$.

4·23. Cos⁻¹ x

(a) *Graph.* The graph of $y = \cos^{-1} x$ is shown in fig. 13.

(b) *Derivative.* If

$$y = \cos^{-1} x, \quad x = \cos y,$$

$$dx/dy = -\sin y = -[\pm \sqrt{(1 - \cos^2 y)}] = \mp \sqrt{(1 - x^2)}.$$

Thus
$$\frac{dy}{dx} = \mp \frac{1}{\sqrt{(1 - x^2)}}. \tag{1}$$

The derivative has the same numerical value as that of $\sin^{-1} x$. This is not surprising, as if $\sin A = x$, then $\cos(\tfrac{1}{2}\pi + A) = -x$. Thus the angles only differ by a constant, $\tfrac{1}{2}\pi$ and when differentiated this constant will vanish.

The *principal value* of $\cos^{-1} x$ is defined to be that value which lies between 0 and π.

This part of the graph is in thick line. It covers all values of x between -1 and $+1$. At any point on it the slope of the graph is seen to be negative. Thus, for the principal value,

$$\frac{d}{dx}(\cos^{-1} x) = -\frac{1}{\sqrt{(1 - x^2)}}. \tag{2}$$

Fig. 13

Unless otherwise stated, the principal value will be intended, e.g. $\cos^{-1}(-\tfrac{1}{2})$ is taken as $\tfrac{2}{3}\pi$.

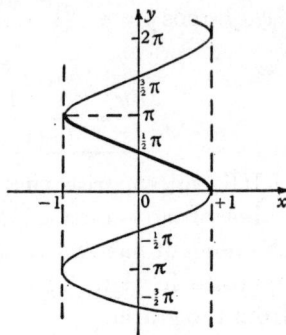

4·24. Tan⁻¹ x

(a) *Graph.* The graph of $y = \tan^{-1} x$ is shown in fig. 14.

(b) *Derivative.* If

$$y = \tan^{-1} x, \quad x = \tan y,$$

$$dx/dy = \sec^2 y = 1 + \tan^2 y = 1 + x^2.$$

Thus
$$\frac{dy}{dx} = \frac{1}{1 + x^2}. \tag{1}$$

Here there is no ambiguity in sign. This can be seen from the graph. For any given value of x the slopes of the possible tangents are all positive and all of the same magnitude.

There is still the question of which of the multi-values of $\tan^{-1} x$ to use. For example, $\tan^{-1} 1$ may be taken as $\frac{1}{4}\pi, \frac{5}{4}\pi, \ldots,$ etc.

Unless otherwise stated the *principal value* will be used. This is the value between $-\frac{1}{2}\pi$ and $+\frac{1}{2}\pi$.

4·25. Differentiation of more complex cases

Any of the general rules for differentiation may be used in dealing with derivatives of more complex inverse trigonometric functions.

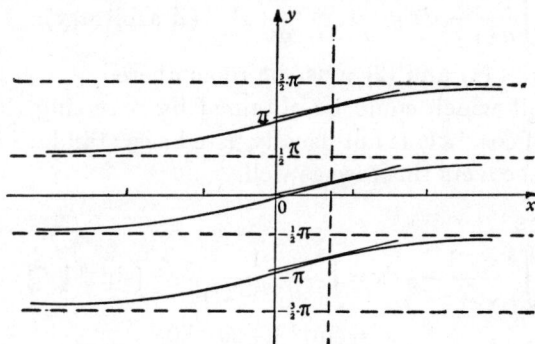

Fig. 14

Example. Differentiate $\tan^{-1} \sqrt{(1 - 2x^2)}$.

$$f'(x) = \frac{1}{1 + \{\sqrt{(1-2x^2)}\}^2} \times \frac{d}{dx}\sqrt{(1-2x^2)}$$

$$= \frac{1}{1 + 1 - 2x^2} \times \frac{(-4x)}{2\sqrt{(1-2x^2)}} = \frac{-x}{(1-x^2)\sqrt{(1-2x^2)}}.$$

Exercise 13

Prove the following:

1. $$D_x \sin^{-1}\frac{x}{a} = \frac{1}{\sqrt{(a^2 - x^2)}}.$$

2. $$D_x \cos^{-1}\frac{x}{a} = -\frac{1}{\sqrt{(a^2 - x^2)}}.$$

3. $$D_x \tan^{-1}\frac{x}{a} = \frac{a}{a^2 + x^2}.$$

4·26. Standard integrals

From the exercise of § 4·25, on reversing the differentiation, the following standard integrals are obtained:

$$\int \frac{1}{\sqrt{(a^2 - x^2)}}\,dx = \sin^{-1}\frac{x}{a} + C, \tag{1}$$

$$\int \frac{a}{a^2 + x^2}\,dx = \tan^{-1}\frac{x}{a} + C.$$

The latter, on dividing both sides by the constant a, gives

$$\int \frac{1}{a^2 + x^2}\,dx = \frac{1}{a}\tan^{-1}\frac{x}{a} + A \quad (A \text{ arbitrary}). \tag{2}$$

The results (1) and (2) must be remembered.

The result which could be obtained by reversing the differentiation of $\cos^{-1}x/a$ is not usually listed as a standard integral, as result (1) covers this case as well.

EXAMPLES

(1) $\displaystyle\int_0^3 \frac{1}{\sqrt{(9 - x^2)}}\,dx = \int_0^3 \frac{1}{\sqrt{(3^2 - x^2)}}\,dx = [\sin^{-1}\tfrac{1}{3}x]_0^3$

$$= \sin^{-1}1 - \sin^{-1}0$$

$$= \tfrac{1}{2}\pi - 0 = \tfrac{1}{2}\pi \simeq 1\cdot571.$$

It will be noticed that the *principal values* of $\sin^{-1}1$ and $\sin^{-1}0$ have been taken. This will be adhered to in all definite integrals.

Actually any *consecutive* pair of values on the graph of $\sin^{-1}x$, which satisfy, could have been taken, e.g.

$$\sin^{-1}1 = \tfrac{5}{2}\pi, \quad \sin^{-1}0 = 2\pi \quad (\text{see fig. 12}).$$

This would give $\tfrac{5}{2}\pi - 2\pi = \tfrac{1}{2}\pi$ for the definite integral above. That is, the same value as before.

(2) $\displaystyle\int_0^{\sqrt5} \frac{1}{5 + x^2}\,dx = \int_0^{\sqrt5} \frac{1}{(\sqrt5)^2 + x^2}\,dx = \frac{1}{\sqrt5}\left[\tan^{-1}\frac{x}{\sqrt5}\right]_0^{\sqrt5}$

$$= \frac{1}{\sqrt5}[\tan^{-1}1 - \tan^{-1}0]$$

$$= \frac{1}{\sqrt5}[\tfrac{1}{4}\pi - 0] = \frac{\pi}{4\sqrt5} \simeq 0\cdot351.$$

Note that the angle is always written as its numerical value in radians. Integration, being the reverse of differentiation, must satisfy the same condition as the latter, viz. when angles occur their value in radians must be used.

(3) $\displaystyle\int \frac{1}{\sqrt{(5-3x^2)}}\,dx = \frac{1}{\sqrt{3}}\int \frac{1}{\sqrt{(\frac{5}{3}-x^2)}}\,dx = \frac{1}{\sqrt{3}}\int \frac{1}{\sqrt{\{(\sqrt{\frac{5}{3}})^2 - x^2\}}}\,dx$

$\displaystyle = \frac{1}{\sqrt{3}}\sin^{-1}\frac{x}{\sqrt{\frac{5}{3}}} + C = \frac{1}{\sqrt{3}}\sin^{-1}\frac{\sqrt{3}\,x}{\sqrt{5}} + C.$

The safest method in this type of example is to make the coefficient of x^2 numerically unity.

EXERCISE 14

1. Differentiate with respect to x:

(i) $\cos^{-1}\sqrt{(3x-2)}$;

(ii) $\tan^{-1}\dfrac{a}{\sqrt{(x^2-a^2)}}$;

(iii) $\sin^{-1}[2x\sqrt{(1-x^2)}]$;

(iv) $\tan^{-1}\dfrac{2x}{1-x^2}$;

(v) $x^3\sin^{-1}x$;

(vi) $x^2\tan^{-1}2x$;

(vii) $\tan^{-1}(\sinh x)$;

(viii) $\cos^{-1}(\sin x)$.

2. Differentiate the following, starting from the definition of inverse trigonometric functions:

(i) $\operatorname{cosec}^{-1}x/a$;   (ii) $\sec^{-1}x/a$;   (iii) $\cot^{-1}x/a$.

3. Find the value of:

(i) $\displaystyle\int \frac{dx}{x^2+16}$;

(ii) $\displaystyle\int \frac{dx}{\sqrt{(9-x^2)}}$;

(iii) $\displaystyle\int \frac{dy}{\sqrt{(9-4y^2)}}$;

(iv) $\displaystyle\int \frac{dx}{\sqrt{(4-(x-1)^2)}}$.

4. Evaluate:

(i) $\displaystyle\int_0^{\frac{1}{2}} \frac{dx}{\sqrt{(1-x^2)}}$;

(ii) $\displaystyle\int_1^2 \frac{dx}{\sqrt{(5-x^2)}}$;

(iii) $\displaystyle\int_{-1}^1 \frac{dx}{1+x^2}$;

(iv) $\displaystyle\int_{-\frac{1}{2}}^1 \frac{dx}{1+3x^2}$.

ANSWERS

1. (i) $-\dfrac{\sqrt{3}}{2\sqrt{(5x-2-3x^2)}}$;

(ii) $-\dfrac{a}{x\sqrt{(x^2-a^2)}}$;

(iii) $\dfrac{2}{\sqrt{(1-x^2)}}$;

(iv) $\dfrac{2}{1+x^2}$;

         B M

(v) $\dfrac{x^3}{\sqrt{(1-x^2)}} + 3x^2 \sin^{-1} x$;   (vi) $\dfrac{2x^2}{(1+4x^2)} + 2x \tan^{-1} 2x$;

(vii) $\operatorname{sech} x$;   (viii) -1.

2. (i) $-\dfrac{a}{x\sqrt{(x^2-a^2)}}$; (ii) $\dfrac{a}{x\sqrt{(x^2-a^2)}}$; (iii) $-\dfrac{a}{a^2+x^2}$.

3. (i) $\tfrac{1}{4}\tan^{-1}\tfrac{1}{4}x + C$;   (ii) $\sin^{-1}\tfrac{1}{3}x + C$;

  (iii) $\tfrac{1}{2}\sin^{-1}\tfrac{2}{3}y + C$;   (iv) $\sin^{-1}\tfrac{1}{2}(x-1) + C$.

4. (i) $\tfrac{1}{6}\pi$;   (ii) 0.643;   (iii) $\tfrac{1}{2}\pi$;   (iv) $\dfrac{\sqrt{3}\,\pi}{6}$.

4·3. Inverse hyperbolic functions

The three that most frequently occur are $\sinh^{-1} x$, $\cosh^{-1} x$ and $\tanh^{-1} x$. These will be dealt with in much the same manner as the inverse trigonometric functions. Their graphs can be quickly sketched in the manner explained in §4·22.

4·31. Sinh$^{-1} x$

(a) *Graph.* The graph of $y = \sinh^{-1} x$ is shown in fig. 15.

(b) *Derivative.* If

$$y = \sinh^{-1} x, \quad x = \sinh y,$$

$$\frac{dx}{dy} = \cosh y = \pm \sqrt{(1 + \sinh^2 y)}$$

$$= \pm \sqrt{(1 + x^2)}.$$

$$\therefore \; \frac{dy}{dx} = \pm \frac{1}{\sqrt{(1 + x^2)}}. \quad (1)$$

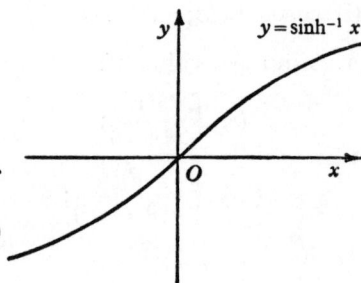

Fig. 15

The negative value is inapplicable, as it can be seen from fig. 15 that the slope of the graph at any point is positive.

Thus

$$\frac{d}{dx}(\sinh^{-1} x) = \frac{1}{\sqrt{(1 + x^2)}}. \quad (2)$$

There is no need of a definition of a principal value as $\sinh^{-1} x$ is single-valued (see fig. 15).

4·32. Cosh⁻¹ x

(a) *Graph.* The graph of $y = \cosh^{-1} x$ is shown in fig. 16.

(b) *Derivative.* If

$$y = \cosh^{-1} x, \quad x = \cosh y,$$

$$dx/dy = \sinh y = \pm \sqrt{(\cosh^2 y - 1)} = \pm \sqrt{(x^2 - 1)}.$$

$$\therefore \frac{dy}{dx} = \pm \frac{1}{\sqrt{(x^2 - 1)}}. \tag{1}$$

From fig. 16 it is seen that there are two values of $\cosh^{-1} x$ for any given value of x ($x > 1$), with the same numerical value but opposite in sign.

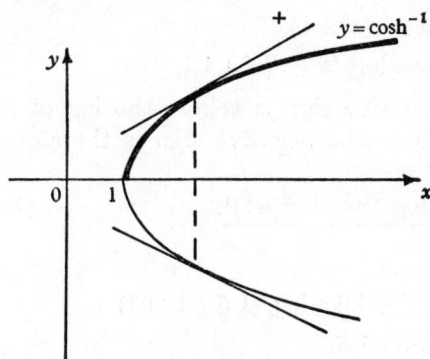
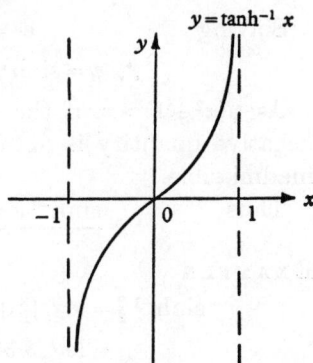

Fig. 16 Fig. 17

The *principal value* of $\cosh^{-1} x$ is defined to be the positive value.

In this case

$$\frac{d}{dx}(\cosh^{-1} x) = \frac{1}{\sqrt{(x^2 - 1)}}. \tag{2}$$

4·33. Tanh⁻¹ x

(a) *Graph.* The graph of $y = \tanh^{-1} x$ is shown in fig. 17.

(b) *Derivative.* $y = \tanh^{-1} x, \quad x = \tanh y,$

$$dx/dy = \operatorname{sech}^2 y = 1 - \tanh^2 y = 1 - x^2,$$

$$\therefore \frac{dy}{dx} = \frac{1}{(1 - x^2)}. \tag{1}$$

As with $\sinh^{-1} x$, there is no need to define a principal value as $\tanh^{-1} x$ is single-valued.

It should also be noted that the derivative is only valid if $x^2 < 1$, i.e. if $-1 < x < 1$.

4·34. Logarithmic forms of the inverse hyperbolic functions

Each of the inverse hyperbolic functions has a logarithmic equivalent which may be used when calculating numerical values.

(a) $\sinh^{-1} x = \log_e \{x + \sqrt{(x^2 + 1)}\}$. If

$$y = \sinh^{-1} x,$$

$$x = \sinh y = \frac{e^y - e^{-y}}{2} = \frac{e^{2y} - 1}{2e^y},$$

$\therefore\ e^{2y} - 2x\, e^y - 1 = 0$, a quadratic in e^y.

Solving $\qquad e^y = x \pm \sqrt{(x^2 + 1)},$

$$\therefore\ y = \sinh^{-1} x = \log_e \{x \pm \sqrt{(x^2 + 1)}\}.$$

As $\sqrt{(x^2 + 1)} > x$, if the negative sign is taken, the log of a negative quantity is obtained. The negative sign is therefore inadmissible.

Thus $\qquad \underline{\sinh^{-1} x = \log_e \{x + \sqrt{(x^2 + 1)}\}}.$ \hfill (1)

EXAMPLE

$$\sinh^{-1} \tfrac{3}{2} = \log_e \{\tfrac{3}{2} + \sqrt{(\tfrac{9}{4} + 1)}\} \simeq \log_e (1·5 + 1·803)$$

$$= \log_e 3·303 \simeq \underline{1·195}.$$

(b) $\cosh^{-1} x = \log_e \{x + \sqrt{(x^2 - 1)}\}$. If

$$y = \cosh^{-1} x,$$

$$x = \cosh y = \frac{e^y + e^{-y}}{2} = \frac{e^{2y} + 1}{2e^y}.$$

$$\therefore\ e^{2y} - 2x\, e^y + 1 = 0.$$

Solving $\qquad e^y = x \pm \sqrt{(x^2 - 1)}.$

$$\therefore\ y = \cosh^{-1} x = \log_e \{x \pm \sqrt{(x^2 - 1)}\}.$$ \hfill (2)

Referring to fig. 16 it is seen that the two possible values should be equal but opposite in sign. This can easily be proved:

$$x - \sqrt{(x^2 - 1)} = \frac{\{x - \sqrt{(x^2 - 1)}\}\{x + \sqrt{(x^2 - 1)}\}}{\{x + \sqrt{(x^2 - 1)}\}} = \frac{x^2 - (x^2 - 1)}{x + \sqrt{(x^2 - 1)}}$$

$$= \frac{1}{x + \sqrt{(x^2 - 1)}}.$$

Thus

$$\log_e \{x - \sqrt{(x^2 - 1)}\} = \log_e \frac{1}{x + \sqrt{(x^2 - 1)}} = -\log_e \{x + \sqrt{(x^2 - 1)}\}.$$

Equation (2) now becomes
$$\cosh^{-1} x = \pm \log_e \{x + \sqrt{(x^2 - 1)}\}.$$

Taking the principal value,
$$\cosh^{-1} x = \log_e \{x + \sqrt{(x^2 - 1)}\}. \tag{3}$$

As mentioned in § 4·32, x must be greater than 1.

(c) $\tanh^{-1} x = \frac{1}{2} \log_e \left(\dfrac{1+x}{1-x}\right)$. If

$$y = \tanh^{-1} x,$$

$$x = \tanh y = \frac{e^y - e^{-y}}{e^y + e^{-y}} = \frac{e^{2y} - 1}{e^{2y} + 1}.$$

Solving for e^{2y}:
$$e^{2y} = \frac{1+x}{1-x}.$$

$$\therefore \ 2y = \log_e \left(\frac{1+x}{1-x}\right),$$

$$y = \tanh^{-1} x = \frac{1}{2} \log_e \left(\frac{1+x}{1-x}\right).$$

As explained in § 4·33, x^2 must be less than 1.

EXAMPLE

Differentiate $\tanh^{-1}(2x + 1)$.

$$D_x \tanh^{-1}(2x + 1) = \frac{1}{1 - (2x+1)^2} \times \frac{d}{dx}(2x + 1)$$

$$= \frac{1}{(-4x^2 - 4x)} \times 2$$

$$= -\frac{1}{2x(x + 1)}.$$

EXERCISE 15

1. Calculate the values, to 3 decimal places, of:

(i) $\cosh^{-1} 2$;   (ii) $\tanh^{-1} \frac{1}{2}$;   (iii) $\sinh^{-1} \frac{1}{2}$,

2. Differentiate with respect to x:

(i) $\sinh^{-1} 2x$;   (ii) $\cosh^{-1} \frac{1}{3}x$;   (iii) $\sinh^{-1} x^2$;

(iv) $\cosh^{-1}(1/x)$;   (v) $\tanh^{-1}(\tan \frac{1}{2}x)$.

3. Differentiate the following, starting with the definition of inverse hyperbolic functions:

(i) $\operatorname{cosech}^{-1} x$;   (ii) $\operatorname{sech}^{-1} x$;   (iii) $\coth^{-1} x$.

ANSWERS

1. (i) 1·317;   (ii) 0·549;   (iii) 0·481.

2. (i) $\dfrac{2}{\sqrt{(1+4x^2)}}$;   (ii) $\dfrac{1}{\sqrt{(x^2-9)}}$;   (iii) $\dfrac{2x}{\sqrt{(1+x^4)}}$;

    (iv) $-\dfrac{1}{x\sqrt{(1-x^2)}}$;

    (v) $\dfrac{\sec^2 \frac{1}{2}x}{2(1-\tan^2 \frac{1}{2}x)}$ which reduces to $\frac{1}{2}\sec x$.

3. (i) $-\dfrac{1}{x\sqrt{(1+x^2)}}$;   (ii) $\dfrac{-1}{x\sqrt{(1-x^2)}}$;   (iii) $\dfrac{-1}{(x^2-1)}$.

4·35. Standard integrals

$$D_x(\sinh^{-1} x/a) = \frac{1}{\sqrt{(1+x^2/a^2)}} \times \frac{1}{a} = \frac{1}{\sqrt{(a^2+x^2)}}.$$

Thus

$$\int \frac{1}{\sqrt{(a^2+x^2)}}\,dx = \underline{\sinh^{-1}\frac{x}{a} + A}$$

$$= \log_e\left\{\frac{x}{a} + \sqrt{\left(1+\frac{x^2}{a^2}\right)}\right\} + \log_e C \quad (C \text{ arbitrary})$$

$$= \underline{\log_e C\left(\frac{x+\sqrt{(a^2+x^2)}}{a}\right)}.$$

In connexion with integration, $\log_e C\left(\dfrac{x+\sqrt{(a^2+x^2)}}{a}\right)$ is often written $\log_e C\{x+\sqrt{(a^2+x^2)}\}$, as if C is arbitrary C/a is also arbitrary.

It is left to the reader to prove in a similar manner that

$$\int \frac{1}{\sqrt{(x^2-a^2)}}\,dx = \cosh^{-1}\frac{x}{a} + A = \log_e C\{x+\sqrt{(x^2-a^2)}\} \quad (x>a),$$

$$\int \frac{1}{a^2-x^2}\,dx = \frac{1}{a}\tanh^{-1}\frac{x}{a} + A = \frac{1}{2a}\log C\left(\frac{a+x}{a-x}\right) \quad (x^2<a^2).$$

EXAMPLES

(1) $\int_3^4 \frac{1}{\sqrt{(x^2+16)}}\, dx = [\sinh^{-1}\tfrac{1}{4}x]_3^4$

$$= [\sinh^{-1} 1 - \sinh^{-1}\tfrac{3}{4}]$$

$$= \log_e(1+\sqrt{2}) - \log_e\{\tfrac{3}{4} + \sqrt{(1+\tfrac{9}{16})}\}$$

$$= \log_e \tfrac{1}{2}(1+\sqrt{2})$$

$$\eqsim \log_e 1\cdot207$$

$$= 0\cdot188.$$

(2) $\int \frac{1}{\sqrt{(b^2x^2-a^2)}}\, dx = \frac{1}{b}\int \frac{1}{\sqrt{\{x^2-(a/b)^2\}}}\, dx = \frac{1}{b}\cosh^{-1}\frac{b}{a}x + C.$

EXERCISE 16

1. $\int \frac{1}{\sqrt{(9+x^2)}}\, dx.$ 2. $\int_2^3 \frac{1}{\sqrt{(x^2-4)}}\, dx.$ 3. $\int_1^2 \frac{1}{(9-x^2)}\, dx.$

4. $\int \frac{1}{\sqrt{(4-x^2)}}\, dx.$ 5. $\int \frac{1}{25+x^2}\, dx.$ 6. $\int \frac{1}{\sqrt{(2x^2+10)}}\, dx.$

7. $\int_{-1}^0 \frac{1}{\sqrt{(x^2+4)}}\, dx.$ 8. $\int_{+3}^{+4} \frac{1}{\sqrt{(x^2-4)}}\, dx.$

9. Find the area between the curve $y^2 = \dfrac{a^4}{a^2+x^2}$, and the ordinates at $x=a$, $x=2a$. Also find the volume formed by rotating this area about the x-axis.

ANSWERS

1. $\sinh^{-1}\tfrac{1}{3}x + C.$ 2. $\cosh^{-1}\tfrac{3}{2} \eqsim 0\cdot962.$

3. $\tfrac{1}{6}\log_e 2\cdot5 \eqsim 0\cdot153.$ 4. $\sin^{-1}\tfrac{1}{2}x + C.$

5. $\tfrac{1}{5}\tan^{-1}\tfrac{1}{5}x + C.$ 6. $\dfrac{1}{\sqrt{2}}\sinh^{-1}(x/\sqrt{5}) + C.$

7. $0\cdot481.$ 8. $0\cdot355.$

9. (i) $2a^2\log_e\left(\dfrac{2+\sqrt{5}}{1+\sqrt{2}}\right) \eqsim 1\cdot125a^2;$

(ii) $\pi a^3(\tan^{-1}2 - \tan^{-1}1) \eqsim 0\cdot322\pi a^3.$

4·4. Formation of differential equations

The inverse trigonometric and hyperbolic functions give rise to important types of differential equations.

EXAMPLE

If $y = (\sin^{-1} x)^2$, show that $(1 - x^2)(dy/dx)^2 = 4y$.

Deduce that
$$(1 - x^2)\frac{d^2y}{dx^2} - x\frac{dy}{dx} - 2 = 0.$$ [L.U.]

$$y = (\sin^{-1} x)^2.$$

$$\therefore \quad \frac{dy}{dx} = 2(\sin^{-1} x)\frac{1}{\sqrt{(1 - x^2)}}.$$

Squaring and cross-multiplying

$$(1 - x^2)(dy/dx)^2 = 4(\sin^{-1} x)^2 = 4y.$$ (1)

Differentiating again:

$$(1 - x^2) \, 2\left(\frac{dy}{dx}\right)\frac{d^2y}{dx^2} + (-2x)\left(\frac{dy}{dx}\right)^2 = 4\left(\frac{dy}{dx}\right).$$

$$\therefore \quad (1 - x^2)\frac{d^2y}{dx^2} - x\frac{dy}{dx} - 2 = 0.$$

EXERCISE 17

1. If $y = \cosh^{-1} x/a$, show that $(x^2 - a^2)\dfrac{d^2y}{dx^2} + x\dfrac{dy}{dx} = 0$.

2. If $y = \sin^{-1} x$, show that $(1 - x^2)\dfrac{d^2y}{dx^2} - x\dfrac{dy}{dx} = 0$.

3. If $y = \tan^{-1}(x^2)$, show that $(1 + x^4)\dfrac{d^2y}{dx^2} + 4x^3\dfrac{dy}{dx} - 2 = 0$.

MISCELLANEOUS EXERCISES ON CHAPTER 4

1. Using principal values, evaluate:

(i) $\cos^{-1}(-\tfrac{1}{2}) + \sin^{-1}\sqrt{3}/2$;   (ii) $\tan^{-1}\sqrt{3} + \tan^{-1} 1/\sqrt{3}$;

(iii) $\tan^{-1}\infty + \cos^{-1}(-1)$;   (iv) $\sec^{-1} 2 + \cot^{-1} 1$;

(v) $\sinh^{-1} 3 + \cosh^{-1} 2$;   (vi) $\tanh^{-1} 0 \cdot 5 + \sinh^{-1} 0 \cdot 5$.

2. Find the principal values of:

(i) $\tan^{-1}\tfrac{1}{2}$;   (ii) $\sin^{-1} 1/\sqrt{3}$;   (iii) $\cos^{-1}\tfrac{5}{12}$.

3. Differentiate:

(i) $x^2 \sin^{-1} x$;   (ii) $\sin^{-1}(\tan x)$;   (iii) $\tan^{-1} \dfrac{x}{1+2x}$;

(iv) $(\sinh^{-1} \sqrt{x})^2$;   (v) $\cosh^{-1} 1/\sqrt{x}$.

4. If $y = \dfrac{\sin^{-1} x}{\sqrt{(1-x^2)}}$, prove that $\dfrac{dy}{dx} = \dfrac{xy+1}{(1-x^2)}$.

5. If $y/x = \sin^{-1} x$, prove that

$$(1-x^2)\frac{d^2y}{dx^2} - x\frac{dy}{dx} + y = 2\sqrt{(1-x^2)}.$$

6. Evaluate:

(i) $\displaystyle\int_a^\infty \frac{dx}{(a^2+x^2)}$;   (ii) $\displaystyle\int_0^{\sqrt{3}} \frac{dx}{\sqrt{(3-x^2)}}$;   (iii) $\displaystyle\int_0^{\sqrt{3}} \frac{dx}{\sqrt{(x^2+1)}}$;

(iv) $\displaystyle\int_0^\infty \frac{dx}{9+x^2}$;   (v) $\displaystyle\int_{\sqrt{3}}^2 \frac{dx}{\sqrt{(x^2-3)}}$.

7. Sketch the graph of $y^2 = \dfrac{1}{1-x^2}$.

Find (i) the area contained between the two branches of the curve from $x=0$ to $x=\frac{1}{2}$; (ii) the volume formed when this area is rotated round the x axis.

8. Find the area between the axis of y, the curve

$$(4+x^2)y^2 = 16 \quad \text{and} \quad x=2.$$

Find also the volume formed when this area is rotated about the x axis.

ANSWERS

1. (i) π;   (ii) $\frac{1}{2}\pi$;   (iii) $\frac{3}{2}\pi$;

 (iv) $\frac{7}{12}\pi$;   (v) $3\cdot135$;   (vi) $1\cdot03$.

2. (i) $0\cdot4637$;   (ii) $0\cdot6152$;   (iii) $1\cdot1412$.

3. (i) $\dfrac{x^2}{\sqrt{(1-x^2)}} + 2x\sin^{-1}x$;   (ii) $\dfrac{\sec^2 x}{\sqrt{(1-\tan^2 x)}}$;

 (iii) $\dfrac{1}{(5x^2+4x+1)}$;   (iv) $\dfrac{\sinh^{-1}\sqrt{x}}{\sqrt{(x+x^2)}}$;

 (v) $-\dfrac{1}{2x\sqrt{(1-x)}}$.

6. (i) $\dfrac{\pi}{4a}$;   (ii) $\frac{1}{2}\pi$;   (iii) $\log_e(2+\sqrt{3})$;

 (iv) $\frac{1}{6}\pi$;   (v) $\frac{1}{2}\log_e 3$.

7. (i) $\frac{1}{3}\pi$;   (ii) $\frac{1}{2}\pi\log_e 3$.

8. $8\log_e(1+\sqrt{2})$; $2\pi^2$.

CHAPTER 5

METHODS OF INTEGRATION

Standard Integrals

A list of standard integrals which have been obtained in the previous chapters, or form part of an Ordinary National Course, is given below. They should be memorized. For brevity the arbitrary constants have been omitted. a, b, n denote any given numbers.

$$\int (ax+b)^n \, dx = \frac{1}{a} \frac{(ax+b)^{n+1}}{(n+1)}. \qquad \int \frac{1}{(ax+b)} \, dx = \frac{1}{a} \log_e (ax+b).$$

$$\int e^{ax+b} \, dx = \frac{1}{a} e^{ax+b}. \qquad \int \sin (ax+b) \, dx = -\frac{1}{a} \cos (ax+b).$$

$$\int \cos (ax+b) \, dx = \frac{1}{a} \sin (ax+b).$$

$$\int \sec^2 (ax+b) \, dx = \frac{1}{a} \tan (ax+b).$$

$$\int \sinh (ax+b) \, dx = \frac{1}{a} \cosh (ax+b).$$

$$\int \cosh (ax+b) \, dx = \frac{1}{a} \sinh (ax+b)$$

$$\int \operatorname{sech}^2 (ax+b) \, dx = \frac{1}{a} \tanh (ax+b).$$

$$\int \frac{1}{\sqrt{(a^2-x^2)}} \, dx = \sin^{-1} \frac{x}{a}.$$

$$\int \frac{1}{a^2+x^2} \, dx = \frac{1}{a} \tan^{-1} \frac{x}{a}.$$

$$\int \frac{1}{\sqrt{(x^2-a^2)}} \, dx = \cosh^{-1} \frac{x}{a} = \log_e \left(\frac{x + \sqrt{(x^2-a^2)}}{a} \right).$$

$$\int \frac{1}{\sqrt{(x^2+a^2)}} \, dx = \sinh^{-1} \frac{x}{a} = \log_e \left(\frac{x + \sqrt{(x^2+a^2)}}{a} \right).$$

$$\int \frac{1}{a^2-x^2} \, dx = \frac{1}{a} \tanh^{-1} \frac{x}{a} = \frac{1}{2a} \log_e \left(\frac{a+x}{a-x} \right) \quad (x^2 < a^2).$$

5·1. Use of $\int \dfrac{f'(x)}{f(x)}\,dx$

$$\frac{d}{dx}(\log_e y) = \frac{1}{y}\frac{dy}{dx} = \frac{dy}{dx}\Big/ y.$$

Thus $\qquad \displaystyle\int\left(\frac{dy}{dx}\Big/ y\right)dx = \log_e y + A \quad (A \text{ arbitrary}).$

Written in functional notation this gives

$$\int \frac{f'(x)}{f(x)}\,dx = \log_e f(x) + A = \log_e Cf(x).$$

$\log_e Cf(x)$ is often a convenient method of introducing the arbitrary constant, when the answer to an integral is a log; for $\log_e Cf(x) = \log_e f(x) + \log_e C$, and if C is arbitrary so is $\log_e C$.

It is always useful to inspect any given integral of a quotient to see whether the numerator is the derivative of the denominator or can be made so by a multiplying *number*. If, so, the integration is easy.

For example, in $\displaystyle\int \frac{(x+1)}{x^2+2x-2}\,dx$, the derivative of the denominator is $2x+2 = 2(x+1)$. The integral can thus be written

$$\int \frac{(x+1)}{x^2+2x-2}\,dx = \frac{1}{2}\int \frac{2(x+1)}{x^2+2x-2}\,dx = \tfrac{1}{2}\log_e C(x^2+2x-2).$$

Again, in the integral $\displaystyle\int \frac{(\cos x + \sin x)}{(\sin x - \cos x)}\,dx$ the numerator is the derivative of the denominator, i.e.

$$I = \int \frac{d(\sin x - \cos x)}{(\sin x - \cos x)} = \log_e C(\sin x - \cos x).$$

Note that the capital letter, I, will often be used to indicate a given integral in order to save space.

Two important standard integrals are obtained by use of the above method. They should be memorized:

$$\int \tan x\,dx = \int \frac{\sin x}{\cos x}\,dx = -\int \frac{(-\sin x)}{\cos x}\,dx = -\int \frac{d(\cos x)}{\cos x}$$
$$= -\log A\,\cos x = \underline{\log C\,\sec x}.$$
$$\int \cot x\,dx = \int \frac{\cos x}{\sin x}\,dx = \int \frac{d(\sin x)}{\sin x} = \underline{\log C\,\sin x}.$$

Often the multiplying number required may be a fraction. The main point is that it *must* be a number, *not* a function of x. It is nonsense to say that

$$\int \frac{x}{x^3+1}\,dx = \frac{1}{3x}\int \frac{3x^2}{x^3+1}\,dx = \frac{1}{3x}\log_e C(x^3+1).$$

If the expression on the right-hand side is differentiated nothing resembling the integrand, $\dfrac{x}{x^3+1}$, is obtained.

This serves to emphasize the fact that only multiplying *constants* can be taken outside the integral sign.

EXERCISE 18

1. $\displaystyle\int \frac{(2x+1)}{(x^2+x-5)}\,dx.$

2. $\displaystyle\int \frac{(4x+4)}{(x^2+2x-1)}\,dx.$

3. $\displaystyle\int \frac{(3x^2-12x+2)}{(x^3-6x^2+2x+1)}\,dx.$

4. $\displaystyle\int \cot 2x\,dx.$

5. $\displaystyle\int \tanh 3x\,dx.$

6. $\displaystyle\int -\frac{(2\sin 2x+\cos x)}{\cos 2x-\sin x}\,dx.$

7. $\displaystyle\int \frac{1}{\sqrt{(1-x^2)}\sin^{-1}x}\,dx.$

8. $\displaystyle\int_0^{\frac{1}{2}\pi} \frac{\cos x}{(1+\sin x)}\,dx.$

ANSWERS

1. $\log_e C(x^2+x-5).$

2. $2\log_e C(x^2+2x-1).$

3. $\log_e C(x^3-6x^2+2x+1).$

4. $\frac{1}{2}\log_e C\sin 2x.$

5. $\frac{1}{3}\log_e C\cosh 3x.$

6. $\log_e C(\cos 2x-\sin x).$

7. $\log_e C(\sin^{-1}x).$

8. $\log_e 2.$

5·2. Simple change of variables

Many integrals may be changed into one of the standard forms by a change of the variable. One of the simplest methods uses the fact that $\displaystyle\int f(u)\,du = \int f(u)\frac{du}{dx}\,dx$, where u is the new variable. This necessitates that (i) one factor of the integrand supplies du/dx, (ii) the rest of the expression is easily put in terms of u.

For example:

$$\int \sin^4 x\,\cos x\,dx.$$

Let $u = \sin x$, then $du/dx = \cos x$.

Thus $\qquad I = \int u^4 du = \dfrac{u^5}{5} + C = \dfrac{\sin^5 x}{5} + C.$

A more convenient way of writing the substitution is:

Let $u = \sin x$, then $du = \cos x\,dx$. du is then substituted for $\cos x\,dx$ and u^4 for $\sin^4 x$.

The thing to watch for is one factor of the integrand being the derivative of another main factor.

EXAMPLES

(1) $\qquad\qquad\qquad \displaystyle\int \dfrac{x^5}{(a^6 + x^6)^2}\,dx.$

Let $u = a^6 + x^6$; then $du = 6x^5\,dx$ and $x^5\,dx = \frac{1}{6}du.$

$$I = \int \frac{1}{6}\frac{1}{u^2}\,du = -\frac{1}{6u} + C = -\frac{1}{6(a^6 + x^6)} + C.$$

(2) If the integral is a definite one, the new variable need not be changed back, as long as the *limits* are changed into corresponding values of the new variable.

$$\int_0^{\frac{1}{2}\pi} \frac{\cos\theta}{(1 + \sin\theta)^2}\,d\theta.$$

Let $u = 1 + \sin\theta$, then $du = \cos\theta\,d\theta$.

When $\theta = 0$, $u = 1$ and when $\theta = \frac{1}{2}\pi$, $u = 2$.

$$\therefore\ I = \int_{u=1}^{u=2} \frac{1}{u^2}\,du = \left[-\frac{1}{u} \right]_1^2 = -[\tfrac{1}{2} - 1] = \tfrac{1}{2}.$$

EXERCISE 19

1. $\displaystyle\int \frac{\cos x}{\sin^4 x}\,dx.$

2. $\displaystyle\int \frac{x}{\sqrt{(x^2 - a^2)}}\,dx.$

3. $\displaystyle\int x^2\,\sqrt{(a^3 + x^3)}\,dx.$

4. $\displaystyle\int \frac{\sec^2 x\,dx}{1 - \tan^2 x}.$

5. $\displaystyle\int_0^a \frac{x\,dx}{\sqrt{(a^4 - x^4)}}.$

6. $\displaystyle\int_0^1 \frac{e^x\,dx}{(1 + e^x)}.$

7. $\displaystyle\int x\,e^{x^2}\,dx.$

8. $\displaystyle\int \frac{(3x^2 + 2)\,dx}{\sqrt{(x^3 + 2x + 1)}}.$

9. $\displaystyle\int \frac{e^x\,dx}{\sqrt{(1 + e^{2x})}}.$

10. $\displaystyle\int \frac{x\,dx}{(1 + x^4)}.$

11. $\displaystyle\int \sec^4 x\,dx.$

ANSWERS

1. $-\frac{1}{3}\operatorname{cosec}^3 x + C.$ **2.** $\sqrt{(x^2 - a^2)} + C.$

3. $\frac{2}{9}(a^3 + x^3)^{\frac{3}{2}} + C.$

4. $\frac{1}{2}\log_e C\left(\dfrac{1 + \tan x}{1 - \tan x}\right)$ [put $u = \tan x$].

5. $\frac{1}{4}\pi$ (put $u = x^2$).   **6.** $\log_e\left(\dfrac{1 + e}{2}\right) \simeq 0\cdot 62.$

7. $\frac{1}{2}e^{x^2} + C.$ **8.** $2\sqrt{(x^3 + 2x + 1)} + C.$

9. $\sinh^{-1}(e^x) + C.$ **10.** $\frac{1}{2}\tan^{-1}(x^2) + C.$

11. $\tan x + \dfrac{\tan^3 x}{3} + C.$

[Put $u = \tan x$, treating $\sec^4 x$ as $(1 + \tan^2 x)\sec^2 x$.]

5·3. Integration of rational algebraic fractions

A rational algebraic fraction is any fraction in which numerator and denominator are polynomials, e.g.

$$\frac{ax^3 + bx^2 + cx + d}{px^2 + qx + r}.$$

5·31. Dividing out

It must be strongly emphasized that if the degree of the numerator equals or is greater than that of the denominator, the fraction must be divided out until this is not so *before* any integration is attempted. This applies to all functions dealt with in this section, e.g.

$$\frac{3x^3 - 2x^2 + x - 5}{x^2 - 2x + 2} = 3x + 4 + \left(\frac{3x - 13}{x^2 - 2x + 2}\right),$$

$$
\begin{array}{r}
3x + 4 \\
x^2 - 2x + 2\overline{)3x^3 - 2x^2 + x - 5} \\
\underline{3x^3 - 6x^2 + 6x} \\
4x^2 - 5x - 5 \\
\underline{4x^2 - 8x + 8} \\
3x - 13.
\end{array}
$$

5·32. Denominator of first degree

These are straightforward.

For example:
$$\int \frac{x^3}{(x-2)}\, dx.$$

By division,
$$I = \int \left(x^2 + 2x + 4 + \frac{8}{(x-2)} \right) dx$$
$$= \tfrac{1}{3}x^3 + x^2 + 4x + 8 \log_e C(x-2).$$

EXERCISE 20

1. $\displaystyle\int \frac{3x-5}{3-2x}\, dx.$

2. $\displaystyle\int \frac{x}{(1-2x)}\, dx.$

3. $\displaystyle\int \frac{x^3}{(2x-1)}\, dx.$

4. $\displaystyle\int \frac{x^2}{px-q}\, dx.$

ANSWERS

1. $-\tfrac{3}{2}x + \tfrac{1}{4}\log_e C(3-2x).$ 2. $-\tfrac{1}{2}x - \tfrac{1}{4}\log_e C(1-2x).$

3. $\tfrac{1}{6}x^3 + \tfrac{1}{8}x^2 + \tfrac{1}{8}x + \tfrac{1}{16}\log_e C(2x-1).$

4. $\dfrac{x^2}{2p} + \dfrac{qx}{p^2} + \dfrac{q^2}{p^3}\log_e C(px-q).$

5·33. Denominator of second degree, factorizable

After dividing out, if necessary, the integrand is split into *partial fractions*.

The method of splitting into partial fractions will be given in the following examples. No proof of the operations will be given; only the technique.

When the degree of the numerator is *less* than that of the denominator
$$\frac{f(x)}{(x-p)(x-q)^2(ax^2+bx+c)}$$
$$\equiv \frac{A}{(x-p)} + \frac{B}{(x-q)} + \frac{C}{(x-q)^2} + \frac{Dx+E}{(ax^2+bx+c)}. \qquad (1)$$

The following rules apply:

(i) To each linear factor, such as $(x-p)$, there corresponds $A/(x-p)$, where A is a number.

(ii) A repeated factor, such as $(x-q)^2$, gives rise to fractions of the form $\dfrac{B}{(x-q)} + \dfrac{C}{(x-q)^2}$ $(B, C$ numbers).

(iii) To a quadratic factor there corresponds a fraction with a linear numerator $(Dx+E)$.

(*a*) *The 'cover-up' rule*. This can only be applied to linear factors. If applied to a repeated factor it only gives the numerator corresponding to the *highest* power of the repeated factor (that is, C in equation (1)).

Consider
$$\frac{3x-2}{(x-2)(x+1)} \equiv \frac{A}{(x-2)} + \frac{B}{(x+1)}.$$

To find A, cover up the factor $(x-2)$ in the left-hand side and put $x=2$ (i.e. the value of x which makes $x-2=0$), in what is left uncovered.

This gives
$$A = \frac{3 \cdot 2 - 2}{2+1} = \frac{4}{3}.$$

For B, $(x+1)$ is covered up and x put equal to -1, giving
$$B = \frac{3 \cdot (-1) - 2}{-1-2} = \frac{5}{3}.$$

Thus
$$\frac{3x-2}{(x-2)(x+1)} \equiv \frac{4}{3(x-2)} + \frac{5}{3(x+1)}.$$

The student can verify this by putting the right-hand side over a common denominator.

(*b*) *Equating coefficients*. When the 'cover-up' rule is not applicable the following method can be used, or the one given in (*c*) below:
$$\frac{3x-2}{(x-2)(x^2+2x-1)} \equiv \frac{A}{(x-2)} + \frac{Bx+C}{(x^2+2x-1)}. \tag{2}$$

The right-hand side is put over a common denominator:
$$\frac{(3x-2)}{(x-2)(x^2+2x-1)} \equiv \frac{A(x^2+2x-1)+(Bx+C)(x-2)}{(x-2)(x^2+2x-1)}.$$

Thus
$$3x-2 \equiv A(x^2+2x-1)+(Bx+C)(x-2). \tag{3}$$

Using the 'cover-up' rule in (2), $A = \frac{4}{7}$. Then (3) becomes
$$(3x-2) \equiv \tfrac{4}{7}(x^2+2x-1)+(Bx+C)(x-2). \tag{4}$$

Equating coefficients of

x^2:
$$0 = \tfrac{4}{7} + B. \quad \therefore \ B = -\tfrac{4}{7}.$$

Numerical terms:
$$-2 = -\tfrac{4}{7} - 2C. \quad \therefore \ C = \tfrac{5}{7}.$$

$$\therefore \ \frac{3x-2}{(x-2)(x^2+2x-1)} \equiv \frac{4}{7(x-2)} - \frac{(4x-5)}{7(x^2+2x-1)}.$$

(c) At stage (4) in paragraph (b) above an alternative method is to give x specially selected values. This can be done as, being an identity, equation (4) is true for all values of x.

The selected values for x are obviously chosen to make the working as short as possible.

For example, in equation (4) put $x = 0$:

$$-2 = -\tfrac{4}{7} + (-2)C. \quad \therefore \; C = \tfrac{5}{7}.$$

Putting $x = 1$, $\qquad 1 = \tfrac{8}{7} + (-1)(B + C).$

$$\therefore \; B = -\tfrac{4}{7}.$$

The same values as before.

The two methods in paragraphs (b) and (c) are equivalent. If convenient they may be intermixed in the same example.

EXAMPLES

Only denominators of the second degree will be given here; denominators of higher degree will be considered in § 5·35.

(1)
$$\int \frac{(5x-4)}{x^2 - 8x + 12}\, dx = \int \frac{(5x-4)}{(x-6)(x-2)}\, dx.$$

Let
$$\frac{5x-4}{(x-6)(x-2)} \equiv \frac{A}{(x-6)} + \frac{B}{(x-2)}.$$

By the 'cover-up' rule: $A = \tfrac{13}{2}$, $B = -\tfrac{3}{2}$.

$$\therefore \; I = \frac{13}{2} \int \frac{1}{(x-6)}\, dx - \frac{3}{2} \int \frac{1}{(x-2)}\, dx$$

$$= \tfrac{13}{2} \log_e (x-6) - \tfrac{3}{2} \log_e C(x-2).$$

(2)
$$\int \frac{(8-x)}{(x-2)^2}\, dx.$$

Let
$$\frac{8-x}{(x-2)^2} \equiv \frac{A}{(x-2)} + \frac{B}{(x-2)^2}.$$

By the 'cover-up' rule $B = 6$.

Thus
$$8 - x \equiv A(x-2) + 6.$$

Put $x = 1$: $\qquad 7 = -A + 6. \quad \therefore \; A = -1.$

Thus
$$I = \int -\frac{1}{x-2}\, dx + 6 \int \frac{1}{(x-2)^2}\, dx$$

$$= -\frac{6}{(x-2)} - \log_e C(x-2).$$

EXERCISE 21

1. Prove (i) $\int \dfrac{1}{x^2 - a^2} dx = \dfrac{1}{2a} \log_e C\left(\dfrac{x-a}{x+a}\right)$ $(x^2 > a^2)$;

(ii) $\int \dfrac{1}{a^2 - x^2} dx = \dfrac{1}{2a} \log_e C\left(\dfrac{a+x}{a-x}\right)$ $(x^2 < a^2)$.

Note. (i) is another standard integral and should be memorized.

Integrate with respect to x:

2. $\dfrac{2x-5}{x^2-5x+6}$.

3. $\dfrac{x^2}{x^2-5x+4}$.

4. $\dfrac{x+1}{(x-1)^2}$.

5. $\dfrac{5+x^2}{9-x^2}$.

6. $\dfrac{x^4}{x^2-5}$.

7. $\dfrac{x^2+1}{5x-2x^2}$.

8. $\dfrac{x^2}{2x^2+x-3}$.

9. $\displaystyle\int_4^5 \dfrac{4x+3}{(x-3)^2} dx$.

10. $\displaystyle\int_3^4 \dfrac{x}{x^2-3x+2} dx$.

11. $\displaystyle\int_5^\infty \dfrac{1}{x^2-7x+12} dx$.

ANSWERS

2. $\log_e C(x^2 - 5x + 6)$.

3. $x + \frac{16}{3} \log_e (x-4) - \frac{1}{3} \log_e C(x-1)$.

4. $\log_e C(x-1) - \dfrac{2}{(x-1)}$.   **5.** $\frac{7}{3} \log_e C\left(\dfrac{3+x}{3-x}\right) - x$.

In QQ. 5 and 6 the results of Q. 1 should be used.

6. $\frac{1}{3}x^3 + 5x + \dfrac{5\sqrt{5}}{2} \log_e C\left(\dfrac{x-\sqrt{5}}{x+\sqrt{5}}\right)$.

7. $-\frac{1}{2}x + \frac{1}{5} \log_e x - \frac{29}{20} \log_e C(5 - 2x)$.

8. $\frac{1}{2}x + \frac{1}{5} \log_e (x-1) - \frac{9}{20} \log_e C(2x+3)$.

9. $4\log_e 2 + 7{\cdot}5 \simeq 10{\cdot}27$.

10. $\log_e \frac{8}{3} \simeq 0{\cdot}981$.

11. $\log_e 2$.

Note. Collect up logs before putting in the limits.

Note. $\underset{x \to \infty}{\text{Lt}} \log\left(\dfrac{x-4}{x-3}\right) = \underset{x \to \infty}{\text{Lt}} \log\left(\dfrac{1-4/x}{1-3/x}\right) = \log 1 = \mathbf{0}.$

5·34. Denominator of second degree, not factorizable

The following standard integrals will be used:

(i) $$\int \frac{1}{x^2 + a^2}\,dx = \frac{1}{a}\tan^{-1}\frac{x}{a} + C.$$

(ii) $$\int \frac{1}{x^2 - a^2}\,dx = \frac{1}{2a}\log_e C\left(\frac{x-a}{x+a}\right) \quad (x^2 > a^2).$$

(iii) $$\int \frac{1}{a^2 - x^2}\,dx = \frac{1}{2a}\log_e C\left(\frac{a+x}{a-x}\right) \quad (x^2 < a^2).$$

The method used is to complete the square of the denominator. Three worked examples, are given.

EXAMPLES

(1) $$\int \frac{1}{2x^2 + 8x + 9}\,dx.$$

In completing the square it is safer to make the coefficient of x^2 unity, if not so already.

$$2x^2 + 8x + 9 = 2(x^2 + 4x + \tfrac{9}{2})$$
$$= 2\{(x+2)^2 - 4 + \tfrac{9}{2}\}$$
$$= 2\{(x+2)^2 + \tfrac{1}{2}\}.$$

$$\therefore I = \frac{1}{2}\int \frac{1}{\{(x+2)^2 + \tfrac{1}{2}\}}\,dx = \frac{1}{2}\int \frac{1}{\left\{(x+2)^2 + \left(\dfrac{1}{\sqrt{2}}\right)^2\right\}}\,dx$$

$$= \tfrac{1}{2}\sqrt{2}\tan^{-1}\sqrt{2}\,(x+2) + C.$$

(2) $$\int \frac{x-4}{x^2 + 3x - 5}\,dx.$$

Here the numerator is of first degree in x (linear). The numerator is made the derivative of the denominator, apart from a constant term.

Derivative of $x^2 + 3x - 5$ is $2x + 3$:

$$x - 4 = \tfrac{1}{2}(2x + 3) - \tfrac{3}{2} - 4 = \tfrac{1}{2}(2x + 3) - \tfrac{11}{2}.$$

The steps are: first write down the derivative $(2x + 3)$; then multiply by the number ($\frac{1}{2}$ in this case) which makes the coefficient of x the same as in the numerator; finally, adjust the constant term to be the same as in the numerator.

Thus

$$I = \int \frac{1}{2} \frac{(2x+3)}{(x^2+3x-5)}\,dx - \int \frac{11}{2(x^2+3x-5)}\,dx$$

$$= \tfrac{1}{2}\log_e C(x^2+3x-5) - \frac{11}{2}\int \frac{1}{\{(x+\tfrac{3}{2})^2-\tfrac{9}{4}-5\}}\,dx$$

$$= \tfrac{1}{2}\log_e C(x^2+3x-5) - \frac{11}{2}\int \frac{1}{\{(x+\tfrac{3}{2})^2-\tfrac{29}{4}\}}\,dx$$

$$= \tfrac{1}{2}\log_e C(x^2+3x-5) - \frac{11}{2\sqrt{(29)}}\log_e\left(\frac{2x+3-\sqrt{(29)}}{2x+3+\sqrt{(29)}}\right).$$

(3)
$$\int \frac{x^3+1}{x^2+4}\,dx.$$

By division:
$$\frac{x^3+1}{x^2+4} = x - \frac{4x-1}{x^2+4}.$$

$$\therefore\; I = \int x\,dx - 2\int \frac{2x-\tfrac{1}{2}}{x^2+4}\,dx$$

$$= \tfrac{1}{2}x^2 - 2\int \frac{2x}{x^2+4}\,dx + \int \frac{1}{x^2+4}\,dx$$

$$= \tfrac{1}{2}x^2 - 2\log_e C(x^2+4) + \tfrac{1}{2}\tan^{-1}\tfrac{1}{2}x.$$

EXERCISE 22

Integrate:

1. $\dfrac{1}{x^2+8x+25}.$ **2.** $\dfrac{1}{x^2+6x-4}.$ **3.** $\dfrac{1}{11-4x-2x^2}.$

4. $\dfrac{1}{x^2-4x+12}.$ **5.** $\dfrac{1}{4-2x-x^2}.$ **6.** $\dfrac{1}{5x^2-17}.$

7. $\dfrac{4x+5}{x^2+2x+2}.$ **8.** $\dfrac{x^2+4x}{x^2+4x-1}.$ **9.** $\dfrac{4x-3}{x^2-5}.$

10. $\dfrac{x^3}{x^2+a^2}.$ **11.** $\dfrac{5x-1}{x^2-3x+5}.$ **12.** $\dfrac{x^2-1}{x^2+5x+3}.$

13. $\displaystyle\int_0^6 \frac{1}{x^2+5x+6}\,dx.$ **14.** $\displaystyle\int_0^1 \frac{x^2-2x+4}{x^2+2x+2}\,dx.$

15. $\displaystyle\int_4^\infty \frac{1}{x^2+4x-5}\,dx.$

Answers

1. $\frac{1}{3}\tan^{-1}\left(\frac{x+4}{3}\right)+C.$ **2.** $\frac{1}{2\sqrt{(13)}}\log_e C\left(\frac{x+3-\sqrt{(13)}}{x+3+\sqrt{(13)}}\right).$

3. $\frac{1}{2\sqrt{(26)}}\log_e C\left(\frac{\sqrt{(26)}+2x+2}{\sqrt{(26)}-2x-2}\right).$

4. $\frac{1}{\sqrt{8}}\tan^{-1}\frac{(x-2)}{\sqrt{8}}+C.$ **5.** $\frac{1}{2\sqrt{5}}\log_e C\left(\frac{\sqrt{5}+1+x}{\sqrt{5}-1-x}\right).$

6. $\frac{1}{2\sqrt{(85)}}\log_e C\left(\frac{\sqrt{5}\,x-\sqrt{(17)}}{\sqrt{5}\,x+\sqrt{(17)}}\right).$

7. $2\log_e C(x^2+2x+2)+\tan^{-1}(x+1).$

8. $x+\frac{1}{2\sqrt{5}}\log_e C\left(\frac{x+2-\sqrt{5}}{x+2+\sqrt{5}}\right).$

9. $2\log_e C(x^2-5)-\frac{3}{10}\sqrt{5}\log_e\left(\frac{x-\sqrt{5}}{x+\sqrt{5}}\right).$

10. $\frac{1}{2}x^2-\frac{1}{2}a^2\log_e C(x^2+a^2).$

11. $\frac{5}{2}\log_e C(x^2-3x+5)+\frac{13}{\sqrt{(11)}}\tan^{-1}\left(\frac{2x-3}{\sqrt{(11)}}\right).$

12. $x-\frac{5}{2}\log_e C(x^2+5x+3)+\frac{17}{2\sqrt{(13)}}\log_e\left(\frac{2x+5-\sqrt{(13)}}{2x+5+\sqrt{(13)}}\right).$

13. $\log_e\frac{4}{3}\simeq 0\cdot 288.$

14. $1-2\log_e\frac{5}{2}+6(\tan^{-1}2-\tan^{-1}1)\simeq 1\cdot 1.$

15. $\frac{1}{6}\log_e 3.$

5·35. Denominator of higher degree than the second

For work of this standard the only types likely to be met with are: (i) where the numerator is a numerical multiple of the derivative of the denominator, (ii) where the denominator easily factorizes into linear and/or quadratic factors. Type (i) has already been dealt with in § 5·1. Type (ii) is dealt with by the method of partial fractions explained in sub-section § 5·33.

Examples

(1) $\displaystyle\int\frac{2x+3}{(x+1)(x^2-5x+6)}\,dx.$

$$\frac{2x+3}{(x+1)(x^2-5x+6)}\equiv\frac{A}{(x+1)}+\frac{B}{(x-2)}+\frac{C}{(x-3)}.$$

By the 'cover-up' rule:

$$A = \tfrac{1}{12}, \quad B = -\tfrac{7}{3}, \quad C = \tfrac{9}{4}.$$

$$\therefore \; I = \frac{1}{12}\int \frac{1}{x+1}\,dx - \frac{7}{3}\int \frac{1}{x-2}\,dx + \frac{9}{4}\int \frac{1}{x-3}\,dx$$

$$= \tfrac{1}{12}\log_e (x+1) - \tfrac{7}{3}\log_e (x-2) + \tfrac{9}{4}\log_e C(x-3).$$

(2) $$\int \frac{x^2}{(x-1)(x^2+4)}\,dx.$$

$$\frac{x^2}{(x-1)(x^2+4)} \equiv \frac{A}{(x-1)} + \frac{Bx+C}{x^2+4}$$

By 'cover-up' rule: $A = \tfrac{1}{5}$.

$$\therefore \; x^2 \equiv \tfrac{1}{5}(x^2+4) + (x-1)(Bx+C).$$

Put $x = 0$: $$0 = \tfrac{4}{5} - C, \quad C = \tfrac{4}{5}.$$

Equate coefficients of x^2:

$$1 = \tfrac{1}{5} + B, \quad B = \tfrac{4}{5}.$$

Thus $$I = \frac{1}{5}\int \frac{1}{x-1}\,dx + \frac{4}{5}\int \frac{x+1}{x^2+4}\,dx$$

$$= \tfrac{1}{5}\log_e C(x-1) + \frac{2}{5}\int \frac{2x}{x^2+4}\,dx + \frac{4}{5}\int \frac{1}{x^2+4}\,dx$$

$$= \tfrac{1}{5}\log_e C(x-1) + \tfrac{2}{5}\log (x^2+4) + \tfrac{2}{5}\tan^{-1}\tfrac{1}{2}x.$$

Exercise 23

Integrate:

1. $$\frac{x^2+1}{x(x^2-4)}.$$

2. $$\frac{8-x}{(x+1)(x-2)^2}.$$

3. $$\frac{1}{x(x^2+1)}.$$

4. $$\frac{6x^2+24x+32}{x^3+6x^2+16x+2}.$$

5. $$\frac{1}{(x-1)^2(x^2+1)}.$$

Answers

1. $\tfrac{5}{8}\log_e (x^2-4) - \tfrac{1}{4}\log_e Cx.$ 2. $\log_e C\left(\dfrac{x+1}{x-2}\right) - \dfrac{2}{x-2}.$

3. $\log_e Cx - \tfrac{1}{2}\log_e (x^2+1).$ 4. $2\log_e C(x^3+6x^2+16x+2).$

5. $\tfrac{1}{4}\log_e (x^2+1) - \tfrac{1}{2}\log_e C(x-1) - \dfrac{1}{2(x-1)}.$

5·4. Integration of irrational fractions of type

$$\frac{px+q}{\sqrt{(ax^2+bx+c)}}.$$

The following standard integrals will be used:

(i) $\displaystyle\int \frac{dx}{\sqrt{(a^2-x^2)}} = \sin^{-1}\frac{x}{a}+C;$ (ii) $\displaystyle\int \frac{dx}{\sqrt{(a^2+x^2)}} = \sinh^{-1}\frac{x}{a}+C;$

(iii) $\displaystyle\int \frac{dx}{\sqrt{(x^2-a^2)}} = \cosh^{-1}\frac{x}{a}+C.$

5·41. Numerator unity

Make the coefficient of x^2 unity, if necessary; then complete the square of the quadratic.

EXAMPLE

$$\int \frac{dx}{\sqrt{(8-5x-3x^2)}} = \frac{1}{\sqrt{3}}\int \frac{dx}{\sqrt{(\frac{8}{3}-\frac{5}{3}x-x^2)}} = \frac{1}{\sqrt{3}}\int \frac{dx}{\sqrt{\{\frac{8}{3}+\frac{25}{36}-(\frac{5}{6}+x)^2\}}}$$

$$= \frac{1}{\sqrt{3}}\int \frac{dx}{\sqrt{\{\frac{121}{36}-(x+\frac{5}{6})^2\}}} = \frac{1}{\sqrt{3}}\int \frac{dx}{\sqrt{\{(\frac{11}{6})^2-(x+\frac{5}{6})^2\}}}$$

$$= \frac{1}{\sqrt{3}}\sin^{-1}\left(\frac{x+\frac{5}{6}}{\frac{11}{6}}\right)+C = \frac{1}{\sqrt{3}}\sin^{-1}\left(\frac{6x+5}{11}\right)+C.$$

In the case of definite integrals, if the answer is an inverse hyperbolic function, the safest way is to put in the limits first and then use the log form. For example,

$$\sinh^{-1}2 - \sinh^{-1}1 = \log_e\{2+\sqrt{(2^2+1)}\} - \log\{1+\sqrt{(1^2+1)}\}$$

$$= \log_e\frac{2+\sqrt{5}}{1+\sqrt{2}}.$$

5·42. Numerator linear

The working is very similar to that in § 5·34.

EXAMPLES

(1) $$\int \frac{x+1}{\sqrt{(x^2-2x)}}\,dx.$$

Derivative of x^2-2x is $2x-2$.

$$x+1 \equiv \tfrac{1}{2}(2x-2)+2.$$

$$\therefore\ I = \frac{1}{2}\int \frac{2x-2}{\sqrt{(x^2-2x)}}\,dx + 2\int \frac{dx}{\sqrt{(x^2-2x)}}$$

$$= \frac{1}{2}\int \frac{du}{\sqrt{u}} + 2\int \frac{dx}{\sqrt{\{(x-1)^2-1\}}}, \quad \text{where} \quad u = x^2 - 2x$$

$$= \sqrt{u} + 2\cosh^{-1}(x-1) + C$$

$$= \sqrt{(x^2-2x)} + 2\cosh^{-1}(x-1) + C.$$

(2) $$\int \sqrt{\left(\frac{x+2}{x-3}\right)}\,dx.$$

The denominator is converted into a quadratic, and the numerator into the form $px+q$, by multiplying by $\sqrt{(x+2)}/\sqrt{(x+2)}$:

$$I = \int \frac{x+2}{\sqrt{\{(x-3)(x+2)\}}}\,dx = \int \frac{x+2}{\sqrt{(x^2-x-6)}}\,dx$$

$$= \frac{1}{2}\int \frac{2x-1}{\sqrt{(x^2-x-6)}}\,dx + \frac{5}{2}\int \frac{dx}{\sqrt{(x^2-x-6)}}$$

$$= \frac{1}{2}\int \frac{du}{\sqrt{u}} + \frac{5}{2}\int \frac{dx}{\sqrt{\{(x-\frac{1}{2})^2-\frac{25}{4}\}}}, \quad \text{where} \quad u = x^2 - x - 6$$

$$= \sqrt{u} + \tfrac{5}{2}\cosh^{-1}\left(\frac{x-\frac{1}{2}}{\frac{5}{2}}\right) + C$$

$$= \sqrt{(x^2-x-6)} + \tfrac{5}{2}\cosh^{-1}\left(\frac{2x-1}{5}\right) + C.$$

Exercise 24

Integrate:

1. $\dfrac{1}{\sqrt{(2x^2+3x)}}.$

2. $\dfrac{1}{\sqrt{(2x^2-7x+5)}}.$

3. $\dfrac{1}{\sqrt{(x^2+10x-11)}}.$

4. $\dfrac{1}{\sqrt{(7-6x-x^2)}}.$

5. $\dfrac{1}{\sqrt{\{(x-1)(x-4)\}}}.$

6. $\dfrac{1}{\sqrt{(9x^2-4ax)}}.$

7. $\dfrac{2x+3}{\sqrt{(x^2+5x+6)}}.$

8. $\sqrt{\left(\dfrac{3-x}{2+x}\right)}.$

9. $\dfrac{3x-4}{\sqrt{(3x^2+4x+7)}}.$

10. $\displaystyle\int_{-1}^{1} \frac{dx}{\sqrt{(x^2+2x+5)}}.$

11. $\displaystyle\int_{0}^{2} \frac{dx}{\sqrt{(3+2x-x^2)}}.$

12. $\displaystyle\int_{-1}^{1} \frac{x\,dx}{\sqrt{(x^2+4x+7)}}.$

ANSWERS

1. $\dfrac{1}{\sqrt{2}}\cosh^{-1}\left(\dfrac{4x+3}{3}\right)+C.$

2. $\dfrac{1}{\sqrt{2}}\cosh^{-1}\left(\dfrac{4x-7}{3}\right)+C.$

3. $\cosh^{-1}\left(\dfrac{x+5}{6}\right)+C.$

4. $\sin^{-1}\left(\dfrac{x+3}{4}\right)+C.$

5. $\cosh^{-1}\left(\dfrac{2x-5}{3}\right)+C.$

6. $\tfrac{1}{3}\cosh^{-1}\left(\dfrac{9x-2a}{2a}\right)+C.$

7. $2\sqrt{(x^2+5x+6)}-2\cosh^{-1}(2x+5)+C.$

8. $\sqrt{(6+x-x^2)}+\tfrac{5}{2}\sin^{-1}\left(\dfrac{2x-1}{5}\right)+C.$

9. $\sqrt{(3x^2+4x+7)}-2\sqrt{3}\sinh^{-1}\left(\dfrac{3x+2}{\sqrt{(17)}}\right)+C.$

10. $\sinh^{-1}1-\sinh^{-1}0=\log_e(1+\sqrt{2})\simeq 0\cdot 881.$

11. $2\sin^{-1}\tfrac{1}{2}=\tfrac{1}{3}\pi.$

12. $2\sqrt{3}-2-2\log_e\left(\dfrac{3+2\sqrt{3}}{3}\right)\simeq -0\cdot 072.$

5·5. Integration of trigonometric functions

5·51. Sec x

Up to the present stage the integration of all the simple trigonometric functions except $\sec x$ and $\csc x$ have been obtained. A method for integrating $\sec x$ is now given:

$$\int \sec x\,dx = \int \frac{1}{\cos x}\,dx = \int \frac{\cos x}{\cos^2 x}\,dx$$

$$= \int \frac{\cos x}{1-\sin^2 x}\,dx = \int \frac{du}{1-u^2}, \quad \text{where} \quad \sin x = u$$

$$= \tfrac{1}{2}\log_e \frac{1+u}{1-u} = \tfrac{1}{2}\log_e C\left(\frac{1+\sin x}{1-\sin x}\right).$$

This standard integral is not usually remembered in this form.

$$\tfrac{1}{2}\log_e\left(\frac{1+\sin x}{1-\sin x}\right) = \tfrac{1}{2}\log_e\frac{(1+\sin x)^2}{(1-\sin^2 x)} = \tfrac{1}{2}\log_e\frac{(1+\sin x)^2}{\cos^2 x}$$

$$= \log_e\left(\frac{1+\sin x}{\cos x}\right) = \log_e(\sec x+\tan x).$$

Also

$$\tfrac{1}{2}\log_e\left(\frac{1+\sin x}{1-\sin x}\right)=\tfrac{1}{2}\log_e\left(\frac{1+\dfrac{2t}{1+t^2}}{1-\dfrac{2t}{1+t^2}}\right),\quad\text{where}\quad t=\tan\tfrac{1}{2}x$$

$$=\tfrac{1}{2}\log_e\frac{(1+t)^2}{(1-t)^2}=\log_e\left(\frac{1+\tan\tfrac{1}{2}x}{1-\tan\tfrac{1}{2}x}\right)$$

$$=\underline{\log_e\tan\left(\tfrac{1}{4}\pi+\tfrac{1}{2}x\right).}$$

Thus, as a standard integral,

$$\int\sec x\,dx=\log_e C(\sec x+\tan x)=\log_e C\tan\left(\tfrac{1}{4}\pi+\tfrac{1}{2}x\right).$$

EXERCISE 25

Prove, by methods similar to those of § 5·51, that

$$\int\operatorname{cosec} x\,dx=-\log_e C(\operatorname{cosec} x+\cot x)=\log_e C\tan\tfrac{1}{2}x.$$

5·52. $\displaystyle\int\sin mx\cos nx\,dx$

Because of their importance in harmonic analysis this type of integral is revised.

The formulae used are:

$$\sin A\cos B=\tfrac{1}{2}[\sin(A+B)+\sin(A-B)],$$
$$\cos A\cos B=\tfrac{1}{2}[\cos(A+B)+\cos(A-B)],$$
$$\sin A\sin B=\tfrac{1}{2}[\cos(A-B)-\cos(A+B)].$$

$\cos A\sin B$ may be treated as $\sin B\cos A$.

EXAMPLES

(1) $\displaystyle\int 3\sin 2x\sin 5x\,dx=\frac{3}{2}\int(\cos 3x-\cos 7x)\,dx$

$$=\underline{\frac{\sin 3x}{2}-\tfrac{3}{14}\sin 7x+C.}$$

(2) $\displaystyle\int 5\cos 3x\sin x\,dx=5\int\sin x\cos 3x\,dx$

$$=\frac{5}{2}\int\{\sin 4x+\sin(-2x)\}\,dx$$

$$=\frac{5}{2}\int(\sin 4x-\sin 2x)\,dx=\underline{\frac{5}{2}\left(\frac{\cos 2x}{2}-\frac{\cos 4x}{4}\right)+C.}$$

5·53. $\int \sin{}^m x \cos{}^n x \, dx$

(a) If one or both powers are odd, 'extract' the *odd* power and use it for du/dx in the substitution method.

(1) $$\int \sin^5 x \, dx = \int \sin^4 x (\sin x \, dx).$$

Let $u = \cos x$, then $du = -\sin x \, dx$.
Converting $\sin^4 x$ into $(1 - \cos^2 x)^2$:

$$I = \int (1 - u^2)^2 (-du) = -u + \tfrac{2}{3}u^3 - \frac{u^5}{5} + C$$

$$= \tfrac{2}{3} \cos^3 x - \cos x - \tfrac{1}{5} \cos^5 x + C.$$

The working may be written

$$\int \sin^5 x \, dx = \int \sin^4 x (\sin x \, dx) = -\int (1 - \cos^2 x)^2 \, d(\cos x)$$

$$= \tfrac{2}{3} \cos^3 x - \cos x - \tfrac{1}{5} \cos^5 x + C.$$

In the last stage the integration is done thinking of '$\cos x$' as a single variable.

(2) $$\int \sin^2 x \cos^3 x \, dx = \int \sin^2 x \cos^2 x (\cos x \, dx)$$

$$= \int \sin^2 x (1 - \sin^2 x) \, d(\sin x)$$

$$= \int (\sin^2 x - \sin^4 x) \, d(\sin x)$$

$$= \frac{\sin^3 x}{3} - \frac{\sin^5 x}{5} + C.$$

(b) If all powers are even, the double-angle formulae of trigonometry are used, repeatedly if necessary, to convert to standard integrals, with the help, sometimes, of types in § 5·52.

EXAMPLE

$$\int \sin^2 x \cos^4 x \, dx = \int (\sin^2 x \cos^2 x) \cos^2 x \, dx$$

$$= \frac{1}{4} \int \sin^2 2x \cos^2 x \, dx = \frac{1}{8} \int \sin^2 2x (1 + \cos 2x) \, dx$$

$$= \frac{1}{16} \int (1 - \cos 4x)(1 + \cos 2x) \, dx$$

$$= \frac{1}{16} \int (1 + \cos 2x - \cos 4x - \cos 2x \cos 4x) \, dx$$

$$= \tfrac{1}{16}x + \tfrac{1}{32}\sin 2x - \tfrac{1}{64}\sin 4x - \frac{1}{32} \int (\cos 6x + \cos 2x) \, dx$$

$$= \tfrac{1}{16}(x + \tfrac{1}{2}\sin 2x - \tfrac{1}{4}\sin 4x) - \frac{1}{32}\left(\frac{\sin 6x}{6} + \frac{\sin 2x}{2}\right) + C$$

$$= \tfrac{1}{16}(x + \tfrac{1}{4}\sin 2x - \tfrac{1}{4}\sin 4x - \tfrac{1}{12}\sin 6x) + C.$$

Note. In vol. II of this book easy methods of dealing with this general type of integral are given if the limits of integration are 0 to $\frac{1}{2}\pi$.

EXERCISE 26

1. $\int \sec 3x \, dx.$

2. $\int \sin 2x \cos x \, dx.$

3. $\int_0^{\frac{1}{4}\pi} \sin 5x \sin 2x \, dx.$

4. $\int \frac{1}{2} \csc 2x \, dx.$

5. $\int 5 \cos 9x \cos 2x \, dx.$

6. $\int_0^{\frac{1}{4}\pi} \sin x \cos^3 x \, dx.$

7. $\int \sin^4 x \, dx.$

8. $\int_0^{\pi} \cos^2 x \sin^3 x \, dx.$

9. $\int \sin^2 2x \cos^2 2x \, dx.$

10. $\int_0^{\frac{1}{4}\pi} \sec (2x - \tfrac{1}{4}\pi) \, dx.$

11. $\int \cos (3 - 2x) \sin (x - 4) \, dx.$

12. $\int \sin^3 (x - \tfrac{1}{8}\pi) \, dx.$

ANSWERS

1. $\frac{1}{3}\log_e C(\sec 3x + \tan 3x).$

2. $-\frac{1}{2}\left(\cos x + \frac{\cos 3x}{3}\right).$

3. $-\frac{2}{21}.$

4. $\frac{1}{4}\log_e C \tan x.$

5. $\frac{5}{2}\left(\frac{\sin 11x}{11} + \frac{\sin 7x}{7}\right) + C.$

6. $\frac{1}{4}.$

7. $\frac{3}{8}x - \frac{1}{4}\sin 2x + \frac{1}{32}\sin 4x + C.$ **8.** $\frac{4}{15}.$

9. $\frac{1}{8}x - \frac{1}{64}\sin 8x + C.$ **10.** $\frac{1}{2}\log_e\left(\dfrac{\sqrt{2}+1}{\sqrt{2}-1}\right) \simeq 0.881.$

11. $\frac{1}{2}\{\cos(x+1) - \frac{1}{3}\cos(3x-7)\} + C.$

12. $\frac{1}{3}\cos^3(x - \frac{1}{8}\pi) - \cos(x - \frac{1}{8}\pi) + C.$

5·6. Trigonometrical and hyperbolic substitutions

Many integrals are made easier by changing the variable to a trigonometric or hyperbolic function.

Some useful substitutions are:

$\sqrt{(a^2 + x^2)}$:   Put $x = a\tan\theta$ or $x = a\sinh u.$

$\sqrt{(x^2 - a^2)}$:   Put $x = a\sec\theta$ or $x = a\cosh u.$

$\sqrt{(a^2 - x^2)}$:   Put $x = a\sin\theta.$

EXAMPLES

(1) $$\int \sqrt{(9 - x^2)}\, dx.$$

Let $x = 3\sin\theta$; then $dx = 3\cos\theta\, d\theta$ and $\sqrt{(9 - x^2)} = 3\cos\theta.$

$$\therefore\ I = \int 3\cos\theta\, 3\cos\theta\, d\theta = \int 9\cos^2\theta\, d\theta$$

$$= \frac{9}{2}\int (1 + \cos 2\theta)\, d\theta = \frac{9}{2}\theta + \frac{9}{4}\sin 2\theta + C.$$

Now $\sin 2\theta = 2\sin\theta\,\cos\theta = 2\dfrac{x}{3}\dfrac{\sqrt{(9 - x^2)}}{3}.$

$$\therefore\ \underline{I = \frac{9}{2}\sin^{-1}\tfrac{1}{3}x + \tfrac{1}{2}x\,\sqrt{(9 - x^2)} + C.}$$

In the case of a definite integral the whole working could be kept in terms of θ, as long as the limits of integration are changed to θ-values.

(2) $$\int_0^a \frac{dx}{(x^2 + a^2)^{\frac{3}{2}}}.$$

Let $x = a\tan\theta$; then $dx = a\sec^2\theta\, d\theta$ and $(x^2 + a^2)^{\frac{3}{2}} = a^3\sec^3\theta.$
When $x = a$, $\theta = \frac{1}{4}\pi$ and when $x = 0$, $\theta = 0.$

$$\therefore\ I = \int_0^{\frac{1}{4}\pi} \frac{a\sec^2\theta\, d\theta}{a^3\sec^3\theta} = \int_0^{\frac{1}{4}\pi} \frac{1}{a^2}\cos\theta\, d\theta = \frac{1}{a^2}[\sin\theta]_0^{\frac{1}{4}\pi}$$

$$= \frac{1}{a^2}\left[\frac{1}{\sqrt{2}} - 0\right] = \underline{\frac{1}{\sqrt{2}\,a^2}}.$$

EXERCISE 27

1. $\int_0^1 \sqrt{(4-x^2)}\, dx.$ 2. $\int \dfrac{x^2}{\sqrt{(x^2+a^2)}}\, dx.$ 3. $\int \dfrac{1}{x^2 \sqrt{(1-x^2)}}\, dx.$

4. $\int_0^1 x^2 \sqrt{(1-x^2)}\, dx.$ 5. $\int \dfrac{1}{x \sqrt{(x^2-1)}}\, dx.$

6. $\int \sqrt{\{9+(x-2)^2\}}\, dx$ [put $x-2 = 3\sinh u$].

7. $\int_0^4 (16-x^2)^{\frac{3}{2}}\, dx.$

ANSWERS

1. $\frac{1}{3}\pi + \frac{1}{2}\sqrt{3}.$ 2. $\frac{1}{2}x\sqrt{(x^2+a^2)} - \frac{1}{2}a^2 \sinh^{-1} x/a + C.$

3. $-\dfrac{\sqrt{(1-x^2)}}{x} + C.$ 4. $\frac{1}{16}\pi.$ 5. $\sec^{-1}x + C.$

6. $\frac{9}{2}\sinh^{-1}\left(\dfrac{x-2}{3}\right) + \dfrac{(x-2)}{2}\sqrt{\{9+(x-2)^2\}} + C.$ 7. $48\pi.$

5·7. Integration by parts

This method of integration comes from the inverse of the formula for the derivative of a product.

As

$$\frac{d}{dx}(uv) = u\frac{dv}{dx} + v\frac{du}{dx},$$

then

$$uv = \int u\left(\frac{dv}{dx}\right) dx + \int v\left(\frac{du}{dx}\right) dx$$

or

$$\int u\left(\frac{dv}{dx}\right) dx = uv - \int v\left(\frac{du}{dx}\right) dx.$$

Note that in order to use this formula effectively the integrand must consist of two factors, one of which (u) is capable of easy differentiation (to give du/dx). and the other (dv/dx) capable of easy integration (to give v). The last condition is usually the decisive one. The success of this method of course depends on the fact that $\int v\left(\dfrac{du}{dx}\right) dx$ should be easier to evaluate than the original integral. It is particularly effective when the expression to be integrated consists of the product of two different types of functions, e.g. $x^2 \log_e x$; $e^x x^3$; $x \sin^{-1} x.$

EXAMPLES

(1) $$\int x^2 \log_e x \, dx.$$

Here, $\log_e x$ is not easily integrated. In consequence x^2 is chosen as 'dv/dx', and $\log_e x$ as 'u'.

Thus

$$\int x^2 \log_e x \, dx = \frac{x^3}{3} \log_e x - \int \frac{x^3}{3} \frac{1}{x} \, dx$$

$$= \frac{x^3}{3} \log_e x - \int \frac{x^2}{3} \, dx$$

$$= \frac{x^3}{3} \log_e x - \frac{x^3}{9} + C.$$

(2) $$\int x^2 e^{2x} \, dx.$$

This is a case where the process has to be repeated until an easy integral is obtained on the right-hand side. Both x^2 and e^{2x} are easily integrated. Note, however, that if x^2 is continually integrated it never vanishes, whilst e^{2x} never vanishes when integrated *or* differentiated. But if x^2 is continually differentiated it gives a number (2) after two differentiations. x^2 is thus chosen as 'u'.

$$\int x^2 e^{2x} \, dx = x^2 \frac{e^{2x}}{2} - \int e^{2x} x \, dx.$$

Continuing the process, choosing x as 'u':

$$I = \frac{x^2 e^{2x}}{2} - \left\{ x \frac{e^{2x}}{2} - \int \frac{e^{2x}}{2} 1 \cdot dx \right\}$$

$$= \frac{x^2 e^{2x}}{2} - \frac{x e^{2x}}{2} + \frac{e^{2x}}{4} + C.$$

The proverb 'never change horses in midstream' applies admirably to integration by parts. Once having chosen one of the functions to be 'u', when differentiated it becomes the new 'u', if the process is to be repeated.

(3) $$\int_0^1 \tan^{-1} x \, dx = \int_0^1 \tan^{-1} x \cdot 1 \cdot dx.$$

Here there is no choice, $\tan^{-1} x$ must be taken as 'u' which leaves 1 as dv/dx:

$$I = [x \tan^{-1} x]_0^1 - \int_0^1 x \frac{1}{1 + x^2} dx$$

$$= [x \tan^{-1} x]_0^1 - \frac{1}{2} \int_0^1 \frac{2x}{1 + x^2} dx$$

$$= [x \tan^{-1} x]_0^1 - \tfrac{1}{2} [\log (1 + x^2)]_0^1 = \tfrac{1}{4}\pi - \tfrac{1}{2} \log_e 2.$$

(4) When each factor gives the same type of function as itself, however many times it be differentiated or integrated, the technique is slightly different.

$$\int e^{3x} \cos 2x \, dx = e^{3x} \frac{\sin 2x}{2} - \int \frac{\sin 2x}{2} 3e^{3x} dx.$$

Integrating by parts a second time:

$$I = \frac{e^{3x} \sin 2x}{2} - \frac{3}{2} \left\{ -\frac{\cos 2x}{2} e^{3x} - \int -\frac{\cos 2x}{2} 3e^{3x} dx \right\},$$

$$I = \frac{e^{3x} \sin 2x}{2} + \tfrac{3}{4} \cos 2x \, e^{3x} - \frac{9}{4} \int e^{3x} \cos 2x \, dx.$$

The original integral has recurred on the right-hand side:

$$I = \frac{e^{3x} \sin 2x}{2} + \tfrac{3}{4} e^{3x} \cos 2x - \tfrac{9}{4} I + C,$$

$$I + \tfrac{9}{4} I = \frac{e^{3x} \sin 2x}{2} + \tfrac{3}{4} e^{3x} \cos 2x + C.$$

$$\therefore \ I = \frac{4}{13} \frac{e^{3x}}{2} (\sin 2x + \tfrac{3}{2} \cos 2x) + A$$

$$= \tfrac{2}{13} e^{3x} (\sin 2x + \tfrac{3}{2} \cos 2x) + A.$$

Obtaining the original integral on the right-hand side of the equation is typical of this type of integrand.

Integration by parts requires long and careful practice before fluency is acquired. The above examples should be followed through carefully two or three times before the exercises are attempted.

EXERCISE 28

Integrate by parts:

1. $x^2 e^{-x}$. 2. $\sin^{-1} x$. 3. $\sqrt{x} \log_e x$.

4. $\dfrac{1}{x^3} \log_e x$. 5. $x \sin x$. 6. $x \tan^{-1} x$.

7. $x^2 \sinh x$. 8. $x \cos 2x$. 9. $x \sec^2 x$.

10. $e^{2x} \sin x.$ **11.** $\displaystyle\int_1^2 x \log_e x \, dx.$ **12.** $\displaystyle\int_0^1 (x+1) e^x \, dx.$

13. $\displaystyle\int_0^{\frac{1}{2}} x \log_e (1+x) \, dx.$

14. Prove that $\displaystyle\int \sqrt{(x^2+a^2)} \, dx = \tfrac{1}{2}x \sqrt{(a^2+x^2)} + \tfrac{1}{2}a^2 \sinh^{-1}(x/a).$

Use the method of §5·7, Example (4), taking, initially, $\sqrt{(x^2+a^2)} = \text{`}u\text{'}$ and $1 = \text{`}dv/dx\text{'}$.

Check the answer by using the substitution method, putting $x = a \sinh u$.

ANSWERS

1. $-(x^2+2x+2) e^{-x} + C.$ **2.** $x \sin^{-1} x + \sqrt{(1-x^2)} + C.$

3. $\tfrac{2}{3}x^{\frac{3}{2}}(\log_e x - \tfrac{2}{3}) + C.$ **4.** $-\dfrac{1}{2x^2}(\log_e x + \tfrac{1}{2}) + C.$

5. $\sin x - x \cos x + C.$ **6.** $\tfrac{1}{2}(x^2+1) \tan^{-1} x - \tfrac{1}{2}x + C.$

7. $(x^2+2) \cosh x - 2x \sinh x + C.$

8. $\tfrac{1}{2}x \sin 2x + \tfrac{1}{4} \cos 2x + C.$ **9.** $x \tan x - \log_e C \sec x.$

10. $\tfrac{1}{5}e^{2x} (2 \sin x - \cos x) + C.$ **11.** $2 \log_e 2 - \tfrac{3}{4} \simeq 0.636.$

12. $e \simeq 2.718.$ **13.** $\tfrac{3}{8}(\tfrac{1}{2} - \log_e \tfrac{3}{2}) \simeq 0.0355.$

MISCELLANEOUS EXERCISES ON INTEGRATION

Integrate:

1. $e^x \sin x.$ **2.** $\dfrac{1}{\sqrt{(16-9x^2)}}.$ **3.** $\dfrac{\sec^2 x}{\tan x}.$

4. $\dfrac{x}{x^2-4}.$ **5.** $\dfrac{1+x}{\sqrt{x}}.$ **6.** $\sin^3 x \cos^4 x.$

7. $\dfrac{\sin 2x}{e^x}.$ **8.** $\dfrac{1}{\sqrt{(x^2+2x+5)}}.$ **9.** $\dfrac{e^x}{1+e^x}.$

10. $\dfrac{5(x+1)}{(x-1)(x+4)}.$ **11.** $\dfrac{x^2-1}{x^2-2x-2}.$ **12.** $\dfrac{x+1}{\sqrt{(x^2+4)}}.$

13. $\sin^4 x.$ **14.** $x \sin x \cos x.$ **15.** $\dfrac{x}{\sqrt{(1+x^4)}}.$

16. $\dfrac{\sinh x}{1+\cosh x}.$ **17.** $\displaystyle\int_1^\infty \dfrac{1}{x^2(x+3)} \, dx.$ **18.** $\dfrac{2x-1}{\sqrt{(x^2+3x+2)}}.$

19. $\dfrac{1}{\sqrt{(1-x-x^2)}}$.   **20.** $7\sin 5x \sin 2x$.   **21.** $\displaystyle\int_1^e \log_e x\,dx$.

22. $\displaystyle\int_0^{\frac{1}{2}\pi} \sin^2 x \cos^4 x\,dx$.   **23.** $\displaystyle\int_0^1 \dfrac{1}{\sqrt{(x^2+x)}}\,dx$.

24. $\displaystyle\int_1^2 \dfrac{1}{x(x^2+1)}\,dx$.   **25.** $\displaystyle\int_0^\pi \tfrac{1}{2}x(1-\cos 2x)\,dx$.

Answers

1. $\tfrac{1}{2}e^x(\sin x - \cos x) + C$.   **2.** $\tfrac{1}{3}\sin^{-1}(\tfrac{3}{4}x) + C$.

3. $\log_e C \tan x$.   **4.** $\tfrac{1}{2}\log C(x^2-4)$.

5. $2\sqrt{x}(1+\tfrac{1}{3}x) + C$.   **6.** $\tfrac{1}{7}\cos^7 x - \tfrac{1}{5}\cos^5 x + C$.

7. $-\tfrac{1}{5}e^{-x}(\sin 2x + 2\cos 2x) + C$.
[Treat integrand as $e^{-x}\sin 2x$.]

8. $\sinh^{-1}\tfrac{1}{2}(x+1) + C$.   **9.** $\log_e C(1+e^x)$.

10. $2\log_e(x-1) + 3\log_e C(x+4)$ or $\log_e C(x-1)^2 (x+4)^3$.

11. $x + \log_e C(x^2-2x-2) + \dfrac{\sqrt{3}}{2}\log_e\left(\dfrac{x-1-\sqrt{3}}{x-1+\sqrt{3}}\right)$.

12. $\sqrt{(x^2+4)} + \sinh^{-1}(\tfrac{1}{2}x) + C$.

13. $\tfrac{1}{4}(\tfrac{3}{2}x - \sin 2x + \tfrac{1}{8}\sin 4x) + C$.

14. $\tfrac{1}{4}(\tfrac{1}{2}\sin 2x - x\cos 2x) + C$. [Treat integrand as $\tfrac{1}{2}x\sin 2x$.]

15. $\tfrac{1}{2}\sinh^{-1}(x^2) + C$. [Substitute $u = x^2$.]

16. $\log_e C(1+\cosh x)$.   **17.** $\tfrac{1}{3} - \tfrac{1}{9}\log_e 4$.

18. $2\sqrt{(x^2+3x+2)} - 4\cosh^{-1}(2x+3) + C$.

19. $\sin^{-1}\left(\dfrac{2x+1}{\sqrt{5}}\right) + C$.   **20.** $\dfrac{7}{2}\left(\dfrac{\sin 3x}{3} - \dfrac{\sin 7x}{7}\right) + C$.

21. 1.   **22.** $\dfrac{\pi}{32}$.

23. $\cosh^{-1} 3 = \log_e(3+2\sqrt{2}) \simeq 1\cdot763$.

24. $\log_e 2 - \tfrac{1}{2}\log_e\tfrac{5}{2} \simeq 0\cdot235$.   **25.** $\tfrac{1}{4}\pi^2$.

CHAPTER 6

SOME APPLICATIONS OF INTEGRATION

6·1. More difficult areas and volumes

In Chapter 2 revision exercises were given on areas, volumes, mean values and root-mean-square values. In this section harder examples will be taken. A summary of formulae to be used, and main points concerning them, which were not detailed in Chapter 2 is now given.

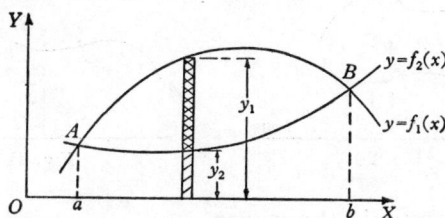

Fig. 18

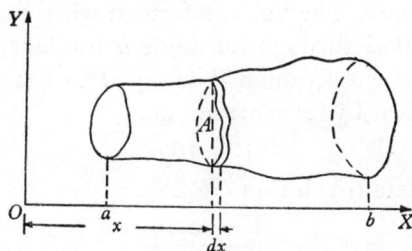

Fig. 19

6·11. Area between two curves

The points of intersection A, B of the curves (fig. 18) are first calculated.

The area completely contained between the two curves is then

$$\int_a^b \{f_1(x) - f_2(x)\}\, dx = \int_a^b (y_1 - y_2)\, dx.$$

6·12. Volumes

The volume of the solid shown in fig. 19 is $\int_a^b A\, dx$, where A is the area of cross-section distance x from O.

If a set of values of A and x only are known, an approximate estimation of the volume can be made using Simpson's rule.

When the solid is one of revolution around the x-axis, A becomes πy^2, as the cross-section is then a circle of radius y.

Note. The student is reminded that the above formulae, as indeed most applications of integration, are obtained from the important theorem

$$\lim_{\delta x \to 0} \sum_{x=a}^{x=b} f(x)\,\delta x = \int_a^b f(x)\,dx,$$

which is demonstrated in a third-year National Certificate Course.

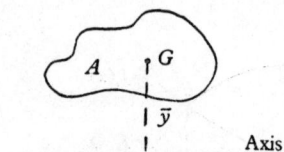
Fig. 20

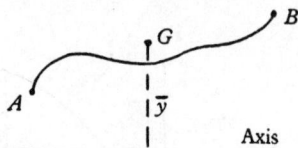
Fig. 21

6·13. Pappus's theorems

(a) *For volumes.* The volume formed when a closed area (A) (fig. 20) is rotated through an angle θ (radians) about an axis which does not cut through it is equal to the area times the distance travelled by its centroid, i.e.

$$V = A\theta\bar{y}.$$

For a complete revolution

$$V = 2\pi A\bar{y}.$$

(b) *For surface areas.* The surface area swept out when a curve (AB) of length L (fig. 21) is rotated through an angle θ (radians) about an axis which does not cut through it is equal to the length of the curve × the distance travelled by its centroid, i.e.

$$S = L\theta\bar{y}.$$

For a complete revolution

$$S = 2\pi L\bar{y}.$$

EXAMPLES

(1) Find the mean value of $\sin pt \sin (pt + \tfrac{1}{3}\pi)$ over a period.

[Sec. A]

$$\sin pt \sin (pt + \tfrac{1}{3}\pi) = \tfrac{1}{2}[\cos \tfrac{1}{3}\pi - \cos (2pt + \tfrac{1}{3}\pi)]$$
$$= \tfrac{1}{4} - \tfrac{1}{2}\cos (2pt + \tfrac{1}{3}\pi).$$

Thus the period is $\dfrac{2\pi}{2p} = \dfrac{\pi}{p}$.

$\therefore$ mean value $= \dfrac{1}{\pi/p - 0} \displaystyle\int_0^{\pi/p} \{\tfrac{1}{4} - \tfrac{1}{2}\cos(2pt + \tfrac{1}{3}\pi)\}\, dt$

$\quad = \dfrac{p}{\pi}\left[\dfrac{t}{4} - \dfrac{1}{2}\dfrac{1}{2p}\sin(2pt + \tfrac{1}{3}\pi)\right]_0^{\pi/p}$

$\quad = \dfrac{p}{\pi}\left[\left\{\dfrac{\pi}{4p} - \dfrac{1}{4p}\sin(2\pi + \tfrac{1}{3}\pi)\right\} - \left\{0 - \dfrac{1}{4p}\sin\tfrac{1}{3}\pi\right\}\right]$

$\quad = \dfrac{p}{\pi}\dfrac{\pi}{4p}, \quad$ as $\quad \sin(2\pi + \tfrac{1}{3}\pi) = \sin\tfrac{1}{3}\pi$

$\quad = \tfrac{1}{4}.$

(2) Find the area completely bounded by the curve $y^2 = 12x$ and the line $3x - 2y + 3 = 0$ (fig. 22).

Curve:
$$y^2 = 12x.$$

Line: $y = \dfrac{3(x+1)}{2}.$

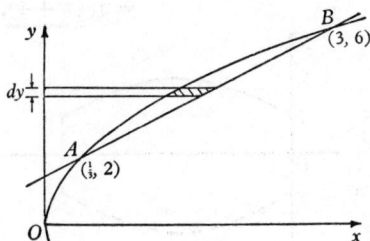

Fig. 22

$\therefore$ At A and B:

$\quad 12x = \tfrac{9}{4}(x+1)^2,$

$\quad 9x^2 - 30x + 9 = 0,$

$\qquad 3(3x-1)(x-3) = 0,$

$\qquad\quad x = \tfrac{1}{3} \quad\text{or}\quad x = 3.$

$\therefore$ A is $(\tfrac{1}{3}, 2)$ and B is $(3, 6)$.

Area of strip is

$(x_{\text{line}} - x_{\text{curve}})\, dy = \{(\tfrac{2}{3}y - 1) - \tfrac{1}{12}y^2\}\, dy = \left(\dfrac{8y - 12 - y^2}{12}\right) dy.$

$\therefore$ Area required $= \dfrac{1}{12}\displaystyle\int_2^6 (8y - 12 - y^2)\, dy$

$\qquad\qquad\qquad = \tfrac{1}{12}[4y^2 - 12y - \tfrac{1}{3}y^3]_2^6 = \tfrac{8}{9}.$

Note. The area could have been found by taking a strip parallel to the y-axis. Its length would have been

$$\sqrt{(12x)} - \tfrac{3}{2}(x+1).$$

Taking a strip parallel to the x-axis avoided square roots.

(3) The curve $y = 1 + \sin x$ is rotated about the x-axis (fig. 23). Prove the volume contained between the surface of revolution so formed and the planes $x = 0$, $x = \pi$ is $\pi(4 + \tfrac{3}{2}\pi)$.

$$\text{Volume} = \int_0^\pi \pi y^2 \, dx$$

$$= \int_0^\pi \pi (1 + \sin x)^2 \, dx$$

$$= \pi \int_0^\pi (1 + 2 \sin x + \sin^2 x) \, dx$$

$$= \pi \int_0^\pi (\tfrac{3}{2} + 2 \sin x - \tfrac{1}{2} \cos 2x) \, dx$$

$$= \pi [\tfrac{3}{2}x - 2 \cos x - \tfrac{1}{4} \sin 2x]_0^\pi$$

$$= \pi [(\tfrac{3}{2}\pi + 2 - 0) - (0 - 2 - 0)]$$

$$= \underline{\pi(\tfrac{3}{2}\pi + 4)}.$$

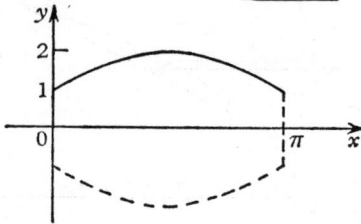

Fig. 23

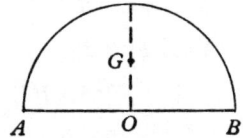

Fig. 24

(4) (i) Find the centroid of a semicircular area.

(ii) A pulley has the shape formed by the rotation of the area shown in fig. 25 about AB as axis. The groove is semicircular. Find the volume of the pulley.

(i) In fig. 24, let r = radius of semicircle. By symmetry, G is on the radius perpendicular to diameter AB. The volume formed if the area is rotated around AB is a sphere of radius r. Area of semicircle is $\tfrac{1}{2}\pi r^2$.

From Pappus's theorem: $\tfrac{4}{3}\pi r^3 = 2\pi . \tfrac{1}{2}\pi r^2 OG$, giving

$$OG = 4r/3\pi. \tag{1}$$

(ii) From (1) $\qquad GM = \dfrac{4}{3\pi}.\dfrac{5}{2} = \dfrac{10}{3\pi} \text{ mm.}$

Volume of cylinder formed by rotation of rectangle $ABCD$ (fig. 25) is $\qquad \pi . 20^2 . 5 = 2000\pi \text{ mm}^3. \tag{2}$

Volume formed by rotation of semicircle about AB is

$$\tfrac{1}{2}\pi(\tfrac{5}{2})^2 . 2\pi OG$$

$$= \tfrac{25}{4}\pi^2(OM - GM)$$

$$= \tfrac{25}{4}\pi^2\left(20 - \frac{10}{3\pi}\right) \text{ mm}^3.$$

From (2) and (3), volume of pulley is

$$2000\pi - \tfrac{25}{4}\pi^2\left(20 - \frac{10}{3\pi}\right)$$

$$\eqsim 5113 \text{ mm}^3.$$

(5) Sketch the curve $x = a\cos^3 t$, $y = a\sin^3 t$. Find the area between the curve and the axes, in the first quadrant.

$$x = a\cos^3 t, \quad y = a\sin^3 t.$$

(i) Writing $-t$ for t does not alter x values, but changes sign of y values.

∴ Graph is symmetrical about the x-axis.

(1)

(ii) Multiples of 2π added to t leave x and y values unaffected.

∴ $t = 0$ to 2π only need be considered.

(2)

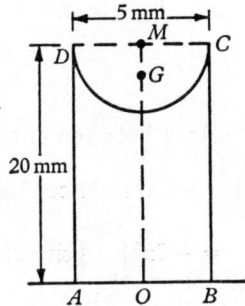

Fig. 25

(iii) For a given value of y, say $a\sin^3 t_1$, putting $t = \pi - t_1$, gives the same y value but changes the sign of the x value, as

$$\cos^3(\pi - t_1) = -\cos^3 t_1.$$

Thus for each y value there is a $\pm x$ value.

∴ Graph is symmetrical about the y-axis. (3)

From (1), (2) and (3) above it is seen that the graph need only be sketched from $t = 0$ to $\tfrac{1}{2}\pi$. After this symmetry enables the sketch to be completed (fig. 26).

t	0	$\tfrac{1}{6}\pi$	$\tfrac{1}{4}\pi$	$\tfrac{1}{2}\pi$
x	a	$\dfrac{3\sqrt{3}}{8}a$	$\dfrac{a}{2\sqrt{2}}$	0
y	0	$\tfrac{1}{8}a$	$\dfrac{a}{2\sqrt{2}}$	a

Area of strip $= y \, dx$.

But $y = a \sin^3 t, \quad dx = -3a \cos^2 t \sin t \, dt.$

$\therefore$ Area of strip $= a \sin^3 t (-3a \cos^2 t \sin t) \, dt$

$$= -3a^2 \sin^4 t \cos^2 t \, dt.$$

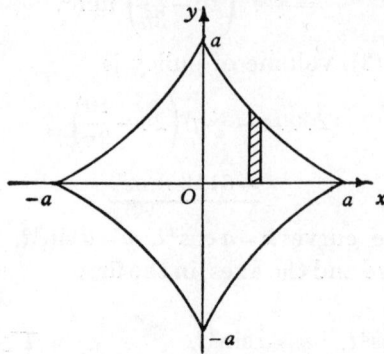

Fig. 26

Area between curve and axis, in first quadrant, is

$$\int_0^{\frac{1}{2}\pi} -3a^2 \sin^4 t \cos^2 t \, dt$$

$$= -3a^2 \int_0^{\frac{1}{2}\pi} \tfrac{1}{4} \sin^2 2t \sin^2 t \, dt = -\tfrac{3}{4} a^2 \int_0^{\frac{1}{2}\pi} \tfrac{1}{2}(1 - \cos 4t) \tfrac{1}{2}(1 - \cos 2t) \, dt$$

$$= -\tfrac{3}{16} a^2 \int_0^{\frac{1}{2}\pi} (1 - \cos 4t - \cos 2t + \cos 4t \cos 2t) \, dt$$

$$= -\tfrac{3}{16} a^2 \int_0^{\frac{1}{2}\pi} \{1 - \cos 4t - \cos 2t + \tfrac{1}{2}(\cos 6t + \cos 2t)\} \, dt$$

$$= -\tfrac{3}{16} a^2 \int_0^{\frac{1}{2}\pi} \{1 - \tfrac{1}{2} \cos 2t - \cos 4t + \tfrac{1}{2} \cos 6t\} \, dt$$

$$= -\tfrac{3}{16} a^2 [t - \tfrac{1}{4} \sin 2t - \tfrac{1}{4} \sin 4t + \tfrac{1}{12} \sin 6t]_0^{\frac{1}{2}\pi}$$

$$= -\tfrac{3}{32} a^2 \pi.$$

The magnitude of the area is $\tfrac{3}{32} \pi a^2$.

The area of the whole figure would be $4 \times \tfrac{3}{32} \pi a^2 = \tfrac{3}{8} \pi a^2$.

The evaluated integral was negative even though the area was above the x-axis.

The reason for this is that in $x = a \cos^3 t$, as t *increases* x *decreases*. Thus the area has been measured in the direction of x *decreasing*. Hence the negative answer.

This example gives a method for evaluating areas under a curve given by parametric co-ordinates.

EXERCISE 29

1. Find the area between the curve $y = x + \sin 2x$ and the x-axis, from $x = 0$ to $x = \frac{1}{2}\pi$.

2. Find the mean value of $2 \cos \omega t \cos (\omega t + \frac{1}{4}\pi)$ over a period.

3. Make a rough sketch of the curve $y^2 = x(x-4)^2$ and find the area of the loop.

4. Find the area of the segment cut off from the curve $y = x(2 - x)$ by the line $y = x$. Find also the volume formed when this segment is rotated about the x-axis.

5. Find the area bounded by the curve $y = e^x$, the y-axis and the line $y = e$.

6. Find the area of the smaller part of the circle $x^2 + y^2 = 12$ which is cut off by the parabola $y = x^2$.

7. Find the volume obtained when the area in Q. 5 is rotated about the y-axis.

8. A semicircular area, of radius a, is rotated about an axis in its plane distant d from the centre of the straight edge and inclined at 60° to it. If the axis of rotation does not cut the area and the area is on the side of the straight edge remote from the axis of rotation, find (a) the volume, (b) the total surface area of the ring formed.   [L.U.]

9. The capping of a stone pillar is a solid with every horizontal cross-section a square. The centres of these squares lie on a vertical axis and the corners lie on the surface of a sphere of radius 0·3 m, whose centre is on the axis, 0·15 m above the plane base of the solid. Calculate the cross-sectional area at any distance x metres from the centre of the sphere, measured along the axis. Hence calculate the volume of the capping.

10. Sketch the curve whose equations are $x = a(t-1)^2, y = at^2$, for values of t in the range $-2 \leqslant t \leqslant 2$. Find the area enclosed by the curve and the line $y = 4a$.

ANSWERS

1. $\frac{1}{8}\pi^2 + 1$.   **2.** $1/\sqrt{2}$.   **3.** $\frac{256}{15}$.   **4.** $\frac{1}{6}$; $\frac{1}{5}\pi$.

5. 1.   **6.** $2\pi + \sqrt{3}$.   **7.** $\pi(e-2) = 0·718\pi$.

8. (a) $\pi^2 a^2 (d + 2a/3\pi)$;   (b) $2\pi(\pi ad + 2ad + a^2)$.

9. $2(0\cdot09 - x^2)\,\text{m}^2$; $0\cdot06075\,\text{m}^3$.

10. $\frac{64}{3}a^2$. Take a strip parallel to the x-axis. Find the length of the strip as the difference in x values for $-t$ and $+t$, both of which give the same y value.

6·2. Lengths of curves

6·21. Definition

No portion of a curve, however small, is identical with a straight line. Before finding a means of calculating the length of a curve it is therefore essential to define exactly what is to be meant by the 'length' of a curve.

Inscribe any n-sided polygon $AP_1P_2\ldots P_{n-1}B$ in the arc of a curve AB (fig. 27). Denote its perimeter $AP_1 + P_1P_2 + \ldots + P_{n-1}B$ by S_n. Keeping A and B fixed, let this polygon be varied in any

Fig. 27

manner such that $n \to \infty$ and the length of every side tends to zero. Under these conditions, if S_n tends to a definite limit, s say, the length of the arc AB is *defined* to be s.

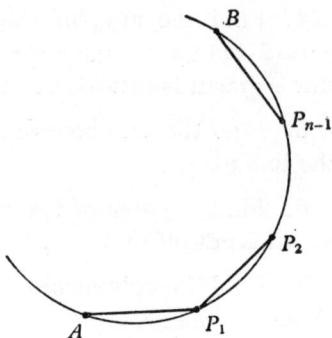

6·22. Sign conventions

What is called a 'right-handed' set of rectangular axes is shown in fig. 28. A convenient way to think of this is that a positive (anti-clockwise) rotation of the x-axis through 90° brings it in line with the y-axis. Clockwise rotations are taken as negative. The positive directions of the axes are, of course, $\overrightarrow{Ox}$ and $\overrightarrow{Oy}$.

The angle between a variable line and a fixed direction is the angle through which the fixed line must be rotated to coincide with the direction of the variable line. If this rotation is anti-clockwise the angle is positive; if clockwise, negative.

It is easily seen that the angles 0 to π and 0 to $-\pi$ cover all possible directions (see. fig. 29).

For rectangular axes the fixed direction is $\overrightarrow{Ox}$.

The angle which a tangent line to a curve makes with $\overrightarrow{Ox}$ is usually denoted by ψ.

If the length of the arc measured from a fixed point A to a point P on the curve is denoted by s, the direction of increase of s can be chosen quite arbitrarily. If a curve is given in the form $y = f(x)$, s *is measured so as to increase as x increases.* If the curve is given parametrically: $x = f_1(t)$, $y = f_2(t)$, s is measured to increase with t.

A line PQ (see fig. 28) drawn along the tangent at P in the direction of s increasing is taken to be *the positive direction of the tangent at P.*

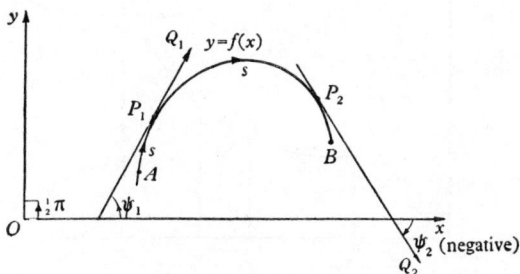

Fig. 28

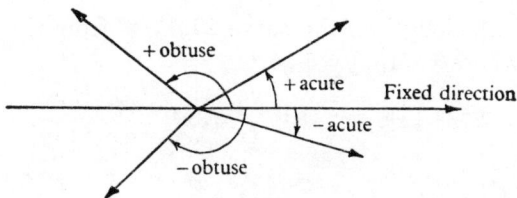

Fig. 29

Thus, in general, the positive direction of the tangent points in the general direction of $\overrightarrow{Ox}$. The angles ψ are then measured as previously described.

6·23. Formula for length of an arc

Let one side of the polygon, mentioned in § 6·21, run from the point (x, y) to the point $(x + \delta x, y + \delta y)$ (fig. 30).

The length of chord PQ is $\sqrt{\{(\delta x)^2 + (\delta y)^2\}}$.

Hence length of arc AB is

$$\lim_{\delta x \to 0} \sum_{x=a}^{x=b} \sqrt{\{(\delta x)^2 + (\delta y)^2\}} = \lim_{\delta x \to 0} \sum_{x=a}^{x=b} \sqrt{\left\{1 + \left(\frac{\delta y}{\delta x}\right)^2\right\}}\, \delta x.$$

Since $\delta y/\delta x \to dy/dx$ as $\delta x \to 0$:

$$\text{Length of arc } AB = \int_a^b \sqrt{\left\{1 + \left(\frac{dy}{dx}\right)^2\right\}}\, dx.$$

If s measures the length of arc from a fixed point (A) to any point (P):

$$\frac{ds}{dx} = \sqrt{\left\{1 + \left(\frac{dy}{dx}\right)^2\right\}}.$$

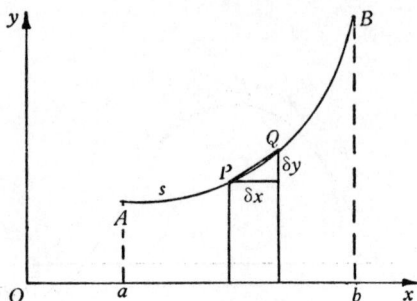

Fig. 30

EXAMPLES

(1) Find the length of the curve $27y^2 = x^3$ from the origin to the point where $x = 15$, $y = 5\sqrt{5}$.

$$27y^2 = x^3. \quad \therefore \quad 54y\frac{dy}{dx} = 3x^2, \quad \frac{dy}{dx} = \frac{x^2}{18y}.$$

Thus $\quad 1 + \left(\frac{dy}{dx}\right)^2 = 1 + \frac{x^4}{324y^2} = 1 + \frac{x^4 \cdot 27}{324x^3} = 1 + \frac{x}{12}.$

Now $\quad s = \int_0^{15} \sqrt{\left\{1 + \left(\frac{dy}{dx}\right)^2\right\}}\, dx = \int_0^{15} \sqrt{\left\{1 + \frac{x}{12}\right\}}\, dx$

$$= \frac{1}{\sqrt{(12)}} \int_0^{15} \sqrt{(12 + x)}\, dx = \frac{1}{2\sqrt{3}} \left[\tfrac{2}{3}(12 + x)^{\frac{3}{2}}\right]_0^{15}$$

$$= \frac{1}{3\sqrt{3}} \left[27^{\frac{3}{2}} - 12^{\frac{3}{2}}\right] = \frac{1}{3\sqrt{3}} \left[3^{\frac{9}{2}} - 8 \cdot 3^{\frac{3}{2}}\right]$$

$$= 3^3 - 8 = \underline{19}.$$

(2) Find the total length of the curve $x = a\cos^3 t$, $y = a\sin^3 t$.

From worked example (5) of §6·13 it is seen that the total length is four times the length of arc in the first quadrant ($t = 0$ to $\frac{1}{2}\pi$).

Now $\qquad x = a\cos^3 t, \quad dx/dt = -3a\cos^2 t\sin t,$

$$y = a\sin^3 t, \quad dy/dt = 3a\sin^2 t\cos t.$$

$$\therefore \ \frac{dy}{dx} = -\tan t \quad \text{and} \quad \sqrt{\left\{1 + \left(\frac{dy}{dx}\right)^2\right\}} = \sqrt{(1 + \tan^2 t)} = \sec t.$$

$$dx = -3a\cos^2 t\sin t\, dt.$$

Thus $\qquad s = 4\int_0^{\frac{1}{2}\pi} -3a\cos^2 t\sin t\sec t\, dt$

$$= -12a\int_0^{\frac{1}{2}\pi} \sin t\cos t\, dt = -6a\int_0^{\frac{1}{2}\pi} \sin 2t\, dt$$

$$= 3a[\cos 2t]_0^{\frac{1}{2}\pi} = 3a(-1-1) = \underline{-6a}.$$

The negative sign appears for the same reason as given in the worked example referred to above.

6·3. Surface areas of solids of revolution

Surface areas of most standard solids may be obtained by use of Pappus's theorem.

Alternatively, referring to fig. 30 of § 6·23, let the arc AB be rotated round the x-axis.

Surface area formed by chord PQ (a frustrum of a cone) is

$$2\pi\left(y + \frac{\delta y}{2}\right)\sqrt{\{(\delta x)^2 + (\delta y^2)\}}.$$

The total surface area is *defined* to be the limit of the sum of the areas formed by rotating the chords. Thus surface area,

$$S = \lim_{\delta x \to 0} \sum_{x=a}^{x=b} 2\pi(y + \tfrac{1}{2}\delta y)\sqrt{\{(\delta x)^2 + (\delta y)^2\}}$$

$$= \lim_{\delta x \to 0} \sum_{x=a}^{x=b} 2\pi(y + \tfrac{1}{2}\delta y)\sqrt{\left\{1 + \left(\frac{\delta y}{\delta x}\right)^2\right\}}\, \delta x,$$

where $\qquad\qquad \delta y \to 0 \ \text{as} \ \delta x \to 0$

$$= \int_a^b 2\pi y\sqrt{\left\{1 + \left(\frac{dy}{dx}\right)^2\right\}}\, dx = \int_{x=a}^{x=b} 2\pi y\, ds.$$

EXAMPLE

Find the area of the surface cut off from a sphere, radius a, by two parallel planes distance h apart. Hence show that this is equal to the surface area cut off by the planes on the circumscribing cylinder.

Fig. 31 shows a section through the centre of the sphere.

The equation of the circle is $x^2 + y^2 = a^2$.

The area required is that formed when the arc AB is rotated round the x-axis. $x^2 + y^2 = a^2$.

Differentiating:

$$2x + 2y\frac{dy}{dx} = 0, \quad \text{giving} \quad \frac{dy}{dx} = -\frac{x}{y}.$$

$$\therefore \quad \sqrt{\left\{1 + \left(\frac{dy}{dx}\right)^2\right\}} = \sqrt{\left(1 + \frac{x^2}{y^2}\right)} = \frac{\sqrt{(x^2 + y^2)}}{y} = \frac{a}{y}.$$

$$\therefore \text{ Area} = \int_b^{b+h} 2\pi y \frac{a}{y} dx = \int_b^{b+h} 2\pi a \, dx = 2\pi a [x]_b^{b+h}$$
$$= 2\pi a h.$$

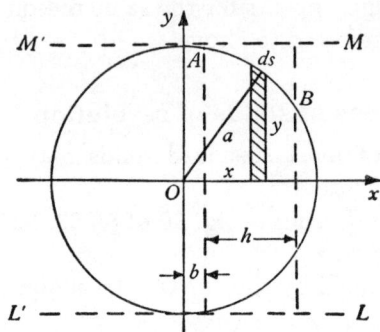

Fig. 31

The radius of the circumscribing cylinder is a. (The figure shows two generators $M'M$ and $L'L$.) The surface area cut off by the two planes distance h apart is $2\pi ah$.

This simple formula for the surface area of the zone of a sphere cut off by two parallel planes should be remembered.

EXERCISE 30

1. Find the length of the arc of the curve $y = \cosh x$ from the point where $x = 1$ to the point where $x = 2$. Also find the surface area swept out if this arc is rotated round the x-axis.

2. Find the length of the arc of the curve $y = \log_e \sec x$ from the point where $x = 0$ to the point where $x = \frac{1}{4}\pi$.

3. Prove that the length of the arc of the parabola given by $x = a \sinh^2 t$, $y = 2a \sinh t$, measured from the vertex $t = 0$ to the point, parameter $t = 1$, is $a(1 + \frac{1}{2}\sinh 2)$.

4. Find the area of the surface of the paraboloid formed by rotating the curve $y^2 = 8x$ about the x-axis from the origin to the point where $x = 16$.

5. The line $4y - 5x = 12$ rotates round the x-axis. Find the surface area of the frustum of the cone generated between the limits $x = 1$ and $x = 5$.

6. Show that the length of the arc of the parabola $y = cx^2$ between $x = 0$ and $x = 2$ is $\int_0^2 \sqrt{(1 + 4c^2x^2)}\, dx$. Hence show that if c is sufficiently small for cubes and higher powers to be neglected, the length of the arc is approximately given by $(2 + \tfrac{16}{3}c^2)$. [Sec. A]

ANSWERS

1. $2 \cdot 45$; $12 \cdot 83\pi$. **2.** $0 \cdot 88$.

4. $277\tfrac{1}{3}\pi \backsimeq 871 \cdot 3$. **5.** $271 \cdot 6$.

6·4. Centroids and centres of gravity

The use of integration in determining centroids of areas under curves and volumes of revolution is done in an Ordinary National Certificate Course, as is also the use of Pappus's theorems in this connexion.

No fundamentally new ideas are introduced when dealing with harder cases. The principle of taking first moments is still followed.

6·41. Centroid of a circular arc

In fig. 32, let a = radius of circle. Let the angle subtended by the arc at the centre O of the circle be 2α. Take the axis of symmetry as x-axis and O as origin. Let G be the centroid. By symmetry G is on Ox.

$$\therefore y_G = 0.$$

Let P be any point on the arc, where $x\hat{O}P = \theta$. Take a small

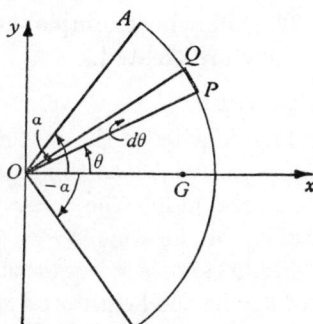

Fig. 32

element PQ, where $\hat{POQ} = d\theta$. Length of $PQ = a\,d\theta$. Distance of PQ from $Oy = a\cos\theta$. Taking first moments round Oy:

$$2a\alpha OG = \int_{-\alpha}^{+\alpha} a\,d\theta\,a\cos\theta = a^2\int_{-\alpha}^{\alpha}\cos\theta\,d\theta = 2a^2\sin\alpha.$$

$$\therefore\ OG = \frac{a\sin\alpha}{\alpha}.$$

Note. In the case of a semicircle $\alpha = \frac{1}{2}\pi$. This gives $OG = 2a/\pi$. This result could also be obtained using Pappus's theorem.

6·42. Centroid of a sector of a circle

Fig. 32 of § 6·41 is used, bearing in mind that it now refers to *areas*.

Area of sector AOB is $\frac{1}{2}a^2 . 2\alpha = a^2\alpha$.

Area of small sector $POQ = \frac{1}{2}a^2\,d\theta$.

As this sector is extremely small it can be treated approximately as a triangle. Its centroid is thus $\frac{2}{3}a$ along the radius. Distance of this centroid from Oy is thus $\frac{2}{3}a\cos\theta$ (approx.).

Taking moments round Oy, in the limit as $d\theta \to 0$:

$$OGa^2\alpha = \int_{-\alpha}^{\alpha}\frac{1}{2}a^2\,d\theta . \frac{2}{3}a\cos\theta = \frac{1}{3}a^3\int_{-\alpha}^{\alpha}\cos\theta\,d\theta = \frac{2}{3}a^3\sin\alpha.$$

$$\therefore\ OG = \frac{2}{3}a\frac{\sin\alpha}{\alpha}.$$

(By symmetry G is again on Ox.)

For a semicircle $\alpha = \frac{1}{2}\pi$ and $OG = 4a/3\pi$, a result already found from Pappus's theorem.

6·43. Variable density

The following example shows how questions involving variable density are treated.

EXAMPLE

The density of the material of which a right circular cone is composed varies as the square of the distance from the vertex, measured along the axis. Find the position of the centre of gravity of the cone.

Fig. 33 shows a cross-section through the axis, Ox of the cone. Let h, r be the height and radius of base of the cone.

Density distance x from vertex is kx^2 (k a constant).

By symmetry, G is on Ox.

Mass of elementary disk PQ is $(\pi y^2 dx)\, kx^2$.

Total mass $= \displaystyle\int_0^h \pi k x^2 y^2\, dx$.

First moment of disk PQ round y-axis is $(\pi y^2 dx)\, kx^2 x$.

Taking moments round Oy for the whole cone

$$\bar{x}\int_0^h \pi k x^2 y^2\, dx = \int_0^h k\pi y^2 x^3\, dx.$$

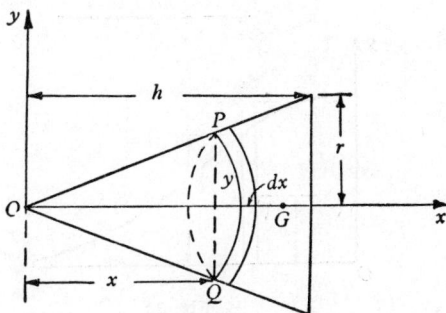

Fig. 33

But $\dfrac{y}{x} = \dfrac{r}{h}$, therefore $y = \dfrac{r}{h} x$. Thus

$$\bar{x}\pi k \int_0^h \frac{r^2}{h^2} x^4\, dx = \pi k \int_0^h \frac{r^2}{h^2} x^5\, dx$$

$$\bar{x}[\tfrac{1}{5}x^5]_0^h = [\tfrac{1}{6}x^6]_0^h,$$

$$\bar{x} = \tfrac{5}{6}h, \quad \bar{y} = 0.$$

6·44. Asymmetrical bodies

If a body is asymmetrical about the axes, the determination of its centroid involves the calculation of both its x and y co-ordinates.

EXAMPLE

Sketch the curve $y = 1 + \cos x$ from $x = 0$ to $x = \pi$. Find the co-ordinates of the centroid of the area between this part of the curve and the x-axis (fig. 34).

Area under curve is

$$\int_0^\pi y\, dx = \int_0^\pi (1 + \cos x)\, dx$$

$$= [x + \sin x]_0^\pi = \pi.$$

Moments round the y-axis:

$$\pi\bar{x} = \int_0^\pi yx\,dx = \int_0^\pi (x + x\cos x)\,dx$$

$$= [\tfrac{1}{2}x^2]_0^\pi + [x\sin x]_0^\pi - \int_0^\pi \sin x\,dx$$

$$= \tfrac{1}{2}\pi^2 + [\cos x]_0^\pi = \tfrac{1}{2}\pi^2 - 2.$$

$$\therefore \quad \bar{x} = \tfrac{1}{2}\pi - \frac{2}{\pi} \simeq 0{\cdot}934.$$

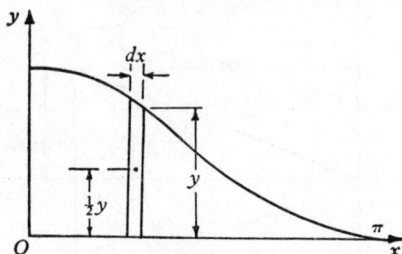

Fig. 34

Moments round the x-axis:

$$\pi\bar{y} = \int_0^\pi y\frac{y}{2}\,dx = \frac{1}{2}\int_0^\pi (1 + \cos x)^2\,dx,$$

$$\pi\bar{y} = \frac{1}{2}\int_0^\pi (1 + 2\cos x + \cos^2 x)\,dx$$

$$= \frac{1}{2}\int_0^\pi \{1 + 2\cos x + \tfrac{1}{2}(1 + \cos 2x)\}\,dx$$

$$= \tfrac{1}{2}[\tfrac{3}{2}x + 2\sin x + \tfrac{1}{4}\sin 2x]_0^\pi$$

$$= \tfrac{3}{4}\pi.$$

$$\therefore \quad \bar{y} = \tfrac{3}{4}.$$

EXERCISE 31

1. The portion of the parabola $y = 2x^2 - 9x$ below the x-axis is rotated round this axis. Find the volume generated and the distance of its centroid from the y-axis.

2. Roughly sketch the curve $a^2y^2 = x^4(a^2 - x^2)$. Find the volume of the solid formed by the rotation of one loop about the x-axis, and the centroid of this solid.

3. Find the area enclosed by the curves $ay = x^2$, $x^3y = a^4$ and the lines $y = 0$, $x = 2a$. Also find the centroid of this area. [L.U.]

4. Find the centroid of a solid hemisphere.

5. Find the centroid of a right circular cone.

6. The linear density of a straight rod AB, length $2a$, is given by $d = d_0(2 - \sqrt{(x/a)})$, where x is the distance from A. Find its mass and the distance of its centre of mass from A.

7. Find the centroid of the area between $y = a^2/x$, the x-axis and the ordinates at $x = a$, $x = 2a$.

8. Find the centroid of the area between $y = \cosh x$, the x-axis and the ordinates at $x = 0$, $x = 2$.

9. A uniform plate is in the form of an equilateral triangle ABC of side 160 mm. Two of the corners have been cut off by circular arcs of radii 20 mm and 60 mm, having their centres at A and C respectively. Find the area of the plate and the distance of its centroid from the edge AB.

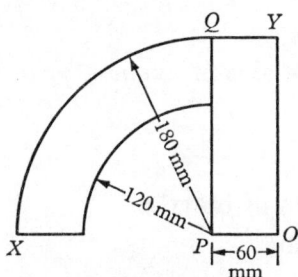

Fig. 35

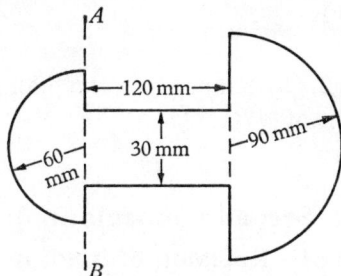

Fig. 36

10. State Pappus's theorem relating to volumes of revolution.

Find the centre of gravity of the uniform sheet of metal shown in fig. 35. The curved edges are quadrants of circles, centre P. QY is straight and parallel to PO.

11. On one side of a diameter of a circular plate of radius $2a$ the surface density is uniform and equal to w. On the other side of the diameter it is uniform and equal to $2w$. Find the distance of the centre of mass of the plate from the centre. Find also the distance the centre of mass is moved if a concentric area of radius a is taken away.

12. A casting, uniform in material and thickness, has the shape and dimensions shown in fig. 36. Find the distance of its centre of gravity from AB.

ANSWERS

1. 773; 2.25.   **2.** $\dfrac{2\pi a^5}{35}$; $\bar{x} = \frac{35}{48}a$, $\bar{y} = 0$.

3. $\frac{17}{24}a^2$; $\bar{x} = \frac{18}{17}a$, $\bar{y} = \frac{189}{680}a$.

4. $\frac{3}{8}$ of radius from base along axis of symmetry.

5. $\frac{3}{4}$ of height from vertex along axis.

6. $\frac{4}{3}d_0 a(3 - \sqrt{2})$; $\frac{3}{35}a(11 - \sqrt{2})$ or $\dfrac{3}{5}\dfrac{(5 - 2\sqrt{2})}{(3 - \sqrt{2})}\,a$.

7. $a/\log_e 2$, $a/4\log_e 2$.   **8.** $(1.24,\ 1.08)$.

9. $9000\ \text{mm}^2$; $35.8\ \text{mm}$.   **10.** $\bar{x} \simeq 102\ \text{mm}$, $\bar{y} \simeq 94\ \text{mm}$.

11. $\dfrac{8a}{9\pi}$; $\dfrac{4a}{27\pi}$.

12. 94.8 mm from AB along the axis of symmetry, to the right of AB.

6·5. Second moments: moments of inertia

6·51. Revision of fundamentals

(i) If a particle, mass m, is distant r from an axis, its moment of inertia (more correctly second moment of mass) about that axis is mr^2. The symbol used is I. For an extended body I is Σmr^2. The summation is usually performed by means of integration.

(ii) If I is written in the form Mk^2, where M is the total mass of the body and k has the dimension of length, k is called the radius of gyration or 'swing radius'.

(iii) For the rotation of a rigid body about a fixed axis:

Kinetic energy $= \frac{1}{2}I\omega^2$, where ω is the angular velocity.

Torque $= I\dfrac{d^2\theta}{dt^2} = I\dfrac{d\omega}{dt}$, where $\dfrac{d^2\theta}{dt^2}$ is the angular acceleration.

When using such equations as those above, absolute units should be used.

For example, M in kilogrammes, r (and k) in metres, $\frac{1}{2}I\omega^2$ in joules and the torque in newton-metres.

(iv) If areas, volumes, etc., are dealt with the corresponding results should properly be called 'second moment of area', etc., although they are often loosely termed moments of inertia.

Second moment of area, or 'moment of inertia of cross-section', is important in the study of the bending of beams.

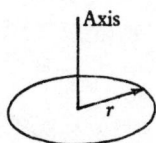

(a)

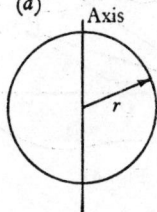

(b)

Fig. 38

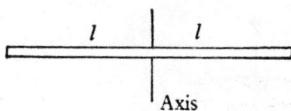

Fig. 37

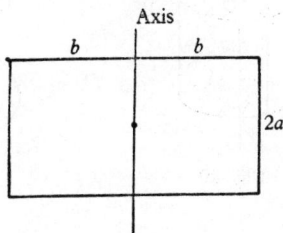

Fig. 39

6·52. Revision of standard results

The following standard results will be assumed known:

Rod. $I = \dfrac{Ml^2}{3}$.

Circular disk. (a) $I = \dfrac{Mr^2}{2}$; (b) $I = \dfrac{Mr^2}{4}$.

Rectangle. $I = \dfrac{Mb^2}{3}$.

6·53. Routh's rule

If a body is *symmetrical* about three perpendicular axes passing through its centre of mass, its moment of inertia about an axis of symmetry is given by

$$M \times \frac{\text{sum of squares of the } other \text{ semi-axes}}{3, \ 4 \text{ or } 5},$$

3 for a thin rod, rectangular lamina or cuboid; 4 for a circular or elliptical disk; 5 for a sphere or ellipsoid.

The results of §6·52 above may be quickly obtained from this rule.

An illustration of the use of the rule in a rather more complicated case is now given.

The moment of inertia of a solid right circular cylinder about an axis through its centre of gravity parallel to an end-face.

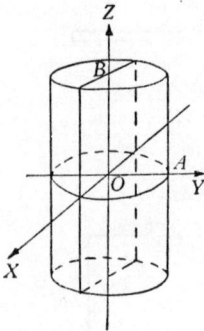

Fig. 40

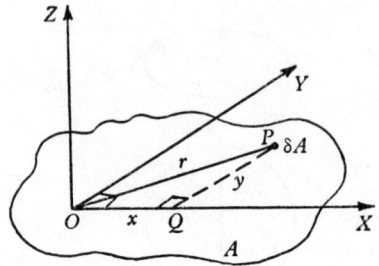

Fig. 41

In fig. 40, let r = radius of an end-face, let h = height of cylinder, let M = total mass.

OX is taken as the required axis parallel to an end-face.

The other semi-axes are:

$OA = r$, associated with a circular section.

$OB = \frac{1}{2}h$, associated with a rectangular section.

Using Routh's rule:

$$I_{OX} = M \times \left[\frac{(\frac{1}{2}h)^2}{3} + \frac{(r)^2}{4} \right]$$

$$= M\left[\frac{h^2}{12} + \frac{r^2}{4} \right].$$

6·54. The perpendicular axes theorem

The moment of inertia of a plane lamina about an axis Oz perpendicular to its plane is equal to the sum of its moments of inertia about any two perpendicular axes Ox, Oy in the plane of the lamina.

In fig. 41, let m = mass per unit area; consider a small element δA at point P distant r from O.

Moment of inertia of δA round Oz is

$$m\delta Ar^2 = m\delta A(x^2 + y^2).$$

$\therefore$ Total moment of inertia of whole lamina

$$= m\Sigma(x^2 + y^2)\,\delta A = m\Sigma x^2\,\delta A + m\Sigma y^2\,\delta A$$

$$= mAk_{Oy}^2 + mAk_{Ox}^2.$$

That is, $$\underline{I_{Oz} = I_{Oy} + I_{Ox}}$$

or $$\underline{k_{Oz}^2 = k_{Oy}^2 + k_{Ox}^2.}$$

EXAMPLE

Given the moment of inertia of a disk, mass M, radius r, about an axis through its centre perpendicular to its plane is $Mr^2/2$, deduce its moment of inertia about a diameter.

In fig. 42 $I_{OX} = I_{OY}$ by symmetry.
But

$$I_{OZ} = I_{OX} + I_{OY}.$$

$$\therefore\ I_{OX} = I_{OY} = \tfrac{1}{2}I_{OZ} = M\,\frac{r^2}{4}.$$

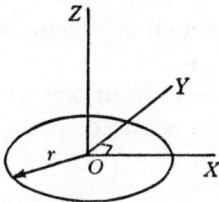

Fig. 42

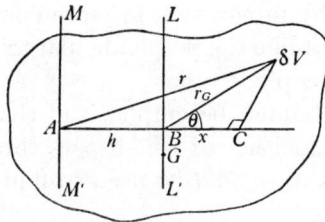

Fig. 43

6·55. The parallel axes theorem

If k_G is the radius of gyration of a body about an axis through its centre of mass G, then k, its radius of gyration about a parallel axis, distant h from the first, is given by

$$k^2 = k_G^2 + h^2.$$

The uneven outline in fig. 43 shows a section of the solid through the plane of the two parallel axes.

Let m = mass per unit volume. Let δV be a small element of volume, *not* necessarily in the plane of the paper. Let the distances of this element from the two axes LL', MM' be r_G and r respectively. Let x be the projection of r_G on the plane through the axes (the plane of the paper). Let θ be the angle between r_G and x.

The cosine rule gives

$$r^2 = r_G + h^2 + 2hr_G \cos\theta = r_G + h^2 + 2hx.$$

Multiply both sides by δVm and sum for all elements δV of the solid:

$$\Sigma m\delta Vr^2 = \Sigma m\delta Vr_G^2 + \Sigma m\delta Vh^2 + \Sigma m\delta V2hx,$$

giving $$Mk^2 = Mk_G^2 + h^2\Sigma m\delta V + 2h\Sigma m\delta Vx$$

or $$Mk^2 = Mk_G^2 + Mh^2.$$

(By definition of the centre of mass $\Sigma m\delta Vx = 0$.)

Thus
$$\underline{k^2 = k_G^2 + h^2.}$$

Points to note are:

(i) This theorem is often used when finding moments of inertia by integration in more complicated cases.

(ii) Unlike the perpendicular axes theorem it is true for solids as well as plane areas.

(iii) It must be emphasized that to 'step' from one axis to another, neither of which pass through the centre of mass, this theorem must *first* be used to find k_G^2.

Examples

(1) Find the moment of inertia of a uniform solid right circular cone about an axis through the vertex parallel to the base. Deduce its moment of inertia about a diameter of the base.

As the density is uniform volumes may be worked with in place of masses.

Divide the cone into thin circular disks, perpendicular to its axis Ox, as shown in fig. 44.

Let h = height of cone and r = base radius.

Second moment of disk PQ about $PQ = (\pi y^2 dx)\frac{1}{4}y^2$.

By the parallel axis theorem,

Second moment of disk PQ about $Oy = (\pi y^2 dx)(\frac{1}{4}y^2 + x^2)$

$$= \frac{\pi r^2}{h^2} x^2 dx \left(\frac{r^2 x^2}{4h^2} + x^2\right) \quad \text{as} \quad \frac{y}{x} = \frac{r}{h}.$$

$$= \frac{\pi r^2}{h^2} \left(\frac{4h^2}{r^2} + 1\right) x^4 dx.$$

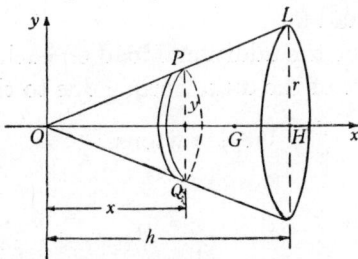

Fig. 44

$\therefore$ Total second moment of volume of cone about Oy is

$$\int_0^h \frac{\pi r^2}{4h^4}(r^2 + 4h^2) x^4 dx = \frac{\pi r^2}{4h^4}(r^2 + 4h^2)\frac{h^5}{5}$$

$$= \frac{\pi r^2 h}{20}(r^2 + 4h^2).$$

But total volume, V, is $\dfrac{\pi r^2 h}{3}$.

$$\therefore \ \underline{I_{Oy} = \tfrac{3}{20} V(r^2 + 4h^2)}.$$

If the cone is of mass M

$$I_{Oy} = \tfrac{3}{20}M(r^2 + 4h^2).$$

If G is the centre of mass of the cone, $OG = \frac{3}{4}h$.

From the parallel axis theorem

$$k_{OY}^2 = k_G^2 + OG^2.$$

Thus $\qquad \tfrac{3}{20}(r^2 + 4h^2) = k_G^2 + (\tfrac{3}{4}h)^2,$

giving $\qquad k_G^2 = \tfrac{3}{20}r^2 + \tfrac{3}{5}h^2 - \tfrac{9}{16}h^2.$

Now $GH = \frac{1}{4}h$ and thus, using the parallel axes theorem again,

$$k_H^2 = k_G^2 + GH^2$$

$$= \tfrac{3}{20}r^2 + \tfrac{3}{5}h^2 - \tfrac{9}{16}h^2 + \tfrac{1}{16}h^2$$

$$= \tfrac{3}{20}r^2 + \tfrac{1}{10}h^2.$$

$$\therefore \ \underline{I_{HL} = \tfrac{1}{20}M(3r^2 + 2h^2)}.$$

(2) A disk 0·4 m in diameter and mass 50 kg is mounted on a shaft which rotates at a speed of 3000 r.p.m. The disk is mounted so that its centre of mass lies on the axis of the shaft, but the normal to the plane of the disk is inclined to the axis of the shaft at an angle of 5 min. Find the extra loads, resulting from this inclination, on bearings placed 0·4 m on each side of the disk.

(Take $\sin 5' = 0 \cdot 00145$.)

Let F newtons be the additional load on each set of bearings. Let O be the centre of the disk. Torque due to the extra loads is

$$0 \cdot 8F \text{ newtons.} \tag{1}$$

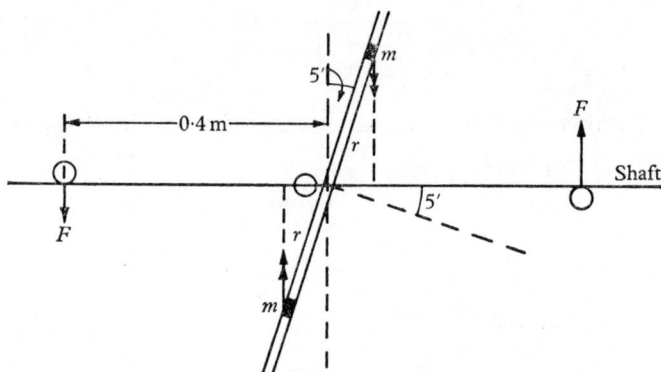

Fig. 45

Consider two small particles of the flywheel, each mass m and distant r from O along a diameter.

Let Ω = angular velocity of disk in radians per sec.

Due to the rotation, each particle has an acceleration $(r \cos 5') \Omega^2$ towards the shaft.

This gives rise to two forces $m(r \cos 5')\Omega^2$, perpendicular to the shaft but in opposite directions.

They form a torque of moment $2mr\Omega^2 \cos 5' (r \sin 5')$.

Summing for all particles of the disk, and using (1),

$$0 \cdot 8F = 2\Omega^2 \sin 5' \cos 5' \, \Sigma mr^2$$

$$\approx 2\Omega^2 \sin 5' \, \Sigma mr^2, \quad \text{taking } \cos 5' \approx 1.$$

$$\therefore \ F = 2 \cdot 5 \Omega^2 \sin 5' \ M k_O^2$$

$$= 2 \cdot 5 \left(\frac{3000 \cdot 2\pi}{60} \right)^2 0 \cdot 00145 \cdot 50 \cdot (0 \cdot 2)^2 \frac{1}{2} \left[k_O^2 = \frac{R^2}{2} \right]$$

$$= 2 \cdot 5 \cdot 14 \cdot 5 \cdot \pi^2 \text{ newtons}$$

$$\simeq 357 \cdot 8 \text{ newtons.}$$

Note that this total load F will be distributed around the axle bearings.

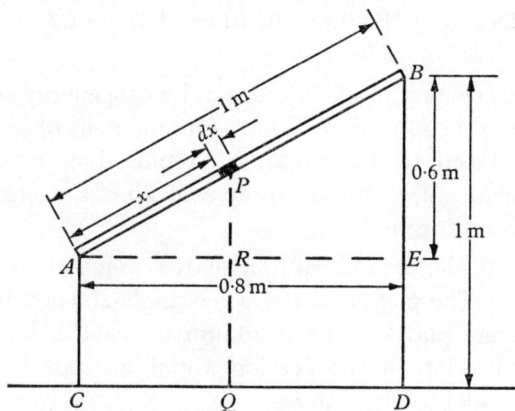

Fig. 46

(3) AB is a uniform bar 1 m long, pivoted at its ends to parallel arms AC and BD, perpendicular to an axis CD, and of lengths 0·4 m and 1 m respectively. Find the radius of gyration of the bar AB about the axis CD.

$\triangle ABE$ is a '3, 4, 5' right-angled triangle (fig. 46). Thus $AE = 0 \cdot 8$ m, $PQ = 0 \cdot 4$ m $+ PR$. But

$$\frac{PR}{x} = \frac{0 \cdot 6}{1}.$$

$$\therefore \ PQ = 0 \cdot 4 + 0 \cdot 6x \tag{1}$$

Let $k =$ radius of gyration of AB round CD. Then

$$1 \times k^2 = \int_0^1 (0 \cdot 4 + 0 \cdot 6x)^2 \, dx = [0 \cdot 16x + 0 \cdot 24x^2 + 0 \cdot 12x^3]_0^1$$

which gives $\qquad k^2 = 0 \cdot 52,$

$$\underline{k \simeq 0 \cdot 72 \text{ m.}}$$

Exercise 32

1. Find in joules the kinetic energy of a flywheel of mass 8 kg, which may be considered a uniform disk of radius 180 mm rotating about its centre at 30 rev/s.

2. Write down the moment of inertia of a rectangle about one edge. A rectangular trap-door, 1·2 m by 0·6 m, can rotate freely about horizontal hinges along one of its shorter sides. If it is released from a horizontal position, find its angular velocity when the door is inclined at 30° to the horizontal, and when it is vertical. (Take g as 9·8 m/s².)

3. A flywheel has 150,000 joules of kinetic energy stored in it when its speed is 300 r.p.m. What is its moment of inertia? The speed is reduced to 150 r.p.m. in 3 min. due to a constant resistive torque. Determine the loss of kinetic energy and the moment of the torque.

4. A box-girder is made of four plates each 200 mm wide and 20 mm thick. The plates enclose a rectangle 200 mm by 120 mm and the flange plates overlap 20 mm on each side. Find the radius of gyration of the section about an axis through the centroid parallel to the flanges.   [L.U.]

5. Show, by integration, that the square of the radius of gyration of a uniform circular disk, of radius r, about an axis perpendicular to its plane and through its centre, is $\frac{1}{2}r^2$.

Also find, by integration, the radius of gyration of a uniform sphere, of diameter d, about a diameter.

6. Obtain, by integration, the 1st and 2nd moments of area about the side of length a of a trapezoidal area with parallel sides of lengths a and b at distance c apart. If k is the radius of gyration about an axis through the centroid parallel to the side of length a, find k^2 in terms of a, b and c.   [L.U.]

7. A rigid body, mass m, whose centre of gravity is G, makes small oscillations about a fixed point O, where $OG = h$. If k is the radius of gyration about the axis of rotation through O, show that the motion is approximately simple harmonic and of period $2\pi\sqrt{(k^2/gh)}$.

(Use the equation: Torque $= I\ddot{\theta}$ and the approximation $\sin\theta \simeq \theta$. This is the compound pendulum formula and will be done in lectures on applied mechanics.)

8. A uniform circular disk of radius r oscillates as a compound pendulum about an axis perpendicular to its plane through a point on its circumference. Find the time period of the oscillation. If the mass of the disk is M, show that the effect of removing a mass m by drilling a circular hole through the centre of the disk is to increase the period in the ratio $\sqrt{\left(\dfrac{3M+m}{3M}\right)} : 1$. [L.U.]

9. Find the moment of inertia of a circular disk, radius r, whose surface density varies as the distance from the centre, about an axis through its centre perpendicular to its plane.

10. The parabola $y^2 = 4x$ is rotated about the x-axis from $x = 0$ to $x = 4$. Find the radius of gyration about Oy of the solid so formed.

ANSWERS

1. 2302 joules.   **2.** 3·5 rad/s.; 4·95 rad/s.

3. 304 kg m^2; 112,500 joules; 59 Nm.

4. 88 mm (approx.).   **5.** $d/\sqrt{10}$.

6. $\frac{1}{6}c^2(a+2b)$; $\frac{1}{12}c^3(a+3b)$; $\dfrac{c^2(a^2+4ab+b^2)}{18(a+b)^2}$.

8. $2\pi\sqrt{(3r/2g)}$.   **9.** $\frac{3}{5}Mr^2$.   **10.** $\sqrt{\dfrac{32}{3}}$.

6·6. Fluid thrusts and centres of pressure

Some of the material in this section is revision of work normally done at Ordinary National Certificate level. On the other hand, it is felt that it may not be too familiar to some of the readers.

All statements made in this section refer to a perfect fluid possessing the following properties:

(i) It has no viscosity. That is, it will not support any shearing force.

(ii) All stress is due to fluid pressure and is normal to every small plane area at a point.

(iii) Pressure is the same in all directions round a point.

(iv) Pressure at any point is instantly transmitted to every other point of the fluid.

(v) The perfect fluid is incompressible.

The pressure at a point at depth h is wh, where w is the weight per unit volume of the fluid. All distances will be assumed measured from the *effective surface level* (E.S.L.).

For example, if atmospheric pressure is to be taken into account the E.S.L. is a distance d above the fluid surface, where $wd = \Pi$, the atmospheric pressure, and w is the weight per unit volume of the fluid.

6·61. Total thrust on a plane area

Fig. 47 shows a plane area A immersed in a liquid and inclined at an angle α to the E.S.L. The right-hand diagram shows a 'side-on' view.

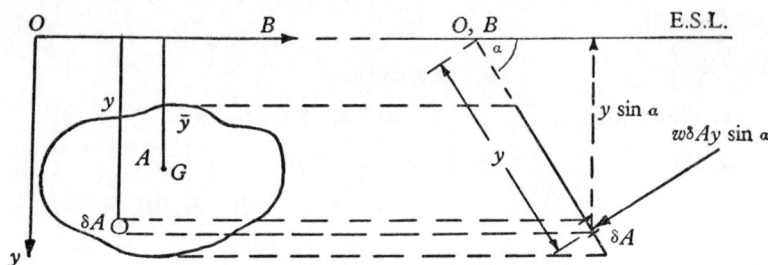

Fig. 47

OB is the line in which the plane of the area cuts the E.S.L. Let w be the weight per unit volume of the liquid.

Take a small element of area δA, distant y from OB. The vertical depth of δA is $y \sin \alpha$.

Thus the thrust on the area δA is $(wy \sin \alpha) \, \delta A$, and is perpendicular to δA.

Total thrust on area A is

$$\lim_{\delta A \to 0} \Sigma wy \sin \alpha \, \delta A = w \sin \alpha \lim_{\delta A \to 0} \Sigma y \, \delta A$$

$$= w \sin \alpha \times \text{sum of first moments of area round } OB$$

$$= w \sin \alpha \, A \bar{y}, \quad \text{where } \bar{y} \text{ is distance of centroid of } A \text{ from } OB.$$

But $\bar{y} \sin \alpha = \bar{z}$, where $\bar{z}$ is the *vertical* depth of the centroid of A.

$\therefore$ Total thrust $= wA\bar{z}$

$$= \text{area} \times \text{pressure at depth of centroid.}$$

6·62. Centre of pressure for a plane area

Fig. 47 will be used again.

Since the total thrust is made up of a system of parallel forces, all perpendicular to the surface, the point at which their resultant (the total thrust) must act can be found by the principle of moments. This point is called the centre of pressure (C.P.). The fluid pressure on the whole area is then equivalent to the total thrust acting at the C.P.

Let $y_p = $ distance of C.P. from OB.

Taking moments round OB:

Moment of thrust on $\delta A = (wy \sin \alpha \, \delta A) \, y$.

$\therefore$ Total thrust $\times y_p = \lim_{\delta A \to 0} \Sigma wy^2 \sin \alpha \, \delta A$.

From § 6·61, total thrust is $w \bar{y} A \sin \alpha$.

$$\therefore \; y_p = \frac{w \sin \alpha \lim \Sigma y^2 \delta A}{w \sin \alpha \, A \bar{y}} = \frac{\lim \Sigma y^2 \delta A}{A \bar{y}}$$

$$= \frac{\int y^2 dA}{A \bar{y}}, \quad \text{where the integration is taken over}$$
$$\text{the whole area.} \quad (1)$$

Now $\int y^2 dA = $ total second moment of area round OB

$$= Ak^2, \quad \text{where } k \text{ is the radius of gyration about } OB.$$

Thus $$y_p = k^2 / \bar{y}. \quad (2)$$

For quite a number of standard cases the formula in equation (2) gives an easy method for calculating the position of the C.P.

Note that:

(i) The position of the C.P. is the same, no matter what the density of the liquid, as long as the liquid is of uniform density.

(ii) The position of the C.P., relative to the immersed area, is not altered if the plane be rotated about OB. This is apparent on noting that α does not appear in equations (1) or (2).

EXAMPLES

(1) Find the total thrust on one face of a vertical circular disk of radius a, immersed in a fluid of weight per unit volume w, when the edge of the disk is in the surface (neglect atmospheric pressure).

The problem will first be worked using the formulae established in §§ 6·61 and 6·62; and then from first principles using integration.

(i) *Using formulae.* Area $= \pi a^2$.

Depth of centre of gravity $= a$.

$$\therefore \text{ Total thrust} = \pi a^2 w a$$
$$= \underline{\pi a^3 w}.$$

$k^2_{XX'} = \tfrac{1}{4}a^2 + a^2 = \tfrac{5}{4}a^2$ (parallel axes theorem).

$$\therefore \text{ Depth of c.p.} = \frac{5a^2}{4a} = \frac{5a}{4}.$$

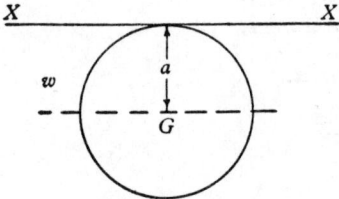

Fig. 48

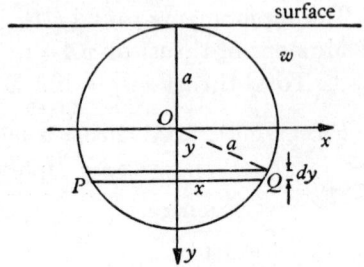

Fig. 49

(ii) *Using integration.* Consider an elementary strip PQ at depth y below the centre O (fig. 49). Pressure at this depth $= w(a+y)$.

Thrust on strip $PQ = w(a+y)\,(2x\,dy)$.

But $x^2 + y^2 = a^2$.

$$\therefore \text{ Thrust on strip } PQ = 2w(a+y)\,\sqrt{(a^2-y^2)}\,dy.$$

Total thrust

$$= \int_{-a}^{+a} 2w(a+y)\,\sqrt{(a^2-y^2)}\,dy$$

$$= 2aw \int_{-a}^{a} \sqrt{(a^2-y^2)}\,dy + 2w \int_{-a}^{a} y\,\sqrt{(a^2-y^2)}\,dy$$

$$= 2aw[\tfrac{1}{2}y\sqrt{(a^2-y^2)} + \tfrac{1}{2}a^2\sin^{-1}y/a]^a_{-a} + 2w[-\tfrac{1}{3}(a^2-y^2)^{\frac{3}{2}}]^a_{-a}$$

$$\text{(see §§5·2 and 5·6)}$$

$$= 2aw[\tfrac{1}{4}\pi a^2 + \tfrac{1}{4}\pi a^2] + 0$$

$$= \underline{\pi a^3 w}.$$

Taking moments around the surface:

Moment of thrust on strip $PQ = \{2w(a+y)\sqrt{(a^2-y^2)}\,dy\}\,(a+y)$
$$= 2w(a+y)^2\sqrt{(a^2+y^2)}\,dy.$$

Thus, if $y_p =$ depth of c.p.:

$$\pi a^3 w y_p = 2w \int_{-a}^{a} (a+y)^2 \sqrt{(a^2-y^2)}\, dy$$

$$= 2w \int_{-a}^{a} a^2 \sqrt{(a^2-y^2)}\, dy + 4w \int_{-a}^{a} ay \sqrt{(a^2-y^2)}\, dy$$

$$+ 2w \int_{-a}^{a} y^2 \sqrt{(a^2-y^2)}\, dy$$

$$= 2wa^2 \left[\frac{\pi^2 a^2}{4} + \frac{\pi a^2}{4} \right] + 0 + 2w \int_{-a}^{a} y^2 \sqrt{(a^2-y^2)}\, dy$$

(see above)

$$= \pi w a^4 + 2w \int_{-a}^{a} y^2 \sqrt{(a^2-y^2)}\, dy. \qquad (1)$$

For the remaining integral, let $y = a \sin\theta$. Then

$$\sqrt{(a^2-y^2)} = a \cos\theta, \quad dy = a \cos\theta\, d\theta.$$

When $y = a$, $\theta = \tfrac{1}{2}\pi$ and when $y = -a$, $\theta = -\tfrac{1}{2}\pi$.

Thus $\quad I = 2w \int_{-\frac{1}{2}\pi}^{\frac{1}{2}\pi} a^2 \sin^2\theta\, a\cos\theta\, a\cos\theta\, d\theta$

$$= \frac{2wa^4}{4} \int_{-\frac{1}{2}\pi}^{\frac{1}{2}\pi} \sin^2 2\theta\, d\theta = \frac{2wa^4}{8} \int_{-\frac{1}{2}\pi}^{\frac{1}{2}\pi} (1 - \cos 4\theta)\, d\theta$$

$$= \frac{2wa^4}{8} \left[\theta - \frac{\sin 4\theta}{4} \right]_{-\frac{1}{2}\pi}^{\frac{1}{2}\pi} = \frac{\pi a^4 w}{4}. \qquad (2)$$

Thus from (1) and (2)

$$\pi a^3 w y_p = \pi w a^4 + \frac{\pi w a^4}{4} = \tfrac{5}{4}\pi w a^4.$$

$$\therefore \underline{y_p = \tfrac{5}{4}a.}$$

From this example it is obvious that the formulae of §§ 6·61 and 6·62 above give by far the quickest method, if applicable.

(2) Find, by any means, the depth of the centre of pressure on one side of a rectangle, immersed in a fluid with one side in the surface (see fig. 50).

Depth of centre of gravity $= b$.

Let depth of c.p. be p.

Now $\qquad\qquad k_{XX'}^2 = \tfrac{1}{3}b^2 + b^2 = \tfrac{4}{3}b^2.$

But $\qquad\qquad p = \dfrac{k_{XX'}^2}{b} = \dfrac{4b^2}{3b} = \tfrac{4}{3}b.$

Centre of pressure is $\tfrac{2}{3}$ the way down the central line.

(3) A vertical canal gate is 4 m broad. On one side the water is 6 m deep and on the other is 3 m deep, measured from the bottom of the gate. Find the magnitude and position of the resultant thrust on the gate (fig. 51).

(Take 1 m^3 of water as 10^3 kg and $g = 9\cdot81$ m/s^2.)

$$\text{Area immersed on one side} = 6 \times 4$$
$$= 24 \text{ m}^2.$$
$$\therefore \text{ Thrust} = 24 \times 3 \times 10^3 \times 9\cdot81 \text{ N}$$
$$= 706\cdot32 \text{ kN}. \tag{1}$$

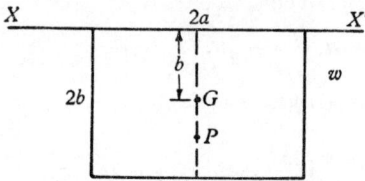

Fig. 50

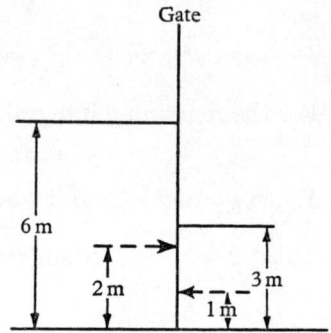

Fig. 51

Similarly, thrust on the other side
$$= 12 \times 1\cdot5 \times 10^3 \times 9\cdot81 \text{ N}$$
$$= 176\cdot58 \text{ kN}. \tag{2}$$
$$\therefore \text{ Resultant thrust} = 529\cdot74 \text{ kN}. \tag{3}$$

Distance of c.p. on left-hand side from bottom $= \frac{1}{3} \times 6 = 2$ m.
Distance of c.p. on right-hand side from bottom $= \frac{1}{3} \times 3 = 1$ m.
Taking moments round bottom of gate:
$$529\cdot74p = 706\cdot32 \times 2 - 176\cdot58$$
$$= 1236\cdot06$$
where p is distance of c.p. from bottom of gate.

This gives $\qquad\qquad p = \frac{7}{3}$ m.

6·63. Thrust on a curved surface

The Principle of Archimedes states that when a solid is wholly or partially, immersed in a fluid at rest the resultant thrust of the fluid on the solid is equal and opposite to the weight of displaced fluid and acts in a vertical line through the centre of gravity of the displaced fluid. The centre of gravity of the displaced fluid is often called the *centre of buoyancy*.

Thrusts on *plane* surfaces have been shown to be equivalent to single forces acting at a specific point, the centre of pressure. In the case of curved surfaces the resultant thrust is quite often not a single force. For a full discussion the student is referred to a book on fluid mechanics. Here, only the simple case of a curved surface bounded by a *plane* curve will be dealt with. In this case the thrust on the curved surface can be found completely and easily.

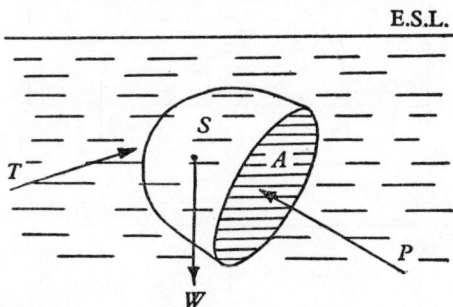

Fig. 52

Let the curved surface be S and let the plane boundary curve circumscribe a *plane* area A (fig. 52). Imagine a thin membrane over the area A and consider the equilibrium of the fluid enclosed by the given surface S and the area A. The thrust P on A can be calculated accurately, as can also its point of application. Let T be the resultant thrust on the curved surface S, and W be the weight of enclosed liquid. Then T, W and P must form three forces in equilibrium. Knowing P and W, T can therefore be calculated.

EXAMPLE

Fig. 53 represents a mould for casting a metal hemisphere of diameter 1 m. Molten metal is poured in through a small aperture. The density of the metal is 8000 kg/m³. (i) Find the upward thrust of molten metal on the mould when $h = 0.25$ m. (ii) If the mould has a mass of 4000 kg, find the value of h when the mould is on the point of being lifted.

The conditions are the same as for a hemispherical surface immersed with vertex at a depth h metres below the free level of the molten metal. This is shown in fig. 54.

The vertically downward thrust, R, on the surface must equal the upward thrust of the molten metal on the mould, for equilibrium.

But $T - R =$ weight of liquid in hemisphere.

Depth of centroid of plane area $ABCD$ is $(h + 0.5)$ m.

$$\therefore \ T = (\pi \cdot 0.5^2)(h + 0.5) \cdot 8000 \cdot 9.81 \text{ N}.$$

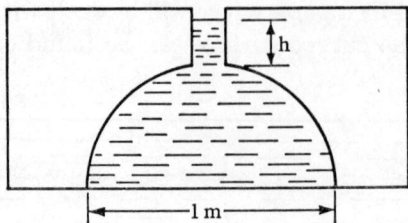

Fig. 53

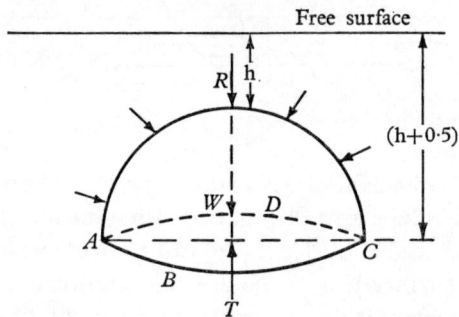

Fig. 54

Weight of liquid in hemisphere $= \frac{2}{3}\pi \cdot (0.5)^3 \cdot 8000 \cdot 9.81$ N.

$$\therefore \ \pi \cdot 2000 \cdot 9.81 \cdot (h + 0.5) - R = \frac{1}{3}\pi \cdot 2000 \cdot 9.81,$$

$$R = \pi \cdot 2000 \cdot 9.81(h + 0.5 - \tfrac{1}{3}),$$

$$R = \pi \cdot 2000 \cdot 9.81(h + \tfrac{1}{6}) \text{ N}.$$

(i) When $h = 0.25$m, $R = \pi \cdot 2000 \cdot 9.81 \cdot \frac{5}{12}$ N

$$\simeq 25{,}686 \text{ N}.$$

(ii) When mould is on the point of being lifted

$$R = 4000 \cdot 9.81 \text{ N}.$$

$$\therefore \ \pi \cdot 2000 \cdot 9.81(h + \tfrac{1}{6}) = 4000 \cdot 9.81,$$

giving $\qquad\qquad h = \dfrac{2}{\pi} - \tfrac{1}{6} \simeq \underline{0.47 \text{ m}.}$

EXERCISE 33

1. A triangular area, height h, is immersed in a liquid with its vertex in the surface, and its base parallel to the surface. Find by integration the depth of the C.P. on one side. Also calculate the depth of the C.P. if the triangle is placed with its base in the free surface.

2. A rectangular plate, immersed in water, has two edges vertical and the other edges, $(x-a)$ metres and $(x+a)$ metres, below the free surface. Prove that the depth of the C.P. below the surface is $(x+a^2/3x)$ metres.

A square trap-door, of side 1 m, in the vertical side of a water tank, has its lower edge horizontal and hinged. It is kept closed by a normal force of 9 kN at its upper edge. Find the greatest depth of water above the lower edge which the tank may contain.

3. A uniform sphere, weight W, radius a, is held in a liquid of weight w per unit volume, with its centre at a height h above the surface of the liquid. Show that the upward force required is $W - \tfrac{1}{3}\pi w(a-h)^2(2a+h)$, if atmospheric pressure is neglected. If the sphere floats in the liquid half immersed, find the work that must be done to move it slowly just clear of the liquid, assuming that the change in the level of the liquid in this movement is negligible.

(*Hint.* For second part find the resultant force, F, acting when the centre is distant x above the surface. Then use, work done
$$= \int F\,dx.)$$

4. An open cubical tank, of edge $2a$, contains a volume $5a^3$ of liquid. If the tank is turned about an edge of the base until the liquid just begins to escape, prove that the liquid pressure on one of the vertical faces is $\tfrac{7}{25}W$, where W is the weight of the liquid in the tank.   [L.U.]

5. A quadrant of a sphere, radius a, is immersed in water so that one of the semicircular plane faces is horizontal and at depth a, and the other is vertical with the lowest point at depth $2a$. Find the magnitude of the thrusts on each of the plane faces and on the curved surface in terms of W, the weight of a volume of water equal to that of the quadrant.   [L.U.]

ANSWERS

1. $\frac{3}{4}h$; $\frac{1}{2}h$.   **2.** 2·5 m.   **3.** $\frac{5}{8}Wa$.

5. $\frac{3}{2}W$ on horizontal face; $\dfrac{W}{2\pi}(3\pi+4)$ on vertical face;

$W\sqrt{\left(\dfrac{(3\pi+4)^2}{4\pi^2}+\dfrac{25}{4}\right)}$ on the curved surface.

6·7. Simple applications to electricity

6·71. Magnetic field intensity near a long straight conductor

The following rule is assumed:

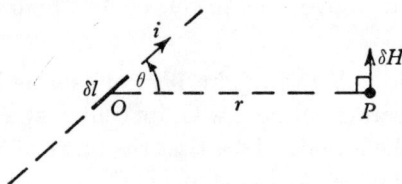

Fig. 55

The strength of the magnetic field due to a circuit of length δl, carrying a current i, at a point distance r, is given by

$$\delta H = \frac{i\,\delta l \sin\theta}{4\pi r^2},$$

where θ is the angle between δl and r (fig. 55).

The field is perpendicular to the plane of δl and the line of $r(OP)$, and in a direction given by Maxwell's corkscrew rule.

In the above formula SI units apply. The field strength is in ampere-turns per metre, the current in amperes and the length in metres.

In fig. 56 the field intensity due to element δl is

$$\delta H = \frac{i\,\delta l \sin\theta}{4\pi x^2}.$$

But $x = R\operatorname{cosec}\theta$, $l = R\cot\theta$.

Thus
$$\delta l \simeq -R\cosec^2\theta\,\delta\theta.$$

$$\therefore\ \delta H \simeq -\frac{i\sin\theta\,\delta\theta}{4\pi R},$$

$$H = \int_{\theta_1}^{\theta_2}\frac{-i\sin\theta\,d\theta}{4\pi R}.$$

If the wire is long, the limits θ_1 and θ_2 are to all intents π and 0. Thus, for a long wire,

$$H := \int_\pi^0 \frac{-i\sin\theta\,d\theta}{4\pi R} = \frac{i}{4\pi R}[\cos\theta]_\pi^0$$

$$= \frac{i}{2\pi R}.$$

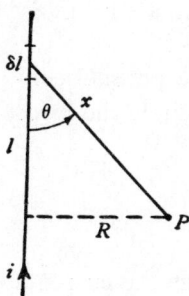

Fig. 56

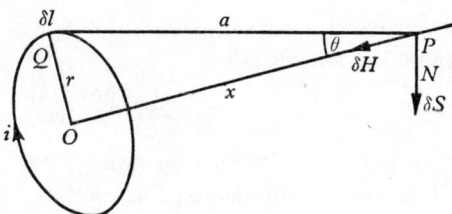

Fig. 57

6·72. Magnetic field intensity due to a circular current

Working in SI units, in fig. 57 let a current i flow in a circular conductor, radius r. The magnetic field intensity at a point distant x from the centre along the axis of the coil will be found. From the 'corkscrew' rule the direction of the field due to an element δl at Q is along $\overrightarrow{PN}$, where PN is perpendicular to PQ.

By symmetry, for the whole circle, the resultant field is along $\overrightarrow{PO}$ as all the components perpendicular to PO cancel out.

From §6·71,
$$\delta S = \frac{i\,\delta l\sin\tfrac12\pi}{4\pi a^2} = \frac{i\,\delta l}{4\pi a^2}.$$

Component along $\overrightarrow{PO} = \delta H = \frac{i\,\delta l\sin\theta}{4\pi a^2}.$

But
$$\sin\theta = \frac{r}{a}, \quad \therefore\ \delta H = \frac{ir\,\delta l}{4\pi a^3}.$$

The total force,

$$H = \int_0^{2\pi r} \frac{ir}{4\pi a^3} dl = \frac{ir}{4\pi a^3} 2\pi r = \frac{ir^2}{2a^3}$$

$$= \frac{ir^2}{2(r^2 + x^2)^{\frac{3}{2}}}.$$

If there are n coils, assuming the coils are thin,

$$H = \frac{nir^2}{2(r^2 + x^2)^{\frac{3}{2}}}.$$

6·73. Force between two long parallel conductors

At a distance R from a long conductor carrying a current i, the field intensity is $i/2\pi R$ (see § 6·71).

Thus, on unit length (1 m in M.K.S. units) of a parallel conductor carrying a current i', distant R from the first, the force per metre length is

$$\frac{ii'\mu_0}{2\pi R} \text{ newtons.}$$

(μ_0 = permeability of free space.)

The force is repulsive or attractive, according as the currents are in opposite or in the same directions.

Note. $\mu_0 = 4\pi \times 10^{-7} H/\text{m}$ for SI units.

EXERCISE 34

1. An indefinitely long straight wire has a charge Q coulombs per metre. Calculate the electric field strength at a point P distant r metres from the wire. (The electric force at distance r metres due to a point charge of q coulombs is $q/4\pi r^2 \epsilon$ volts per metre, where ϵ is the absolute permittivity of the surrounding medium.)

2. Two parallel wires carry equal and opposite high-frequency currents $\pm i$ amperes. All the current may be assumed concentrated evenly on the surface of the wires. The radius, a metres, of the wire is small compared with the distance, d metres, between them. Using the result of § 6·71, show that the total flux between the wires, per metre length, is given by

$$\int_a^{d-a} \frac{\mu_0 i}{2\pi} \left(\frac{1}{x} + \frac{1}{d-x} \right) dx \text{ webers.}$$

Deduce that the inductance per metre length of line is

$$\frac{\mu_0}{\pi} \log_e \left(\frac{d-a}{a}\right) \text{ henrys} \simeq \frac{\mu_0}{\pi} \log_e \frac{d}{a} \text{ henrys.}$$

3. Two coils of inductances L_1, L_2 are in inductive relationship to each other. M is their mutual inductance. If i_1, i_2 are the currents flowing at time t due to a voltage V applied across L_1, then

$$i_2 = -\frac{M}{L_2}\left(\frac{V}{L_1 - M^2/L_2}\right)t = -\frac{M}{L_2}i_1.$$

The total electromagnetic energy stored in the system at any time t is given by $U = \int_0^t Vi_1\,dt$. Show that $U = \frac{1}{2}\frac{V^2}{L_1 - M^2/L_2}t^2$ and prove this is equal to $\frac{1}{2}i_1^2 L_1 + \frac{1}{2}i_2^2 L_2 + i_1 i_2 M$.

4. Two spheres of radius a metres have their centres distant d metres apart ($d > 2a$). They are given equal and opposite charges of Q coulombs. Find the p.d. between the spheres (this is the work done when unit charge is moved from one sphere to the other), and the capacity of the condenser formed by the two spheres.

5. A spherical surface of radius a metres, with centre at O, carries an electric charge of surface density σ coulombs/metre², symmetrically distributed with respect to an axis OX. P is a point on OX such that $OP = x$ (metres). Prove that the potential at P is given by

$$\int_0^\pi \frac{a^2\sigma \sin\theta\,d\theta}{2\epsilon_0(a^2 + x^2 - 2ax\cos\theta)^{\frac{1}{2}}} \text{ volts.}$$

Evaluate this integral, with ϵ_0 and σ constant, in the case when $x > a$.

(*Note.* Potential of a point charge q coulombs at distance r metres is $\dfrac{q}{4\pi\epsilon_0 r}$ volts.)

ANSWERS

1. $\dfrac{Q}{2\pi r\epsilon}$ volts per metre.

4. $\dfrac{Q(d-2a)}{2\pi\epsilon a(d-a)}$ volts; $\dfrac{2\pi\epsilon a(d-a)}{(d-2a)}$.   **5.** $\dfrac{a^2\sigma}{x\epsilon_0}$.

CHAPTER 7

CURVATURE. EVOLUTES AND INVOLUTES

7·1. Definitions

The sign conventions adopted in § 6·22 will be used. Fig. 58 shows a tangent TP to a curve at any point $P(x, y)$.

As P moves along the curve, s and ψ vary.

It is seen from a figure that for a *given* change in the arc length s, the larger the change in the angle ψ, the larger will be the 'bending' of the curve.

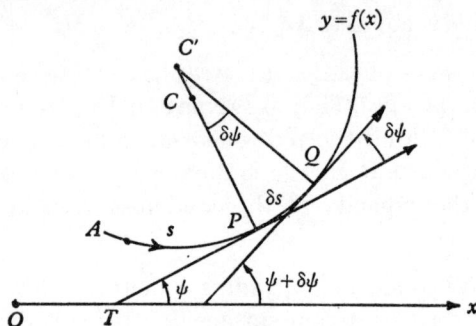

Fig. 58

Curvature at a point on a curve is defined to fit this impression. $\delta\psi/\delta s$ is defined as the *average curvature* over the interval δs.

$$\lim_{\delta s \to 0} \frac{\delta\psi}{\delta s} = \frac{d\psi}{ds} \text{ is defined as the } \textit{curvature at a point.}$$

If ψ increases with s the curvature is positive and if ψ decreases with s it is negative.

Let the normals at P and a neighbouring point Q intersect at C'. As Q approaches P, that is $\delta s \to 0$, C' tends to a point C, on the normal at P, such that $CP = ds/d\psi$, for

$$\frac{C'P}{\sin \hat{P}QC'} = \frac{\text{chord } PQ}{\sin P\hat{C}'Q} \quad \text{(sine rule for } \triangle PQC')$$

$$= \frac{\text{chord } PQ}{\delta s} \times \frac{\delta s}{\delta\psi} \times \frac{\delta\psi}{\sin \delta\psi}.$$

Now as $\delta s \to 0$, $C' \to C$, $\dfrac{\text{chord } PQ}{\delta s} \to 1$, $\dfrac{\delta s}{\delta \psi} \to \dfrac{ds}{d\psi}$ and $\dfrac{\delta \psi}{\sin \delta \psi} \to 1$.

$$\therefore \ \frac{CP}{1} = 1 \times \frac{ds}{d\psi} \times 1,$$

$$CP = \frac{ds}{d\psi} = \frac{1}{\text{curvature}}.$$

The circle, centre C, radius CP, is called the *circle of curvature*. It has the same curvature as the curve at the point of contact P, and obviously the same tangent.

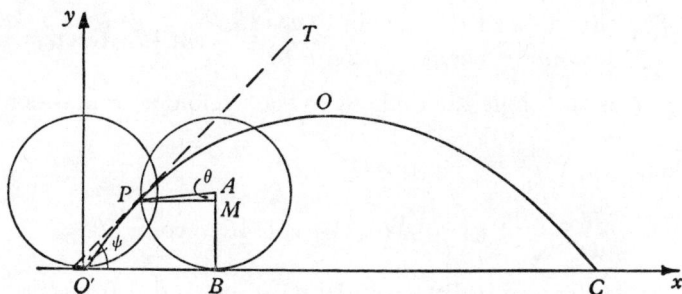

Fig. 59

CP is called the *radius of curvature* and denoted by R (some books denote it by ρ).

Thus, radius of curvature $= R = \dfrac{ds}{d\psi}$

$$= \frac{1}{\text{curvature}}.$$

7·2. Intrinsic equations. The cycloid

An intrinsic equation is one in which the arc length s is given as a function of ψ, i.e. $s = f(\psi)$.

Some curves have their simplest equation in this form. In this case the radius of curvature is easily found as $ds/d\psi$.

One important curve which has a simple intrinsic equation is the cycloid. This is the curve traced out by a point on the circumference of a circle which rolls on a fixed straight line, as shown in fig. 59.

Take O', the position when the point is on the line, as origin, and the line as x-axis. Let the circle be of radius a. At time t let the circle have turned through the angle $BAP = \theta$ say. As the circle rolls, without sliding,

$$O'B = \text{arc } BP = a\theta.$$

Let P have co-ordinates (x, y).

Then

$$x = O'B - PM = a\theta - a\sin\theta = a(\theta - \sin\theta),$$
$$y = BA - MA = a - a\cos\theta = a(1 - \cos\theta). \tag{1}$$

These are the parametric equations of the cycloid.

Now

$$dx/d\theta = a(1 - \cos\theta), \quad dy/d\theta = a\sin\theta. \tag{2}$$

$$\therefore \frac{dy}{dx} = \frac{a\sin\theta}{a(1 - \cos\theta)} = \frac{2\sin\tfrac{1}{2}\theta \cos\tfrac{1}{2}\theta}{2\sin^2\tfrac{1}{2}\theta} = \cot\tfrac{1}{2}\theta = \tan\psi,$$

where ψ is the angle the tangent to the cycloid at P makes with $O'x$.

Thus

$$\psi = (\tfrac{1}{2}\pi - \tfrac{1}{2}\theta) \tag{3}$$

$$\frac{ds}{dx} = \sqrt{\left\{1 + \left(\frac{dy}{dx}\right)^2\right\}} = \sqrt{(1 + \cot^2\tfrac{1}{2}\theta)} = \operatorname{cosec}\tfrac{1}{2}\theta.$$

$$\therefore ds = \operatorname{cosec}\tfrac{1}{2}\theta \, dx = \operatorname{cosec}\tfrac{1}{2}\theta \, a(1 - \cos\theta) \, d\theta, \quad \text{from (2)}$$

$$= 2a \operatorname{cosec}\tfrac{1}{2}\theta \sin^2\tfrac{1}{2}\theta \, d\theta$$

$$= 2a \sin\tfrac{1}{2}\theta \, d\theta.$$

Thus up to the point P

$$s = \int_0^\theta 2a \sin\tfrac{1}{2}\theta \, d\theta$$

$$= -4a[\cos\tfrac{1}{2}\theta]_0^\theta = 4a(1 - \cos\tfrac{1}{2}\theta).$$

But from (3), $\psi = \tfrac{1}{2}\pi - \tfrac{1}{2}\theta$ and therefore $\sin\psi = \cos\tfrac{1}{2}\theta$.

Thus

$$s = 4a(1 - \sin\psi). \tag{4}$$

This is the intrinsic equation of the cycloid referred to O' as origin.

If O is the point where $\psi = 0$, then from (4), $s = 4a$. That is, arc OO' of the cycloid is of length $4a$.

If the origin is chosen at O, the equation becomes

$$s = 4a \sin\psi. \tag{5}$$

O is called a *vertex* of the cycloid. C is called a *cusp*.

Another important curve which has a simple intrinsic equation is the catenary (see Chapter 10).

For a cycloid, radius of curvature,

$$R = \frac{ds}{d\psi} = \frac{d}{d\psi}(4a \sin \psi)$$

$$= \underline{4a \cos \psi}.$$

7·3. Radius of curvature in Cartesian co-ordinates

Several important relationships between the co-ordinates x, y and ψ and s are first obtained.

$$\frac{ds}{dx} = \sqrt{\left\{1 + \left(\frac{dy}{dx}\right)^2\right\}}.$$

But $\qquad\qquad dy/dx = \tan \psi.$

Thus $\qquad\quad ds/dx = \sqrt{(1 + \tan^2 \psi)} = \sec \psi,$

and $\qquad\qquad\quad \underline{dx/ds = \cos \psi}.$

Now $\sin \psi = \tan \psi \cos \psi = \dfrac{dy}{dx}\dfrac{dx}{ds} = \dfrac{dy}{ds},$

$$\underline{\sin \psi = dy/ds}.$$

These results can easily be remembered by using the following diagram (fig. 60), which, however, should *not* be used as a proof.

$$\frac{dy}{dx} = \tan \psi,$$

$$\frac{dy}{ds} = \sin \psi,$$

$$\frac{dx}{ds} = \cos \psi.$$

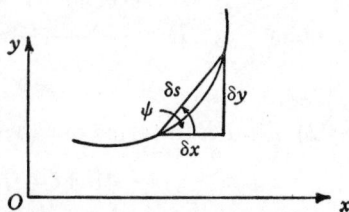

Fig. 60

The formula for R can now be established:

$$\tan \psi = dy/dx.$$

$$\therefore \frac{d}{d\psi}(\tan \psi)\frac{d\psi}{ds} = \frac{d}{ds}(\tan \psi) = \frac{d}{ds}\left(\frac{dy}{dx}\right) = \frac{d}{dx}\left(\frac{dy}{dx}\right)\frac{dx}{ds},$$

giving $\qquad\qquad \sec^2 \psi \dfrac{d\psi}{ds} = \dfrac{d^2y}{dx^2}\dfrac{dx}{ds} = \dfrac{d^2y}{dx^2}\cos \psi.$

$$\therefore \frac{d\psi}{ds} = \frac{d^2y}{dx^2}\bigg/ \sec^3 \psi = \frac{d^2y}{dx^2}\bigg/ (1 + \tan^2 \psi)^{\frac{3}{2}}.$$

$$\therefore \frac{d\psi}{ds} = \frac{d^2y}{dx^2}\bigg/ \left\{1 + \left(\frac{dy}{dx}\right)^2\right\}^{\frac{3}{2}}.$$

Thus
$$R = \frac{ds}{d\psi} = \frac{\{1 + (dy/dx)^2\}^{\frac{3}{2}}}{d^2y/dx^2}.$$

If the curve is given in parametric form

$$x = f_1(t), \quad y = f_2(t).$$

Then
$$\frac{dy}{dx} = \frac{dy}{dt} \bigg/ \frac{dx}{dt}$$

and
$$\frac{d^2y}{dx^2} = \frac{d}{dt}\left(\frac{dy}{dx}\right) \bigg/ \frac{dx}{dt}.$$

Radius of curvature enters into the theory of the bending of beams.

EXAMPLES

(1) Find the value of R at any point on the curve $y = c \cosh x/c$.

$$\frac{dy}{dx} = \sinh\frac{x}{c}. \quad \therefore \; 1 + \left(\frac{dy}{dx}\right)^2 = 1 + \sinh^2\frac{x}{c} = \cosh^2\frac{x}{c}.$$

$$\frac{d^2y}{dx^2} = \frac{1}{c}\cosh\frac{x}{c}.$$

Thus
$$R = \frac{c\left(\cosh^2\dfrac{x}{c}\right)^{\frac{3}{2}}}{\cosh\dfrac{x}{c}} = c\cosh^2\frac{x}{c} = \frac{cy^2}{c^2} = \underline{\frac{y^2}{c}}.$$

(2) Find R in terms of t for any point on the cycloid

$$x = a(t + \sin t), \quad y = a(1 - \cos t).$$

$$\frac{dx}{dt} = a(1 + \cos t) = 2a\cos^2 \tfrac{1}{2}t.$$

$$\frac{dy}{dt} = a\sin t = 2a\sin\tfrac{1}{2}t\cos\tfrac{1}{2}t.$$

$$\therefore \; \frac{dy}{dx} = \frac{dy}{dt}\bigg/\frac{dx}{dt} = \tan\tfrac{1}{2}t.$$

$$\frac{d^2y}{dx^2} = \frac{d}{dt}\left(\frac{dy}{dx}\right)\bigg/\frac{dx}{dt} = \tfrac{1}{2}\sec^2\tfrac{1}{2}t\,\frac{1}{2a\cos^2\tfrac{1}{2}t} = \frac{1}{4a}\sec^4\tfrac{1}{2}t.$$

$$\therefore \; R = \frac{(1 + \tan^2\tfrac{1}{2}t)^{\frac{3}{2}}\,4a}{\sec^4\tfrac{1}{2}t} = \frac{\sec^3\tfrac{1}{2}t\,4a}{\sec^4\tfrac{1}{2}t} = \underline{4a\cos\tfrac{1}{2}t.}$$

7·4. Newton's formula for the value of R at the origin, when a curve passes through the origin and has the x-axis as a tangent there

It is assumed that if a circle is drawn to touch the curve at O and pass through another point $P(x, y)$ of the curve, then the limiting position of this circle as P approaches O is the circle of curvature at O. Its radius will be the radius of curvature, R (fig. 61).

Let OA be the diameter of such a circle. Draw PN perpendicular to AO.

Then
$$ON . NA = NP^2$$

(product of segments of chords of a circle).

$$\therefore NA = NP^2/ON,$$

giving $\quad NA = x^2/y.$

Now as $P \to O$, $NA \to$ diameter of circle of curvature.

Thus

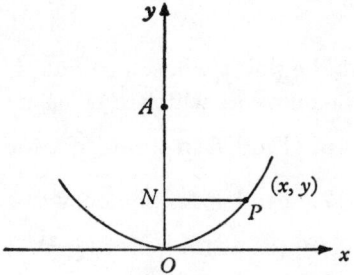

Fig. 61

$$2R = \lim_{x \to 0} \frac{x^2}{y},$$

$$R = \tfrac{1}{2} \lim_{x \to 0} \frac{x^2}{y} = \lim_{x \to 0} \frac{x^2}{2y}.$$

Similarly, if the y-axis is a tangent at the origin

$$R = \lim_{y \to 0} \frac{y^2}{2x}.$$

EXAMPLE

Find the radius of curvature at the origin of the curve

$$y = 2x^2 + 3x^3.$$

The curve obviously goes through the origin

$$dy/dx = 4x + 9x^2; \quad \text{when} \quad x = 0, \quad dy/dx = 0.$$

Thus the x-axis is a tangent to the curve at the origin.

$$\therefore R = \lim_{x \to 0} \frac{x^2}{2(2x^2 + 3x^3)} = \lim_{x \to 0} \frac{x^2/x^2}{4x^2/x^2 + 6x^3/x^2}$$

$$= \lim_{x \to 0} \frac{1}{4 + 6x} = \frac{1}{4}.$$

EXERCISE 35

1. Prove that the radius of curvature at any point on the hyperbola $xy = c^2$ is $\dfrac{(x^2 + y^2)^{\frac{3}{2}}}{2c^2}$.

2. Find R at any point on the curve $x = a \cos^3 t$, $y = a \sin^3 t$.

3. Find R at any point on the curve $x = t - \frac{4}{3}t^3$, $y = 2t^2$.

4. Find the radius of curvature at the point where $x = 4$ on the curve $y = \frac{1}{6}x^3 - 8x + 6$.

5. Show that the radius of curvature of the curve

$$y = \tfrac{1}{2}x^2 - x + 3$$

at the point whose ordinate is k is $(2k - 4)^{\frac{3}{2}}$. Find the point on the curve at which R is least. [Sec. A]

6. Find R in terms of x for the curve $y = a \log_e \sin (x/a)$.

7. Find R for the curve $x = \dfrac{3t}{1 + t^3}$, $y = \dfrac{3t^2}{(1 + t^3)}$; (i) at the point $(0, 0)$; (ii) at the point $(\frac{3}{2}, \frac{3}{2})$.

8. Find the radius of curvature at the origin for the curve $y = x^2 - x^3$.

ANSWERS

2. $\frac{3}{2}a \sin 2t$.   **3.** $\frac{1}{4}(1 + 4t^2)^2$.   **4.** $\frac{1}{4}$.

5. $(1, \frac{5}{2})$.   **6.** $-a \operatorname{cosec} x/a$.

7. (i) $\frac{3}{2}$;   (ii) $\dfrac{3\sqrt{2}}{16}$.   **8.** $\frac{1}{2}$.

7·5. Evolutes

7·51. Definitions

The centre of the circle of curvature is called the centre of curvature. The locus of the centre of curvature corresponding to different points on the curve is called the *evolute* of the curve.

If the co-ordinates of any point on the curve are (x, y), the centre of curvature, C (fig. 62), is given by

$$x_c = x - R \sin \psi, \quad y_c = y + R \cos \psi.$$

In fig. 62:

$$x_c = OM = OL - ML = OL - NP = x - R \sin \psi,$$
$$y_c = MC = MN + NC = LP + NC = y + R \cos \psi.$$

7·52. Construction of the evolute

From § 7·51:

$$d(x_c) = dx - R\cos\psi\,d\psi - \sin\psi\,dR$$

$$= dx - \frac{ds}{d\psi}\frac{dx}{ds}\,d\psi - \sin\psi\,dR \quad \left(\text{as } R = \frac{ds}{d\psi} \text{ and } \cos\psi = \frac{dx}{ds}\right)$$

$$= dx - dx - \sin\psi\,dR.$$

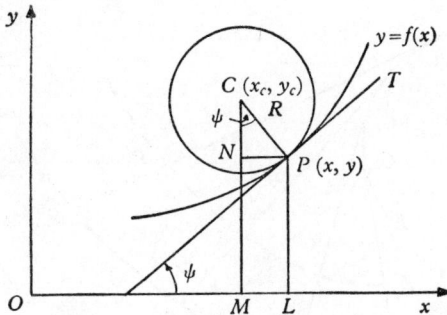

Fig. 62

Thus $\qquad\qquad d(x_c) = -\sin\psi\,dR.$ $\hfill(1)$

$$d(y_c) = dy - R\sin\psi\,d\psi + \cos\psi\,dR$$

$$= dy - \frac{ds}{d\psi}\frac{dy}{ds}\,d\psi + \cos\psi\,dR$$

$$= dy - dy + \cos\psi\,dR.$$

Thus $\qquad\qquad d(y_c) = \cos\psi\,dR.$ $\hfill(2)$

From equations (1) and (2):

$$dy_c/dx_c = -\cot\psi = \text{gradient of } PC \text{ (as } PC \perp PT).$$

But dy_c/dx_c is the slope of the tangent to the locus of C, the evolute.

Thus *the normal to the curve at any point is a tangent to the evolute.*

If normals are drawn at every point on a curve they all touch the evolute. They will, in fact, show its shape.

In such a case the evolute is said to be the *envelope* of these normals. A diagram showing this is seen in fig. 63.

7·53. Arc of the evolute

In fig. 63, let A be the point on the curve, $y = f(x)$, from which arc length, s, is measured. Let C_0, C_1, C_2, C be the centres of curvature for the points A, P_1, P_2, P of the curve.

The evolute is the curve EE_1.

If the co-ordinates of C are (x_c, y_c), from 7·52:

$$dx_c = -\sin \psi\, dR, \quad dy_c = \cos \psi\, dR.$$

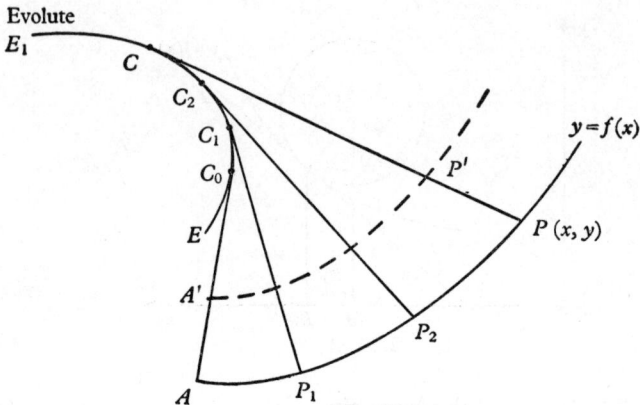

Fig. 63

Let e be the arc length of the evolute measured from C_0.

Then
$$(de)^2 = (dx_c)^2 + (dy_c)^2,$$

i.e.
$$(de)^2 = \sin^2 \psi (dR)^2 + \cos^2 \psi (dR)^2$$
$$= (dR)^2.$$

Thus
$$de = dR.$$

Integrating,
$$e = R - R_0,$$

where R_0 is the radius of curvature (AC_0) at the point A.

Thus, *the length of the arc of the evolute is equal to the difference between the radii of curvature corresponding to its ends.*

In fig. 63 arc $C_0 C = CP - C_0 A$.

7·6. Involutes

If a curve, S, is the evolute of a curve Q, then Q is called an involute of the curve S.

It is easily seen that a given evolute can have any number of involutes.

For example, in fig. 63 the curve AP is an involute to the evolute C_0C; but *any* curve parallel to AP is also an involute. For example, curve $A'P'$.

Owing to the property of the length of the arc of the evolute given in § 7·53, the physical description of an involute of a given curve is easily visualized.

Imagine a fixed length of string fastened at C and wrapped round the curve $(C_0C_1C_2C)$. Keeping the string taut, gradually unwind it. Then the other end (A) will sweep out an involute to the curve, as the condition regarding the length of the arc of the evolute in § 7·53 is automatically satisfied.

The profiles of the teeth of many types of gear wheels are involutes of circles. They have practical advantages over other shapes.

EXAMPLES

(1) Find the co-ordinates of the centre of curvature at the point t on the parabola $x = at^2$, $y = 2at$. Hence find the evolute and sketch it.

$$x = at^2, \quad \frac{dx}{dt} = 2at,$$

$$y = 2at, \quad \frac{dy}{dt} = 2a.$$

$$\therefore \frac{dy}{dx} = \frac{1}{t} \quad \text{and} \quad \frac{d^2y}{dx^2} = -\frac{1}{t^2}\frac{1}{2at} = -\frac{1}{2at^3}.$$

Thus $\qquad R = -2at^3\left(1 + \frac{1}{t^2}\right)^{\frac{3}{2}} = \underline{-2a(1+t^2)^{\frac{3}{2}}}.$

Now $\tan\psi = 1/t$; thus

$$\sin\psi = \frac{1}{\sqrt{(1+t^2)}} \quad \text{and} \quad \cos\psi = \frac{t}{\sqrt{(1+t^2)}}.$$

$$x_c = x - R\sin\psi = at^2 + \frac{2a(1+t^2)^{\frac{3}{2}}}{\sqrt{(1+t^2)}} = a(2 + 3t^2).$$

$$y_c = y + R\cos\psi = 2at - \frac{2a(1+t^2)^{\frac{3}{2}}t}{\sqrt{(1+t^2)}} = -2at^3.$$

To find the locus of C, the evolute:

$$x_c - 2a = 3at^2, \quad y_c = -2at^3.$$

$$\therefore t^6 = \frac{(x_c - 2a)^3}{27a^3} = \frac{y_c^2}{4a^2}.$$

The evolute has, therefore, the equation

$$27ay^2 = 4(x - 2a)^3.$$

For the sketch (fig. 64):

    (i) Only term in y is y^2. $\therefore$ Graph is symmetrical about x-axis.

    (ii) When $x < 2a$, y^2 is negative. $\therefore$ No part of graph for $x < 2a$.

    (iii) When $x = 2a$, $y = 0$.

    (iv) When $x = 3a$, $y \simeq a/5$; when $x = 4a$, $y \simeq 1 \cdot 1a$; when $x = 5a$, $y = 2a$.

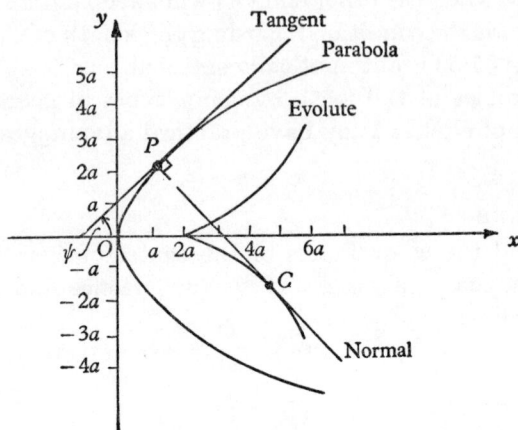

Fig. 64

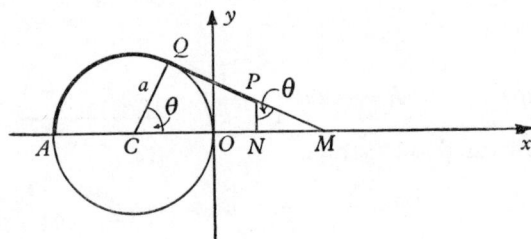

Fig. 65

Note that in the sketch, normals to the part of the parabola in the 4th quadrant are tangents to the part of the evolute in the 1st quadrant, and vice versa.

The simplest way of avoiding awkward ambiguity of signs when calculating $\sin \psi$ and $\cos \psi$ from $\tan \psi$ is to sketch a graph of the curve and draw in a typical tangent and normal. Substituting the values of R, $\sin \psi$, etc., into the formulae for x_c and y_c to give the correct results should then be easy.

(2) Find a set of parametric co-ordinates for the involute of a circle. (This will be the shape of the teeth of a gear-wheel.)

The principle of the 'wrapped string' will be used. It is assumed tied at A, wrapped round to O, and hence of fixed length πa. It is then unwound from O. Let the point P on the involute correspond to a length QO of string unwrapped. Let $O\hat{C}Q = \theta$. Take axes as shown in fig. 65.

$$PQ = OQ = a\theta, \quad QM = a\tan\theta \quad (\triangle CQM).$$
$$\therefore \; PM = a(\tan\theta - \theta).$$
$$NP = PM\cos\theta = a(\tan\theta - \theta)\cos\theta.$$

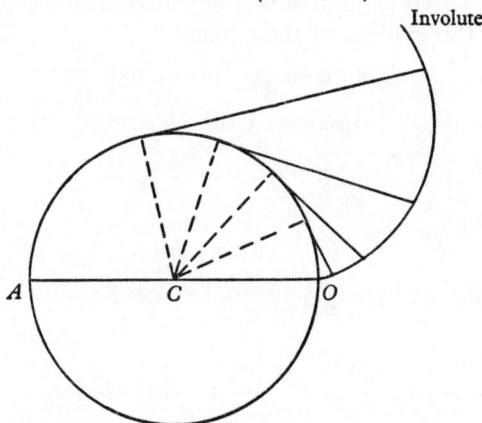

Involute

Fig. 66

$$\therefore \; y_P = a(\tan\theta - \theta)\cos\theta = a(\sin\theta - \theta\cos\theta).$$
$$NM = PM\sin\theta = a(\tan\theta - \theta)\sin\theta.$$
$$CM = \frac{CQ}{\cos\theta} = \frac{a}{\cos\theta}.$$
$$\therefore \; x_P = ON = CM - NM - CO$$
$$= \frac{a}{\cos\theta} - a - a(\tan\theta - \theta)\sin\theta$$
$$= a\left[\frac{(1 - \sin^2\theta)}{\cos\theta} - 1 + \theta\sin\theta\right]$$
$$= a[\cos\theta + \theta\sin\theta - 1].$$

Parametric co-ordinates for an involute are thus:
$$x = a(\cos\theta + \theta\sin\theta - 1), \quad y = a(\sin\theta - \theta\cos\theta).$$

A sketch of a circle and part of an involute to it is shown in fig. 66.

EXERCISE 36

1. Find the co-ordinates of the centre of curvature at any point 'θ' on the cycloid

$$x = a(\theta - \sin\theta), \quad y = a(1 + \cos\theta).$$

2. Find dy/dx and d^2y/dx^2 at the point whose parameter is 't' on the curve $x = a(t - \tanh t)$, $y = a\operatorname{sech} t$. Deduce the length of the radius of curvature. Show that the equation of the evolute is $y = a\cosh x/a$.   [L.U.]

3. Find the co-ordinates of the centre of curvature and the equation of the evolute of the ellipse

$$x = a\cos\theta, \quad y = b\sin\theta.$$

4. Find the co-ordinates of the centre of curvature at any point on the curve $\quad x = 2t^2, \quad y = 2t^3.$

ANSWERS

1. $x = a(\theta + \sin\theta), \quad y = 4a - a(1 + \cos\theta).$ This is another cycloid.

2. $\dfrac{dy}{dx} = -\dfrac{1}{\sinh t}, \ \dfrac{d^2y}{dx^2} = \dfrac{\cosh^3 t}{a\sinh^4 t}; \quad R = a\sinh t.$

3. $x_c = \dfrac{a^2 - b^2}{a}\cos^3\theta, \ y_c = -\dfrac{(a^2 - b^2)}{b}\sin^3\theta;$

$$(ax)^{\frac{2}{3}} + (by)^{\frac{2}{3}} = (a^2 - b^2)^{\frac{2}{3}}.$$

4. $x_c = -2t^2 - 9t^4, \ y_c = 8t^3 + \tfrac{8}{3}t.$

CHAPTER 8

INFINITE SERIES. CONVERGENCE

A full treatment of this subject is beyond the scope of this book. Here, an intuitive discussion is given, together with the simpler tests for convergence.

8·1 Some types of infinite series. Definitions of convergence and divergence

An infinite series is the name given to a series whose terms go on indefinitely. At least two examples will have already been met with; the infinite geometric series, and certain expansions using the binomial theorem.

The case of the infinite G.P. (geometric progression) is recalled:
$$S = a + ar + ar^2 + \ldots + ar^n + \ldots.$$
The sum to n terms, S_n, is
$$\frac{a(1-r^n)}{(1-r)} = \frac{a}{(1-r)} - \left(\frac{a}{1-r}\right)r^n.$$
As long as $-1 < r < 1, r^n \to 0$ as $n \to \infty$.

Thus, if $-1 < r < 1$, $\qquad S_\infty = \dfrac{a}{(1-r)}.$

Some of the symbols may not be well known to the reader.

$n \to \infty$ means that n is made as large as we please. S is the symbol to be used for an infinite series, S_n for the sum to n terms and S_∞ for the 'sum to infinity'. The last statement needs further explanation. If, as n is made as large as we please, the sum to n terms of a series can be made as close as we please to a certain value, then the series is said to be *convergent* and the value is called the sum to infinity of the series.

The case of the infinite G.P. should now be read through again.

The letter u, with suffix denoting the number of the term, will be used to denote any term of a series. Thus,
$$S_n = u_1 + u_2 + u_3 + \ldots + u_n = \sum_1^n u_r, \quad \text{shortly.}$$
A broken line at the *end* will denote an infinite series:
$$S = u_1 + u_2 + u_3 + \ldots + u_n + \ldots = \sum_1^\infty u_r.$$

The definition of convergence given above may be written symbolically:

If, as $n \to \infty$, $S_n \to l$ (l finite), then $S_\infty = l$. Or

$$\lim_{n \to \infty} S_n = l.$$

A series which does not converge is said to *diverge*.

Different types of series can be illustrated graphically by supposing the sum to n terms, S_n, plotted against the number of terms, n.

Different types are shown in figs. 67 and 68.

Note. The diagrams have been drawn full-line to make the demonstrations clearer. As the number of terms, n, is clearly an integer the diagrams should really be sets of isolated points.

8·2. Tests for convergence

8·21. Determination of S_n in terms of n

This is not often possible. When it is, $\lim\limits_{n \to \infty} S_n$ is calculated. The geometric series is a good example.

8·22. Determination of $\lim\limits_{n \to \infty} u_n$

It is obvious that any *finite* number of terms can be neglected when considering whether a series is convergent; they could only add or subtract a finite quantity to the sum of any number of terms taken.
$$u_n = S_{n+1} - S_n.$$

$$\therefore \; \lim_{n \to \infty} u_n = \lim_{n \to \infty} (S_{n+1} - S_n).$$

Now, *if* the series is convergent S_{n+1} and S_n approach the same value and thus
$$\lim_{n \to \infty} (S_{n+1} - S_n) = 0.$$

Thus, *if* a series converges
$$\lim_{n \to \infty} u_n = 0.$$

This must never be used as a test for convergence. All that has been proved is that a series *cannot* converge unless $\lim\limits_{n \to \infty} u_n = 0$; that is, the terms are ultimately as small as we please.

If $\lim\limits_{n \to \infty} u_n \neq 0$, the series diverges.

If $\lim\limits_{n \to \infty} u_n = 0$ the series may converge or may *not* converge.

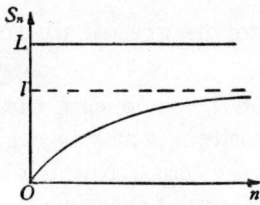

(a) All terms positive, e.g.

$$1+\frac{1}{2}+\frac{1}{2^2}+\frac{1}{2^3}+\ldots$$

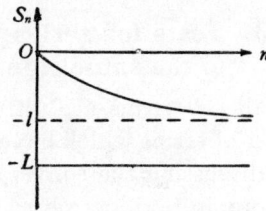

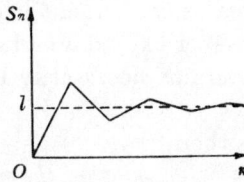

(b) All terms negative, e.g.

$$-1-\frac{1}{2}-\frac{1}{2^2}-\frac{1}{2^3}-\ldots$$

(c) Containing positive and negative terms, e.g.

$$1-\frac{1}{2}+\frac{1}{2^2}-\frac{1}{2^3}+\frac{1}{2^4}-\ldots$$

Fig. 67. Convergent series.

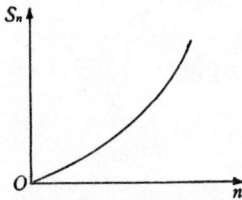

(a) Diverges to $+\infty$, e.g.

$$1+2+3+4+\ldots$$

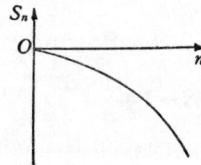

(b) Diverges to $-\infty$, e.g.

$$-1-2-3-4-\ldots$$

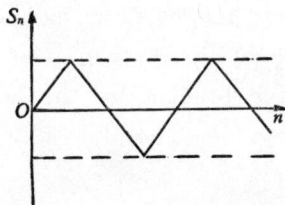

(c) Finitely oscillating series, e.g.

$$1-1-1+1+1-1-1+\ldots$$

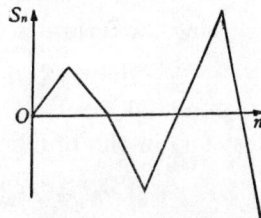

(d) Infinitely oscillating series, e.g.

$$1-2+3-4+5\ldots$$

Fig. 68. Divergent series.

8·23. Tests for series whose terms are all ultimately of the same sign

As all the terms are ultimately of the same sign, the finite number of terms which have the opposite sign may be neglected when discussing the convergency of the series. Nothing is lost, therefore, in taking a series with all terms of the same sign in a general discussion. This will be done.

(i) As all the terms are of the same sign the series cannot oscillate. It must either be convergent or diverge to $+\infty$ or to $-\infty$. To prove such a series convergent it is only necessary to prove that its sum can never exceed a certain numerical value, L say. L, when found, is not necessarily the sum to infinity. (See fig. 67 (a) and (b).)

For this type of series then:

If $|S_n| < L$ for all n sufficiently large, the series converges.

If $|S_n| > L$, however large L is taken, as long as n is taken large enough, then the series diverges.

$|S_n|$, called the 'modulus' of S_n, denotes the numerical value of S_n; that is, the value of S_n independent of its sign, e.g.

$$|-5| = 5.$$

8·231. Some important examples

(1) $S = 1 + \dfrac{1}{2} + \dfrac{1}{3} + \dfrac{1}{4} + \dots + \dfrac{1}{n} + \dots.$

This is called the *harmonic* series.

At first sight this series seems likely to converge. At least,

$$\lim_{n \to \infty} u_n = \lim_{n \to \infty} \frac{1}{n} = 0.$$

Bracketing the terms successively in twos, fours, eights, etc.,

$$S = 1 + \tfrac{1}{2} + (\tfrac{1}{3} + \tfrac{1}{4}) + (\tfrac{1}{5} + \tfrac{1}{6} + \tfrac{1}{7} + \tfrac{1}{8}) + \dots.$$

Replacing all fractions in each bracket by the last, and smallest one in the bracket:

$$S > 1 + \tfrac{1}{2} + (\tfrac{1}{4} + \tfrac{1}{4}) + (\tfrac{1}{8} + \tfrac{1}{8} + \tfrac{1}{8} + \tfrac{1}{8})$$
$$+ (\tfrac{1}{16} + \tfrac{1}{16} + \tfrac{1}{16} + \tfrac{1}{16} + \tfrac{1}{16} + \tfrac{1}{16} + \tfrac{1}{16} + \tfrac{1}{16}) + \dots,$$
$$S > 1 + \tfrac{1}{2} + \tfrac{1}{2} + \tfrac{1}{2} + \tfrac{1}{2} + \dots.$$

Thus S can be made as large as we please.

Therefore the series is divergent.

This is a standard series and occurs frequently. It should be memorized.

(2) $S = 1 + \dfrac{1}{2^2} + \dfrac{1}{3^2} + \dfrac{1}{4^2} + \ldots + \dfrac{1}{n^2} + \ldots.$

Bracketing the terms in twos, fours, eights, etc., as shown

$$S = 1 + \left(\frac{1}{2^2} + \frac{1}{3^2}\right) + \left(\frac{1}{4^2} + \frac{1}{5^2} + \frac{1}{6^2} + \frac{1}{7^2}\right) + \ldots.$$

$$\therefore \ S < 1 + \left(\frac{1}{2^2} + \frac{1}{2^2}\right) + \left(\frac{1}{4^2} + \frac{1}{4^2} + \frac{1}{4^2} + \frac{1}{4^2}\right) + \ldots$$

$$S < 1 + \tfrac{1}{2} + \tfrac{1}{4} + \tfrac{1}{8} + \ldots.$$

The right-hand side is a G.P. whose sum to infinity is $\dfrac{1}{1 - \frac{1}{2}} = 2.$

$$\therefore \ S < 2.$$

Thus the series converges.

In a similar manner it can be shown that the general series

$$S = 1 + \frac{1}{2^p} + \frac{1}{3^p} + \ldots + \frac{1}{n^p} + \ldots$$

converges if $p > 1$ and diverges if $p \leqslant 1$.

8·232. Comparison tests

Let $\qquad U = u_1 + u_2 + u_3 + \ldots + u_n + \ldots$

be any series to be tested.

Let $\qquad V = v_1 + v_2 + v_3 + \ldots + v_n + \ldots$

be a known standard series.

Then *if $u_r \leqslant cv_r$, where c is a constant, and V converges, then so does U.*

If $u_r \geqslant cv_r$ and V diverges, then so does U.

Note that the suffix 'r' denotes any general term of a series.

The first statement, in words, reads: 'If each term of the series to be tested is less than or equal to a constant multiple of the corresponding term of a standard series, then if the standard series converges, so does the one under test.'

For: $\qquad U_n = u_1 + u_2 + \ldots + u_n.$

$$\therefore \ U_n \leqslant c(v_1 + v_2 + \ldots + v_n),$$

i.e. $\qquad U_n \leqslant cV_n.$

But as $n \to \infty$, V_n approaches a definite number, V say.

Therefore U_n must also approach a definite limit.

The second statement can be demonstrated in a similar manner.

It should be noted that series whose terms are all ultimately of the same sign are still being dealt with.

EXAMPLES

(1)
$$U = 1 + \frac{1}{2!} + \frac{1}{3!} + \frac{1}{4!} + \ldots$$

As the standard series take the G.P.

$$V = 1 + \frac{1}{2} + \frac{1}{2^2} + \frac{1}{2^3} + \ldots$$

Then $\qquad\qquad u_r \leqslant 1 v_r.$

But V converges, *therefore U converges.*

(2) $\qquad\qquad U = \frac{1}{2} + \frac{1}{4} + \frac{1}{6} + \frac{1}{8} + \ldots$

Take $\qquad\qquad V = 1 + \frac{1}{2} + \frac{1}{3} + \frac{1}{4} + \ldots$

Then $\qquad\qquad u_r = \frac{1}{2} v_r.$

But V diverges, *therefore U diverges.*

8·24. Test for an alternating series

An alternating series is one in which the terms are alternately positive and negative, e.g.

$$1 - \tfrac{1}{2} + \tfrac{1}{3} - \tfrac{1}{4} + \ldots$$

In this case there is a simple test.

Let $\qquad\qquad S = u_1 - u_2 + u_3 - u_4 + \ldots$

Then, if $\lim\limits_{n \to \infty} u_n \to 0$ *and* $u_n < u_{n-1}$ *for all* $n > N$ *(N a certain number), the series converges.*

The fact that $\lim\limits_{n \to \infty} u_n \to 0$ is necessary for a series to be convergent has already been discussed.

The statement '$u_n < u_{n-1}$ for all $n > N$' means that each term must be less than the previous one from the $(N+1)$th term onwards. (Usually the condition is satisfied from the first term onwards for series likely to be met with at this stage.)

As it has already been shown that neglecting a finite number of terms does not affect the convergency or divergency of a series, the following proof loses nothing in generality:

$$S = u_1 - u_2 + u_3 - u_4 + \ldots$$
$$= (u_1 - u_2) + (u_3 - u_4) + (u_5 - u_6) + \ldots$$

All the brackets are positive as $u_n < u_{n-1}$.
Therefore S is positive and increases steadily.
Also
$$S > (u_1 - u_2).$$
But
$$S = u_1 - (u_2 - u_3) - (u_4 - u_5) - \dots,$$
where all the brackets are again positive.
$$\therefore \ S < u_1.$$

The series thus satisfies the conditions of 8·23. Therefore the series converges.

EXAMPLE
$$S = 1 - \tfrac{1}{2} + \tfrac{1}{3} - \tfrac{1}{4} + \tfrac{1}{5} - \dots.$$

(i) $\lim\limits_{n\to\infty} u_n = \lim\limits_{n\to\infty} \dfrac{1}{n} = 0.$

(ii) $u_n < u_{n-1}$ for all n.

$\therefore$ *The series converges.*

This is another standard series that should be memorized.

8·241. Full treatment of the geometric progression

Consider $\quad S = a + ar + ar^2 + ar^3 + \dots + ar^n + \dots.$

It has already been proved *convergent if* $|r| < 1$.
If $|r| > 1$, then
$$\lim_{n\to\infty} u_n = \lim_{n\to\infty} ar^n = \pm\infty$$

($\pm$ to cover the case when r is negative).
Thus S *diverges if* $|r| > 1$.
If $r = 1$, $\qquad S = a(1 + 1 + 1 + \dots + 1 + \dots),$
which is obviously divergent.
If $r = -1$, $\qquad S = a(1 - 1 + 1 - 1 + 1 - 1 + \dots),$
which again is divergent (oscillatory).
Thus the G.P. *converges if* $|r| < 1$ *and diverges if* $|r| \geqslant 1$.

8·25. Absolute convergence

It has been seen that the series
$$1 + \tfrac{1}{2} + \tfrac{1}{3} + \tfrac{1}{4} + \dots \text{ diverges,}$$
whilst $\qquad 1 - \tfrac{1}{2} + \tfrac{1}{3} - \tfrac{1}{4} + \dots \text{ converges.}$

The latter series is known as *semi-convergent* or *conditionally convergent*. Such a series is convergent, but loses its convergency and becomes divergent if all its terms are taken with the same sign.

A series which is convergent, and *retains* its convergency when all its terms are taken with the same sign, is called an *absolutely* or *unconditionally convergent series*. This gives a method of investigating a series which has mixed positive and negative terms, not necessarily alternating. The corresponding series, taking all terms positive, is tested. If this proves to be convergent, so is the original series.

8·26. Ratio test for convergence

In this test, often called D'Alembert's ratio test after its instigator, only the numerical values (moduli) of the terms are taken.

Let $$S = u_1 + u_2 + u_3 + \ldots + u_n + \ldots$$

be the series under test, where u_r denotes the rth term, *including its sign*.

The series with all its terms positive is denoted by

$$|S| = |u_1| + |u_2| + |u_3| + \ldots + |u_n| + \ldots.$$

Suppose $$\left|\frac{u_2}{u_1}\right| < r, \quad \left|\frac{u_3}{u_2}\right| < r \ldots \left|\frac{u_n}{u_{n-1}}\right| < r \ldots \tag{1}$$

That is, the numerical value of the ratio of each term to the preceding one is less than a certain positive number r.

From (1) above:

$$|u_2| < |u_1|r, \quad |u_3| = \frac{|u_3|}{|u_2|}\frac{|u_2|}{|u_1|}|u_1| < |u_1|r^2 \quad \text{and so on.}$$

Thus $$|S| < |u_1|(1 + r + r^2 + \ldots + r^n + \ldots).$$

Now the series in the bracket, a G.P., converges if $r < 1$.

Therefore, *if* $\left|\dfrac{u_n}{u_{n-1}}\right| < r < 1$, *the original series is absolutely convergent and thus convergent.*

If $\left|\dfrac{u_n}{u_{n-1}}\right| > r \geqslant 1$ then the numerical value of the terms cannot be decreasing. Thus $\lim\limits_{n \to \infty} u_n$ cannot be zero. *Thus the series diverges if* $\left|\dfrac{u_n}{u_{n-1}}\right| > r \geqslant 1.$

In the practical applications of this test it is often convenient to find the limit of $|u_n/u_{n-1}|$ as $n\to\infty$. If this limit is λ, that is

$$\lim_{n\to\infty}\left|\frac{u_n}{u_{n-1}}\right|=\lambda:$$

If $\lambda < 1$ the series is absolutely convergent and thus convergent.
If $\lambda > 1$ the series diverges.
If $\lambda = 1$ the test fails and another test should be tried.

8·27. Power series

A series of the form $a_0 + a_1 x + a_2 x^2 + \ldots + a_n x^n + \ldots$, where the coefficients are constants, is called a power series in x. If any of the tests for convergency are applied, the result of the test may depend on the value of x. For example, it may happen that the series is only convergent if $|x| < 1$.

Practice must be obtained in testing power series, as the next chapter concerns itself with this type of series.

EXAMPLES

(1)
$$1 + x + \frac{x^2}{2!} + \frac{x^3}{3!} + \ldots + \frac{x^n}{n!} + \ldots,$$

$$\frac{u_n}{u_{n-1}} = \frac{x^{n-1}}{(n-1)!}\frac{(n-2)!}{x^{n-2}} = \frac{x}{(n-1)}.$$

$$\therefore \lim_{n\to\infty}\left|\frac{u_n}{u_{n-1}}\right| = \lim_{n\to\infty}\frac{|x|}{(n-1)} = 0 \quad \text{for all } x.$$

The series converges for all finite x.

(2)
$$x - \frac{x^2}{2} + \frac{x^3}{3} - \frac{x^4}{4} + \ldots$$

$$\left|\frac{u_n}{u_{n-1}}\right| = \left|\frac{x^n}{n}\frac{n-1}{x^{n-1}}\right| = |x|\frac{(n-1)}{n} = |x|\left(1-\frac{1}{n}\right).$$

$$\therefore \lim_{n\to\infty}\left|\frac{u_n}{u_{n-1}}\right| = \lim_{n\to\infty}|x|\left(1-\frac{1}{n}\right) = |x|.$$

The series converges if $|x| < 1$ and diverges if $|x| > 1$.
If $|x| = 1$ the test fails.
For $x = 1$:
$$S = 1 - \tfrac{1}{2} + \tfrac{1}{3} - \tfrac{1}{4} + \ldots.$$

This converges (see example in 8·24).

For $x = -1$: $S = -1 - \frac{1}{2} - \frac{1}{3} - \frac{1}{4} - \dots$.

This diverges (see example (1) in 8·231).

Collecting the results:

$$x - \frac{x^2}{2} + \frac{x^3}{3} - \frac{x^4}{4} + \dots$$

converges if $-1 < x \leqslant 1$ *and diverges if* $x > 1$ *or* $x \leqslant -1$.

EXERCISE 37

Test the following series for convergence:

1. $\frac{2}{3} + \frac{4}{9} + \frac{8}{27} + \dots$ (also find the sum to infinity).

2. $\frac{1}{100} + \frac{1}{200} + \frac{1}{300} + \dots$.   **3.** $1 - \frac{1}{3} + \frac{1}{5} - \frac{1}{7} + \dots$.

4. $1 + \frac{2}{2} + \frac{2^2}{3} + \frac{2^3}{4} + \dots$.   **5.** $1 + \frac{3}{2!} + \frac{3^2}{3!} + \frac{3^3}{4!} + \dots$.

6. $x + \frac{x^3}{3!} + \frac{x^5}{5!} + \dots$.   **7.** $1 + \frac{x^2}{2!} + \frac{x^4}{4!} + \dots$.

8. $x - \frac{x^3}{3!} + \frac{x^5}{5!} - \dots$.   **9.** $1 - \frac{x^2}{2!} + \frac{x^4}{4!} - \dots$.

10. $-x - \frac{x^2}{2} - \frac{x^3}{3} - \frac{x^4}{4} - \dots$.   **11.** $1^2 + 2^2 x + 3^2 x^2 + 4^2 x^3 + \dots$.

12. $1 + nx + \frac{n(n-1)}{2!} x^2 + \frac{n(n-1)(n-2)}{3!} x^3 + \dots$.

ANSWERS

(D. = divergent, C. = convergent)

1. G.P., convergent. Sum is 2.   **2.** D.

3. C.   **4.** D.   **5.** C.

6. C. for all x.   **7.** C. for all x.   **8.** C. for all x.

9. C. for all x.

10. C. $-1 \leqslant x < 1$; D. for $x < -1$ or $x \geqslant 1$.

11. C. $|x| < 1$; D. $|x| \geqslant 1$.

12. C. $|x| < 1$ from ratio test. This is the binomial series. Further treatment is beyond the scope of this book.

CHAPTER 9

EXPANSION OF FUNCTIONS AS POWER SERIES

9·1. General discussion

Fig. 69 shows the graph of some function of x, $y = f(x)$.

Take any two points on the graph, say A, B. If their co-ordinates are substituted into the equation of the line $y = a_0 + a_1 x$, two equations are obtained from which coefficients a_0, a_1 could be found. The line $y = a_0 + a_1 x$ would then coincide with the curve at two points.

Take any three points on the graph, say A, B, C. Substituting their co-ordinates into the equation $y = a_0 + a_1 x + a_2 x^2$, values for a_0, a_1, a_2 could be found so that the parabola $y = a_0 + a_1 x + a_2 x^2$ 'fitted' the curve at three points.

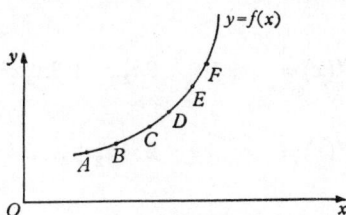

Fig. 69

With patience, this method could be extended. Coefficients $a_0, a_1, ..., a_n$ could be found such that the curve

$$y = a_0 + a_1 x + a_2 x^2 + ... + a_n x^n$$

fitted the given curve for any $(n + 1)$ points.

If the original curve has *not* an equation of the form

$$y = a_0 + a_1 x + a_2 x^2 + ... + a_n x^n \quad (n \text{ finite}),$$

the curve obtained by the method described above could never fit exactly.

In this chapter a method of finding an infinite power series to represent a given function of x is explained.

9·2. Importance of convergence

If an infinite power series in x is found, by any means, to represent any function of x, then in general it only represents the function accurately in the range of values of x for which the series is *convergent*.

Most of the power series tested in Chapter 8, either in the text or exercises, are series representing standard functions such as e^x, $\sin x$, etc. Their ranges of convergency should be memorized.

9·3. Maclaurin's expansion

It is assumed $f(x)$ can be expanded in the form of a power series:

$$f(x) = a_0 + a_1 x + a_2 x^2 + \ldots + a_n x^n + \ldots \tag{1}$$

It is also assumed that an infinite power series can be differentiated term by term. This is true for values of x for which the series is convergent.

Then

$$f(x) = a_0 + a_1 x + a_2 x^2 \quad + a_3 x^3 \quad + a_4 x^4 + \ldots$$
$$+ a_n x^n + \ldots,$$
$$f'(x) = \quad 1a_1 + 2a_2 x \quad + 3a_3 x^2 \quad + 4a_4 x^3 + \ldots$$
$$+ n a_n x^{n-1} + \ldots,$$
$$f''(x) = \quad 2.1a_2 + 3.2a_3 x \quad + 4.3a_4 x^2 + \ldots$$
$$+ n(n-1) a_n x^{n-2} + \ldots,$$
$$f'''(x) = \quad 3.2.1a_3 + \ldots$$
$$+ n(n-1)(n-2) a_n x^{n-3} + \ldots,$$

$$\cdots\cdots\cdots\cdots\cdots\cdots\cdots\cdots\cdots\cdots\cdots\cdots\cdots\cdots\cdots\cdots\cdots\cdots$$

$$f^n(x) = \quad + n! a_n + \ldots \text{ powers of } x.$$

Putting $x = 0$ in each stage in turn:

$$f(0) = a_0, \qquad a_0 = f(0),$$
$$f'(0) = a_1, \qquad a_1 = f'(0),$$
$$f''(0) = 2! a_2, \qquad a_2 = \frac{1}{2!} f''(0),$$
$$f'''(0) = 3! a_3, \qquad a_3 = \frac{1}{3!} f'''(0).$$

$$\cdots\cdots\cdots\cdots \qquad \cdots\cdots\cdots\cdots$$

$$f^n(0) = n! a_n, \qquad a_n = \frac{1}{n!} f^n(0),$$
$$\text{etc.}$$

This gives Maclaurin's expansion, on substitution in (1):

$$f(x) = f(0) + x f'(0) + \frac{x^2}{2!} f''(0) + \frac{x^3}{3!} f'''(0) + \ldots + \frac{x^n}{n!} f^n(0) + \ldots.$$

9·4. Some expansions of standard functions

The ranges of x for which the series are convergent will be given. They will be found in the text or exercises of Chapter VIII.

9·41. e^x

$$f(x) = e^x, \qquad f(0) = 1,$$
$$f'(x) = e^x, \qquad f'(0) = 1,$$
$$f^n(x) = e^x, \qquad f^n(0) = 1.$$

Substituting in Maclaurin's expansion:

$$e^x = 1 + x + \frac{x^2}{2!} + \frac{x^3}{3!} + \dots + \frac{x^n}{n!} + \dots.$$

The series is convergent for all values of x. Putting $-x$ for x:

$$e^{-x} = 1 - x + \frac{x^2}{2!} - \frac{x^3}{3!} + \frac{x^4}{4!} + \dots + (-1)^n \frac{x^n}{n!} + \dots.$$

9·42. Sinh x and cosh x

$$f(x) = \sinh x, \qquad f(0) = 0,$$
$$f'(x) = \cosh x, \qquad f'(0) = 1,$$
$$f''(x) = \sinh x, \qquad f''(0) = 0,$$
$$f'''(x) = \cosh x, \qquad f'''(0) = 1, \quad \text{etc.}$$
$$f^n(0) = 0, \quad n \text{ even},$$
$$= 1, \quad n \text{ odd}.$$

Thus
$$\sinh x = x + \frac{x^3}{3!} + \frac{x^5}{5!} + \dots.$$

It is left to the reader to show in like manner that

$$\cosh x = 1 + \frac{x^2}{2!} + \frac{x^4}{4!} + \dots.$$

Both series are convergent for all values of x.

9·43. Sin x and cos x

$$f(x) = \sin x, \qquad f(0) = 0,$$
$$f'(x) = \cos x, \qquad f'(0) = 1,$$
$$f''(x) = -\sin x, \qquad f''(0) = 0,$$
$$f'''(x) = -\cos x, \qquad f'''(0) = -1,$$
$$f^{iv}(x) = \sin x, \qquad f^{iv}(0) = 0.$$

Hereafter the values are periodically $0, 1, -1, 0$. Hence

$$\sin x = x - \frac{x^3}{3!} + \frac{x^5}{5!} - \frac{x^7}{7!} + \dots$$

The series for $\cos x$ is

$$\cos x = 1 - \frac{x^2}{2!} + \frac{x^4}{4!} - \frac{x^6}{6!} + \frac{x^8}{8!} - \dots$$

Both series are convergent for all values of x.

9·44. $\log_e (1 + x)$

It may seem surprising that $\log_e x$ is not chosen, but this would give $f(0) = \log_e 0 \to -\infty$, for the first term.

$$f(x) = \log_e (1 + x), \qquad f(0) = \log_e 1 = 0,$$

$$f'(x) = \frac{1}{(1+x)}, \qquad f'(0) = 1,$$

$$f''(x) = \frac{-1}{(1+x)^2}, \qquad f''(0) = -1$$

$$f'''(x) = \frac{-1 \cdot -2}{(1+x)^3} = \frac{1 \cdot 2}{(1+x)^3}, \quad f'''(0) = 2!,$$

$$f^{iv}(x) = \frac{-1 \cdot -2 \cdot -3}{(1+x)^4}, \qquad f^{iv}(0) = -3!.$$

From here on the sequence is obvious.

Hence

$$\log_e (1 + x) = 0 + x - \frac{x^2}{2!} + \frac{2! \, x^3}{3!} - \frac{3! \, x^4}{4!} + \frac{4! \, x^5}{5!} - \dots$$

$$= x - \frac{x^2}{2} + \frac{x^3}{3} - \frac{x^4}{4} + \frac{x^5}{5} - \dots + \frac{(-1)^{n+1} x^n}{n} + \dots$$

The series is convergent for $-1 < x \leqslant 1$ (see example (2) of § 8·27).

Writing $-x$ for x:

$$\log_e (1 - x) = -x - \frac{x^2}{2} - \frac{x^3}{3} - \frac{x^4}{4} - \dots$$

This series converges for $-1 \leqslant x < 1$ (Q. 10, Exercise 37).

9·5. Odd and even functions

If $f(x) = f(-x)$ the function is said to be an even function, e.g. $\cos x$ is an even function.

The definition implies that the graph of the function is symmetrical about the y-axis, as equal but opposite x values give the same y value.

As all odd powers of x change sign if $-x$ is written for x, it follows that the power series for an even function can only contain *even* powers of x.

Similarly, if $f(x) = -f(-x)$ the function is said to be odd. Its power series will only contain odd powers of x. Its graph is symmetrical about the origin.

Thus in finding a power series for $\tan x$ it can be said at once that the series will only contain odd powers of x, as

$$\tan x = -\tan(-x)$$

and is therefore an odd function.

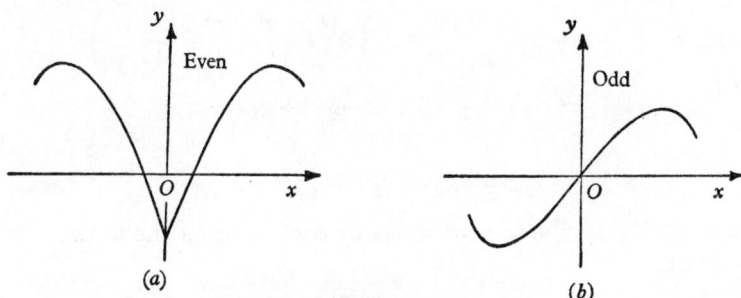

Fig. 70

Most functions are neither odd or even, e.g. $x - x^2$, e^x, etc.

In manipulating odd and even functions, note that:

(i) The product, quotient, sum and difference of two even functions is even.

(ii) The product and quotient of two odd functions is even.

(iii) The sum and difference of two odd functions is odd.

(iv) The product and quotient of an odd and even function is odd; their sum and difference is neither even nor odd.

All these rules are easily proved. One is proved below; the rest may be proved by the reader.

Let $f(x)$, $g(x)$ be two odd functions.

Then if $\phi(x) = f(x) g(x)$,

$$\phi(-x) = f(-x) g(-x) = \{-f(x)\}\{-g(x)\} = f(x) g(x)$$
$$= \phi(x).$$

Hence $\phi(x)$, their product, is an even function.

Fig. 70 shows sketches of an even and an odd function.

9·6. Other methods of expansion

In general, for the range of values for which two series are absolutely convergent, they may be added, subtracted, multiplied, divided, differentiated and integrated term by term. The proof of this statement is beyond the scope of this book.

Many new series may be obtained from standard ones by operating on them by one of the above processes. In the case of integration care has to be taken to evaluate the 'constant of integration' by giving x a specific value.

Examples are given below.

9·61. Multiplication

$$e^x \cos x = \left(1 + x + \frac{x^2}{2!} + \frac{x^3}{3!} + \frac{x^4}{4!} + \ldots\right) \times \left(1 - \frac{x^2}{2!} + \frac{x^4}{4!} - \frac{x^6}{6!} + \ldots\right)$$

$$= (1 - \tfrac{1}{2}x^2 + \tfrac{1}{24}x^4 - \ldots) + (x - \tfrac{1}{2}x^3 + \tfrac{1}{24}x^5 - \ldots)$$
$$+ (\tfrac{1}{2}x^2 - \tfrac{1}{4}x^4 + \ldots)$$
$$+ (\tfrac{1}{6}x^3 - \tfrac{1}{12}x^5 + \ldots) + (\tfrac{1}{24}x^4 - \tfrac{1}{48}x^6 + \ldots) + \ldots$$

on multiplying the second series by each term in the first.
Thus
$$e^x \cos x = 1 + x - \tfrac{1}{3}x^3 - \tfrac{1}{6}x^4 - \ldots.$$

9·62. Division

It is best to enumerate factorials.

$$\tan x = \frac{\sin x}{\cos x} = \frac{x - \tfrac{1}{6}x^3 + \tfrac{1}{120}x^5 - \ldots}{1 - \tfrac{1}{2}x^2 + \tfrac{1}{24}x^4 - \ldots}.$$

$$1 - \tfrac{1}{2}x^2 + \tfrac{1}{24}x^4 - \ldots \overline{)x - \tfrac{1}{6}x^3 + \tfrac{1}{120}x^5 - \ldots} (x + \tfrac{1}{3}x^3 + \tfrac{2}{15}x^5 - \ldots$$
$$\underline{x - \tfrac{1}{2}x^3 + \tfrac{1}{24}x^5 - \ldots}$$
$$\tfrac{1}{3}x^3 - \tfrac{1}{30}x^5 - \ldots$$
$$\underline{\tfrac{1}{3}x^3 - \tfrac{1}{6}x^5 - \ldots}$$
$$\tfrac{2}{15}x^5 - \ldots$$
$$\underline{\tfrac{2}{15}x^5 - \ldots.}$$

Thus
$$\tan x = x + \tfrac{1}{3}x^3 + \tfrac{2}{15}x^5 - \ldots.$$

9·63. Differentiation

$$\sin x = x - \frac{x^3}{3!} + \frac{x^5}{5!} - \frac{x^7}{7!} + \ldots.$$

Differentiating
$$\cos x = 1 - \frac{3x^2}{3!} + \frac{5x^4}{5!} - \frac{7x^6}{7!} + \dots$$

$$= 1 - \frac{x^2}{2!} + \frac{x^4}{4!} - \frac{x^6}{6!} + \dots.$$

9·64. Integration

Important series which can be obtained by this method are those for any function having an algebraic derivative which may be expanded by the binomial expansion.

EXAMPLES

(1) $\log_e (1+x) = \int \frac{1}{(1+x)} \, dx = \int (1+x)^{-1} \, dx$

$$= \int (1 - x + x^2 - x^3 + \dots) \, dx$$

$$= x - \tfrac{1}{2}x^2 + \tfrac{1}{3}x^3 - \tfrac{1}{4}x^4 + \dots + C.$$

When $x = 0$, $\log_e (1+x) = 0$. Therefore $C = 0$.

$$\therefore \ \underline{\log_e (1+x) = x - \tfrac{1}{2}x^2 + \tfrac{1}{3}x^3 - \dots.}$$

(2) $\sin^{-1} x = \int \frac{1}{\sqrt{(1-x^2)}} \, dx = \int (1 - x^2)^{-\frac{1}{2}} \, dx$

$$= \int (1 + \tfrac{1}{2}x^2 + \tfrac{3}{8}x^4 + \tfrac{5}{16}x^6 + \dots) \, dx,$$

from the binomial expansion.

Thus $\sin^{-1} x = C + x + \tfrac{1}{6}x^3 + \tfrac{3}{40}x^5 + \tfrac{5}{112}x^7 + \dots.$

When $x = 0$, $\sin^{-1} x = 0$. Thus $C = 0$.

$$\therefore \ \underline{\sin^{-1} x = x + \tfrac{1}{6}x^3 + \tfrac{3}{40}x^5 + \tfrac{5}{112}x^7 + \dots.}$$

(3) $\cosh x = 1 + \dfrac{x^2}{2!} + \dfrac{x^4}{4!} + \dots.$

Integrating $\sinh x = C + x + \dfrac{x^3}{3 \cdot 2!} + \dfrac{x^5}{5 \cdot 4!} + \dots.$

When $x = 0$, $\sinh x = 0$. Thus $C = 0$.

$$\therefore \ \underline{\sinh x = x + \dfrac{x^3}{3!} + \dfrac{x^5}{5!} + \dots.}$$

9·7. Approximate values of definite integrals

As long as the limits of integration are within the range of convergency of the expansion of a function, an approximate value for a definite integral may be obtained by expanding the function and integrating term by term.

This method is usually reserved for integrals which cannot be calculated accurately. Unless the terms rapidly get smaller the method may be too long to be of much use.

EXAMPLE

$$\int_0^1 \sqrt{x}\cos x\,dx$$

$$= \int_0^1 \sqrt{x}\,(1 - \tfrac{1}{2}x^2 + \tfrac{1}{24}x^4 - \tfrac{1}{720}x^6 + \ldots)\,dx$$

$$= \int_0^1 (x^{\frac{1}{2}} - \tfrac{1}{2}x^{\frac{5}{2}} + \tfrac{1}{24}x^{\frac{9}{2}} - \tfrac{1}{720}x^{\frac{13}{2}} + \ldots)\,dx$$

$$\simeq [\tfrac{2}{3}x^{\frac{3}{2}} - \tfrac{1}{7}x^{\frac{7}{2}} + \tfrac{1}{132}x^{\frac{11}{2}} - \tfrac{1}{5400}x^{\frac{15}{2}}]_0^1$$

$$\simeq 0\!\cdot\!6667 - 0\!\cdot\!1429 + 0\!\cdot\!0076 - 0\!\cdot\!0002.$$

$$= 0\!\cdot\!5312.$$

Correct to 3 decimal places $\displaystyle\int_0^1 \sqrt{x}\cos x\,dx = 0\!\cdot\!531.$

9·8. Evaluation of limits using power series

Expansion of the functions as power series is often successful in evaluating the limit of the quotient of two functions.

EXAMPLE

Find $\displaystyle\lim_{x\to 0}\frac{\sin^2 x - x^2\cos x}{x^3\tan x}.$ [L.U.]

$$\frac{\sin^2 x - x^2\cos x}{x^3\tan x} = \frac{\tfrac{1}{2}(1 - \cos 2x) - x^2\cos x}{x^3\tan x}$$

$$= \frac{\tfrac{1}{2} - \tfrac{1}{2}\cos 2x - x^2\cos x}{x^3\tan x}$$

$$= \frac{\tfrac{1}{2} - \tfrac{1}{2}(1 - \tfrac{4}{2}x^2 + \tfrac{16}{24}x^4 - \ldots) - x^2(1 - \tfrac{1}{2}x^2 + \tfrac{1}{24}x^4 - \ldots)}{x^3(x + \tfrac{1}{3}x^3 + \tfrac{2}{15}x^5 + \ldots)}$$

$$= \frac{\tfrac{1}{2} - (\tfrac{1}{2} - x^2 + \tfrac{1}{3}x^4 - \ldots) - (x^2 - \tfrac{1}{2}x^4 - \ldots)}{(x^4 + \tfrac{1}{3}x^6 + \ldots)}$$

$$= \frac{+\tfrac{1}{6}x^4 + \text{higher powers of } x}{x^4 + \text{higher powers of } x}$$

$$= \frac{\tfrac{1}{6} + \text{powers of } x}{1 + \text{powers of } x} \quad \begin{array}{l}\text{(dividing numerator and}\\ \text{denominator by } x^4\text{).}\end{array}$$

$$\therefore \lim_{x\to 0}\frac{\sin^2 x - x^2\cos x}{x^3\tan x} = \frac{1}{6}.$$

9·9. Examples

(1) Determine, by Maclaurin's series, the expansions of $\sinh x$ and $\cosh x$ in ascending powers of x. Show that the expansion of $x \coth x$ in ascending powers of x contains only even powers, and obtain the expansion as far as the term in x^6. Use this expansion to calculate $\coth 0·2$ to five significant figures. [L.U.]

$$\sinh x = x + \frac{x^3}{3!} + \frac{x^5}{5!} + \dots, \quad \cosh x = 1 + \frac{x^2}{2!} + \frac{x^4}{4!} + \dots.$$

These series have been derived in the text and will not be worked through again.

Now $x = -(-x)$. Thus x is an odd function,

$$\coth x = -\coth(-x).$$

Thus $\coth x$ is an odd function. $x \coth x$ is therefore the product of two odd functions and is thus an even function.

The expansion of $x \coth x$ can therefore only contain even powers:

$$x \coth x = x \frac{\cosh x}{\sinh x}$$

$$= \frac{x(1 + \frac{1}{2}x^2 + \frac{1}{24}x^4 + \frac{1}{720}x^6 + \dots)}{x + \frac{1}{6}x^3 + \frac{1}{120}x^5 + \frac{1}{5040}x^7},$$

$$
\begin{array}{r}
1 + \frac{1}{3}x^2 - \frac{1}{45}x^4 + \frac{2}{945}x^6 - \dots \\
x + \frac{1}{6}x^3 + \frac{1}{120}x^5 + \frac{1}{5040}x^7 + \dots \overline{)\, x + \frac{1}{2}x^3 + \frac{1}{24}x^5 + \frac{1}{720}x^7 + \dots} \\
x + \frac{1}{6}x^3 + \frac{1}{120}x^5 + \frac{1}{5040}x^7 + \dots \\
\hline
\frac{1}{3}x^3 + \frac{1}{30}x^5 + \frac{1}{840}x^7 + \dots \\
\frac{1}{3}x^3 + \frac{1}{18}x^5 + \frac{1}{360}x^7 + \dots \\
\hline
-\frac{1}{45}x^5 - \frac{1}{630}x^7 \\
-\frac{1}{45}x^5 - \frac{1}{270}x^7 \\
\hline
\frac{2}{945}x^7
\end{array}
$$

$$\underline{x \coth x = 1 + \tfrac{1}{3}x^2 - \tfrac{1}{45}x^4 + \tfrac{2}{945}x^6 - \dots.}$$

Put $x = 0·2$:

$$\tfrac{1}{5} \coth 0·2 \simeq 1 + \frac{0·04}{3} - \frac{0·0016}{45} + \frac{2}{945}(0·000064).$$

$$\coth 0·2 \simeq 5 + \frac{0·2}{3} - \frac{0·008}{45} + \frac{2}{945}(0·00032)$$

$$\simeq 5 + 0·06667 - 0·00018 + 0·000001$$

$$= 5·06649.$$

Correct to five significant figures

$$\coth 0 \cdot 2 = 5 \cdot 0665.$$

(2) If $\sin y = (1 + x) \sin \alpha$, evaluate dy/dx, d^2y/dx^2 and d^3y/dx^3, when $x = 0$, in terms of $\tan \alpha$. Hence expand y in ascending powers of x as far as the term in x^3, given $y = \alpha$ when $x = 0$.

[L.U.]

Suffixes denote values when $x = 0$.

$$\sin y = (1 + x) \sin \alpha. \quad \therefore \ y_0 = \alpha.$$

$$\cos y \frac{dy}{dx} = \sin \alpha. \quad \therefore \ \cos y_0 \left(\frac{dy}{dx}\right)_0 = \sin \alpha.$$

But $\cos y_0 = \cos \alpha$, therefore $\left(\dfrac{dy}{dx}\right)_0 = \tan \alpha$.

Differentiating again:

$$\cos y \left(\frac{d^2y}{dx^2}\right) - \sin y \left(\frac{dy}{dx}\right)^2 = 0.$$

$$\therefore \ \left(\frac{d^2y}{dx^2}\right)_0 = \tan y_0 \left\{ \left(\frac{dy}{dx}\right)_0 \right\}^2 = \tan \alpha \tan^2 \alpha = \tan^3 \alpha.$$

Differentiating a third time:

$$\cos y \left(\frac{d^3y}{dx^3}\right) - \sin y \frac{dy}{dx}\frac{d^2y}{dx^2} - 2 \sin y \frac{dy}{dx}\frac{d^2y}{dx^2} - \cos y \left(\frac{dy}{dx}\right)^3 = 0.$$

$$\therefore \ \cos y_0 \left(\frac{d^3y}{dx^3}\right)_0 - 3 \sin y_0 \left(\frac{dy}{dx}\right)_0 \left(\frac{d^2y}{dx^2}\right)_0 - \cos y_0 \left(\frac{dy}{dx}\right)_0^3 = 0.$$

$$\therefore \ \left(\frac{d^3y}{dx^3}\right)_0 = 3 \tan \alpha \tan \alpha \tan^3 \alpha + \tan^3 \alpha = 3 \tan^5 \alpha + \tan^3 \alpha.$$

Hence, from Maclaurin's expansion,

$$y = \alpha + x \tan \alpha + \frac{x^2}{2!} \tan^3 \alpha + \frac{x^3}{3!}(3 \tan^5 \alpha + \tan^3 \alpha) + \dots.$$

(3) Find the expansion of $\log_e (1 + x)$ in ascending powers of x when $-1 < x < 1$. Give the general term.

The expansion of $(a + bx + cx^2) \log_e (1 - 3x)$ begins with the terms $6x - \frac{3}{2}x^3 + dx^4$. Find a, b, c, d. [Sec. A]

The series for $\log_e (1 + x)$ was obtained in § 9·44.

$$\log_e (1 - 3x) = -3x - \tfrac{9}{2}x^2 - \tfrac{27}{3}x^3 - \tfrac{81}{4}x^4 - \dots.$$

$$\therefore \ (a + bx + cx^2) \log_e (1 - 3x)$$

$$= (a + bx + cx^2)(-3x - \tfrac{9}{2}x^2 - 9x^3 - \tfrac{81}{4}x^4 - \dots)$$

$$= -3ax - (3b + \tfrac{9}{2}a)x^2 - (9a + \tfrac{9}{2}b + 3c)x^3 - (\tfrac{81}{4}a + 9b + \tfrac{9}{2}c)x^4 - \dots.$$

Thus, if the first terms are $6x - \frac{3}{2}x^3 + dx^4$:

$$-3a = 6, \quad \therefore \ a = -2.$$

$$3b + \tfrac{9}{2}a = 0, \quad \therefore \ b = -\tfrac{3}{2}a = 3.$$

$$9a + \tfrac{9}{2}b + 3c = \tfrac{3}{2}, \quad \therefore \ c = -\tfrac{3}{2}b - 3a + \tfrac{1}{2} = -\tfrac{9}{2} + 6 + \tfrac{1}{2} = 2.$$

$$-\left(\tfrac{81}{4}a + 9b + \tfrac{9}{2}c\right) = d, \quad \therefore \ d = -\left[-\tfrac{81}{2} + 27 + 9\right] = 4\tfrac{1}{2}.$$

$$\underline{a = -2, \quad b = 3, \quad c = 2, \quad d = 4\tfrac{1}{2}.}$$

EXERCISE 38

1. Find the first and second derivatives of $e^{\sin x}$. Hence, or otherwise, write down the first three terms in the expansion of this function in a series of ascending powers of x.   [Sec. A]

2. Using the result of Q. 1 and the expansion of e^x, find the first three terms of the expansion of $e^{x+\sin x}$.   [Sec. A]

3. Find, by any means, an expansion for $\tan^{-1} x$.

4. Find the value of $\int_0^{\frac{1}{2}} x \log_e(1+x)\,dx$ correct to three decimal places. Check the result by evaluating the integral, using integration by parts.

5. Find a value for $\int_0^1 x \tan^{-1} x\, dx$ using the expansion of $\tan^{-1} x$. Check the result, using integration by parts.

6. Find a value for $\int_0^1 e^{-x^2}\, dx$ correct to three decimal places.

7. Expand $\sin^2 x$ by Maclaurin's theorem as far as the term in x^6.

8. Show that $s = c \sinh x/c$ may be replaced approximately by $s = x + x^3/6c^2$ if c is large compared with x.

9. Prove $\dfrac{d}{dx}\tan^{-1}(x^2) = \dfrac{2x}{1+x^4}$. Hence obtain a power series for $\tan^{-1}(x^2)$.

10. (i) State the expansions for $\log_e(1+x)$ and $\log_e(1-x)$. (ii) Deduce an expansion for $\log_e\left(\dfrac{1+x}{1-x}\right)$. (iii) Use this to find a value for $\log_e 1{\cdot}5$ correct to four decimal places.

11. Show that $\dfrac{x}{(e^x - 1)} \simeq 1 - \frac{1}{2}x + \frac{1}{12}x^2$ when x is small.

12. Evaluate: (i) $\lim\limits_{x \to 0} \dfrac{(1 - \cos x)}{x^2}$; (ii) $\lim\limits_{x \to 0} \dfrac{e^x - 1}{x}$.

13. Evaluate: $\lim\limits_{m \to 0} \dfrac{A}{m^2}\left(\dfrac{\sin mx}{\sin ml} - \dfrac{x}{l}\right)$.

ANSWERS

1. $1 + x + \dfrac{x^2}{2!}$.

2. $1 + 2x + 2x^2$.

3. $x - \frac{1}{3}x^3 + \frac{1}{5}x^5 - \frac{1}{7}x^7 + \dots$

4. $0 \cdot 035$.

5. $0 \cdot 285$.

6. $0 \cdot 747$.

7. $x^2 - \frac{1}{3}x^4 + \frac{2}{45}x^6$.

9. $2(\frac{1}{2}x^2 - \frac{1}{6}x^6 + \frac{1}{10}x^{10} - \frac{1}{14}x^{14} - \dots)$.

10. (ii) $2(x + \frac{1}{3}x^3 + \frac{1}{5}x^5 + \dots)$; (iii) $0 \cdot 4055$.

12. (i) $\frac{1}{2}$; (ii) 1.

13. $\dfrac{Ax(l^2 - x^2)}{6l}$.

CHAPTER 10

THE CATENARY

The catenary is the name give to the shape of the curve taken up by a heavy uniform, perfectly flexible chain or string when suspended from two points.

10·1. Equations of a catenary

Let A be the lowest point of the catenary, shown in fig. 71.
Take the y-axis as the vertical through A.
Let $w = $ weight per unit length of the chain, or string.

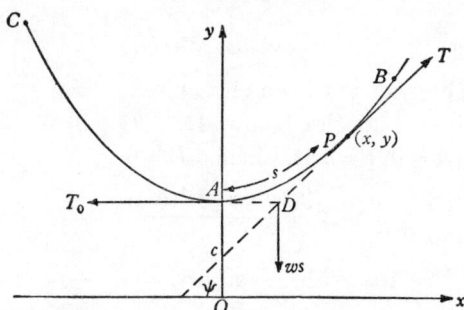

Fig. 71

Let $s = $ length of arc from A to any point P on the catenary.
Let $T_0 = $ tension at point A, acting along the tangent at A.
Let $T = $ tension at point P acting along the tangent at P.

The portion AP of the catenary is in equilibrium under three forces; its weight ws, T_0 and T. They must therefore be concurrent, at D.

Resolving horizontally and vertically for the portion AP:

$$T \cos \psi = T_0 = cw, \tag{1}$$

$$T \sin \psi = ws. \tag{2}$$

Note that T_0 has been replaced by a constant c (peculiar to any individual catenary) times w. c is often called the parameter of the catenary.

Dividing (2) by (1): $\quad \dfrac{s}{c} = \tan \psi = \dfrac{dy}{dx}. \tag{3}$

This gives the intrinsic equation

$$s = c \tan \psi. \tag{4}$$

Now $\dfrac{ds}{dx} = \dfrac{1}{\cos \psi} = \sec \psi = \surd(1 + \tan^2 \psi) = \dfrac{\surd(c^2 + s^2)}{c}$, from (4).

$$\therefore \quad \frac{ds}{\surd(c^2 + s^2)} = \frac{1}{c} dx.$$

Integrating $\qquad \sinh^{-1} \dfrac{s}{c} = \dfrac{x}{c} + C.$

When $x = 0$, $s = 0$ as the y-axis was chosen through A. Hence $C = 0$.

$$\therefore \quad s = c \sinh x/c. \tag{5}$$

From (3) $\qquad\qquad \dfrac{dy}{dx} = \dfrac{s}{c} = \sinh \dfrac{x}{c}.$

Integrating $\qquad\qquad y = c \cosh x/c + D.$

The origin has not yet been chosen.
It is chosen at a depth c below A.
Thus when $x = 0$, $y = c$ and hence $D = 0$.

$$\therefore \quad y = c \cosh x/c. \tag{6}$$

From (5) and (6):

$$1 = \cosh^2 \frac{x}{c} - \sinh^2 \frac{x}{c} = \frac{y^2}{c^2} - \frac{s^2}{c^2}.$$

$$\therefore \quad y^2 = c^2 + s^2. \tag{7}$$

From (1) and (2): $\qquad T^2 = w^2(c^2 + s^2).$

$\therefore$ From (7): $\qquad\qquad T = wy. \tag{8}$

The inverse forms of the functions in equations (5) and (6) should be noted:

$$\frac{x}{c} = \sinh^{-1} \frac{s}{c} = \log_e \left\{ \frac{s + \surd(c^2 + s^2)}{c} \right\}.$$

$$\therefore \quad x = c \log_e \frac{(s + y)}{c}. \tag{9}$$

Equation (6) would give the same.
From (8) and (1), $\qquad y = T/w = c \sec \psi.$

Using this, and $s = c \tan \psi$, in (9):

$$x = c \log_e (\sec \psi + \tan \psi). \tag{10}$$

The ten formulae derived above serve to connect any two variables of the catenary together.

In exercises on the catenary it is usually best to find the parameter c as early as possible.

EXAMPLE

A heavy uniform chain, length l, lies in a straight line on a horizontal floor. One end of the chain is slowly raised vertically until half the chain is clear of the floor, the remainder being on the point of slipping (fig. 72). If the coefficient of friction between the chain and the floor is given by $\mu = \frac{1}{3}$, show that the height above the floor of the raised end is $\frac{1}{6}l(\sqrt{10}-1)$, and the horizontal distance of the midpoint of the chain from that of the raised end is $\frac{1}{6}l \log_e(\sqrt{10}+3)$. [L.U.]

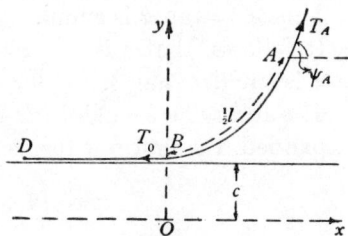

Fig. 72

The raised part of the chain, AB, will form part of a catenary with B as its lowest point.

Let $w =$ weight per unit length.

Weight of part $BD = \frac{1}{2}wl$.

$$\therefore \ T_0 = \mu \frac{wl}{2} = \frac{wl}{6} = cw.$$

Thus
$$c = \tfrac{1}{6}l. \tag{1}$$

At point A:
$$T_A \cos \psi_A = T_0 = \tfrac{1}{6}wl,$$
$$T_A \sin \psi_A = ws_A = \tfrac{1}{2}wl.$$

Squaring and adding:
$$T_A^2 = \frac{w^2 l^2 10}{36}.$$

$$\therefore \ T_A = \frac{wl \sqrt{10}}{6}.$$

But $T_A = wy_A$, therefore
$$y_A = \frac{\sqrt{10}}{6}l. \tag{2}$$

Now the height of A above the table is $y_A - c$
$$= \frac{\sqrt{10}}{6}l - \frac{l}{6}$$
$$= \tfrac{1}{6}l(\sqrt{10}-1).$$

As $y = c \cosh x/c$, then $x_A = c \cosh^{-1} y_A/c$.

$$\therefore \quad x_A = \frac{l}{6} \cosh^{-1}\left(\frac{\sqrt{10}}{6} l \frac{6}{l}\right), \quad \text{using (1) and (2)}$$

$$x_A = \tfrac{1}{6} l \log_e\{\sqrt{10} + \sqrt{(10-1)}\} = \tfrac{1}{6} l \log_e(\sqrt{10} + 3).$$

10·2. Approximations for a tightly stretched wire

If the wire is tightly stretched, $\tan \psi$ will be small along the whole length.

Thus $s/c = \tan \psi$ is small.

It follows, that c is large compared with s, and also with x, as x is smaller than s.

The formulae $s = c \sinh x/c$ and $y = c \cosh x/c$ can therefore be expanded, the first few terms being taken as an approximation:

$$y \simeq c\left(1 + \frac{x^2}{2c^2}\right) = c + \frac{x^2}{2c}. \tag{1}$$

It is seen that the curve is approximately a parabola.

Note that $y - c$ is a measure of the 'sag'.

EXAMPLE

A chain 30·4 m long is suspended from two points A and B 30 m apart. Find the approximate depth of the mid-point of the chain below AB.

$$s = c \sinh x/c.$$

$$\therefore \quad 15\cdot2 = c \sinh 15/c.$$

$$15\cdot2 \simeq c\left(\frac{15}{c} + \frac{15^3}{6c^3}\right),$$

$$0\cdot2 \simeq \frac{15^3}{6c^2},$$

giving
$$c \simeq \frac{75}{\sqrt{2}}. \tag{i}$$

Now the sag is $y - c \simeq x^2/2c$.

$$\therefore \quad \text{Sag} \simeq \frac{15^2 \cdot \sqrt{2}}{150} = \frac{3}{2}\sqrt{2}.$$

$$\underline{\text{Sag} \simeq 2\cdot12 \text{ m.}}$$

10·3. Suspension bridge

The suspension chains, cables, etc., are taken as light compared with the weight of the bridge.

The lowest point of the suspension cable is taken as the origin, and axes as shown in fig. 73. The weight suspended from the portion OP is wx, where w is the weight per unit length of the bridge. Resolving horizontally and vertically:

$$T_0 = T \cos \psi = cw \text{ say.}$$

$$T \sin \psi = wx.$$

$\therefore \tan \psi = dy/dx = x/c$. Integrating,

$$y = x^2/2c \quad (x = 0 \text{ when } y = 0).$$

The shape of the suspension cable is thus *parabolic*.

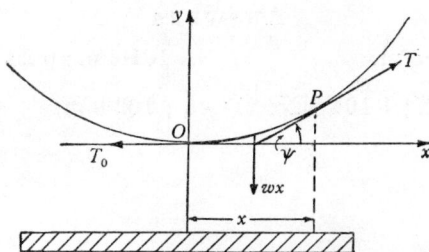

Fig. 73

Exercise 39

1. An electrical transmission line is stretched between two towers 300 m apart. If the tension at the lowest point of the wire is equal to the weight of 1500 m of wire, find the sag midway between the two towers.

2. From any point P on a catenary the ordinate PQ is drawn, meeting the x-axis at Q. From Q a perpendicular QR is drawn to the tangent at P, meeting the latter at R. Prove that the length of QR is constant and equal to c, and the length of PR is s.

(The symbols c, s, have their usual meanings.) [L.U.]

3. A cable hangs in the catenary $y = 200 \cosh{(0 \cdot 005x)}$, x, y in metres. Find the length of wire and the sag at the middle if the points of suspension are 100 m apart.

4. A uniform wire, tightly stretched between two points in a horizontal line 30 m apart, has a maximum sag of 1 m. Show that the length of the wire is less than $0 \cdot 15$ m in excess of the 30 m span.

5. In a suspension bridge the span is 100 m and the sag of the two supporting cables at their lowest point is 10 m. The bridge has a mass of 900 kg per metre. Prove that the curve has the equation $y = \dfrac{x^2}{250}$. Calculate the tensions at the ends and the middle, the slope of a cable at its ends, and the length of cable.

6. A kite is flown at the end of 40 m of string. The tension at the hand is equal to the weight of 16 m of string and is inclined at 30° to the horizontal. Show that the kite is about 34 m above the hand.

ANSWERS

1. About 7·5 m.   **3.** 101·4 m; 6·28 m.

5. 1·189 MN; 1·104 MN; 21° 48′; 102·6 m.

CHAPTER 11

COMPLEX NUMBERS

11·1. Introduction

Throughout the history of mathematics the number system has had to be gradually extended. Even the solution of apparently easy equations in algebra has meant either the introduction of a new set of numbers, or calling such equations insoluble. For example, unless negative numbers are allowed the equation $3x + 4 = 0$ is insoluble. A brief, incomplete, review is given below. It is not in chronological order of adoption of various sets of numbers.

Positive integers, e.g. $1, 2, 3, \ldots$
Fractions, e.g. $\frac{1}{3}, \frac{3}{4}$ [Decimals are, of course, fractions]
Negative integers and negative fractions

RATIONAL NUMBERS

The symbol for zero

REAL NUMBERS

Surds { A surd, or *irrational* number, is a root which cannot be expressed as a rational number, e.g. $\sqrt{2}$

Transcendental numbers, e.g. $\pi, e, \sin 31°$ { Numbers which cannot be found as solutions to algebraic equations

Complex numbers

11·11. A simple origin of complex numbers

If an attempt is made to solve the equation $x^2 + 2x + 2 = 0$ by means of the usual formula, the result is

$$x = \frac{-2 \pm \sqrt{(4-8)}}{2} = -1 \pm \frac{\sqrt{-4}}{2}.$$

It is obvious that $\sqrt{-4}$ is not a member of the real number system. After much argument the number system was extended to include such roots of negative numbers. Unfortunately they were christened 'imaginary numbers'. In fact they are just as 'real', and no more so, than any other form of number. To the engineer, the imaginary part of a complex number, representing a certain electric current, is as 'real' as the number three in the phrase 'these three chairs'.

11·12. The symbol j. Definition of a complex number

The symbol j is used to denote $\sqrt{-1}$. Thus $j^2 = -1$. The symbol i is also often used, but many engineers prefer j, as i is used to denote electric current.

The rigorous definition and treatment of complex numbers is difficult. It must suffice to state that they are defined in such a manner that the usual rules of algebra are obeyed, if it is always remembered that j^2 is -1.

Thus:
$$j^5 = j^4 j = j^2 j^2 j = (-1)(-1)j = j.$$

In time, statements such as $j^7 = -j$ should be capable of being verified mentally.

Note that only *one* new symbol is required for the whole range of imaginary numbers. For example

$$\sqrt{-25} = \sqrt{25}\,\sqrt{-1} = j5 \quad \text{and} \quad \sqrt{-31} = \sqrt{31}\,\sqrt{-1} = j\,\sqrt{31}.$$

A complex number is a number of the form $a + jb$, where a and b are real numbers.

a is called the *real part*.

b is called the *imaginary part*.

The '$a + jb$' form of a complex number is called the *algebraic* form to distinguish it from another form to be met with presently.

Note that all real numbers may be said to be part of the complex number system with zero imaginary part. Thus,

$$3 = 3 + j0.$$

A number such as $\sqrt{-4} = j2$ is often said to be purely imaginary.

11·13. Addition, subtraction, and multiplication of complex numbers

In addition and subtraction the real and imaginary parts are dealt with separately:

$$(3+j4)+(5-j2)+(2+j9)=(3+5+2)+j(4-2+9)$$
$$=10+j11,$$
$$(11-j2)-(4+j3)=(11-4)-j(2+3)$$
$$=7-j5.$$

In multiplication the normal rules for multiplying out brackets hold:

$$(3+j4)(9-j2)=27-j6+j36-j^28$$
$$=(27+8)+j(-6+36)$$
$$=35+j30.$$

With practice multiplication should be done in one step:

$$(3+j4)(9-j2)=(27+8)+j(36-6)=35+j30.$$

11·14. Conjugate complex numbers

Numbers of the form $a+jb$ and $a-jb$ are said to be conjugate. Their product is a real number, the *sum* of two squares:

$$(a+jb)(a-jb)=a^2+jab-jab-j^2b^2$$
$$=a^2+b^2.$$

Henceforth, the product of two conjugates should be written down on sight: $\quad (3-j4)(3+j4)=3^2+4^2=25.$

Note that to form a conjugate all that is necessary is to change the sign of the imaginary part.

11·15. Division of complex numbers

The numerator and denominator are each multiplied by the conjugate of the *denominator*. This makes the new denominator a real number.

For example

$$\frac{2-j3}{3+j4}=\frac{(2-j3)}{(3+j4)}\frac{(3-j4)}{(3-j4)}=\frac{-6-j17}{(3^2+4^2)}=-\frac{6}{25}-j\frac{17}{25}.$$

EXAMPLES

Simplify: (1) $(2-j)(4+j3)(5-j2)$.

(2) $(\cos x+j\sin x)(\cos 2x+j\sin 2x)$.

(3) $\dfrac{1}{2-j3}+\dfrac{5-j}{6+j2}$.

(1) $(2-j)(4+j3)(5-j2)=(11+j2)(5-j2)$
$$=59-j12.$$

(2) $(\cos x+j\sin x)(\cos 2x+j\sin 2x)$
$$=(\cos x\cos 2x-\sin x\sin 2x)+j(\sin x\cos 2x+\cos x\sin 2x)$$
$$=\cos 3x+j\sin 3x.$$

(3) $\dfrac{1}{2-j3}+\dfrac{5-j}{6+j2}=\dfrac{2+j3}{2^2+3^2}+\dfrac{(5-j)(6-j2)}{6^2+2^2}$
$$=(\tfrac{2}{13}+j\tfrac{3}{13})+(\tfrac{28}{40}-j\tfrac{16}{40})$$
$$=\tfrac{111}{130}-j\tfrac{11}{65}.$$

EXERCISE 40

Simplify the following:

1. $(2+j)^2$.   **2.** $(3-j)^4$.

3. $(1+j)(2+j)(3-j4)$.   **4.** $\dfrac{5-j}{5+j}$.

5. $\dfrac{1}{2}-\left(\dfrac{1+j}{2+j}\right)$.   **6.** $\dfrac{(3+j2)}{(4-j)(1+j4)}$.

7. $(\cos 2x+j\sin 2x)^2$.   **8.** $\dfrac{\cos 3x+j\sin 3x}{\cos x+j\sin x}$.

9. Find the roots of $x^2+x+1=0$ in $a+jb$ form.

10. If $z=\dfrac{2+j}{1-j}$, find the real and imaginary parts of the complex number $z+z^{-1}$.   [L.U.]

ANSWERS

1. $3+j4$.   **2.** $28-j96$.   **3.** $15+j5$.

4. $\tfrac{12}{13}-j\tfrac{5}{13}$.   **5.** $-\tfrac{1}{10}-j\tfrac{1}{5}$.   **6.** $\tfrac{54}{289}-j\tfrac{29}{289}$.

7. $\cos 4x+j\sin 4x$.   **8.** $\cos 2x+j\sin 2x$.

9. $-0.5\pm j0.866$.   **10.** $\tfrac{7}{10}$ R.P.; $\tfrac{9}{10}$ I.P.

11·2. Complex numbers as vectors

11·21. The Argand diagram

The reader will already be familiar with the representation of
real numbers by points on a straight line (see fig. 74). It is easy
to convince oneself that to every real number there corresponds
one point on the line. Harder to show, but equally true, is the
fact that to *every* point on the line there corresponds a real
number. There are no points left by which imaginary and complex numbers may be represented.

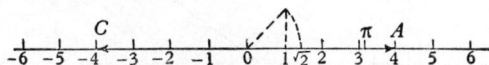

Fig. 74

To represent imaginary numbers another line must be chosen.
This other line must have one point in common with the 'real
number line', the point for zero.

A number, say 4, can be represented by the *point A*, or just
as well by the *vector* $\overrightarrow{OA}$. For the rest of this discussion the
vector method will be chosen. Now $j^2 4 = j . j . 4 = -4$. Thus,
operating by j *twice* on the real number 4 has the effect of turning
the vector $\overrightarrow{OA}$ through 180° into the position $\overrightarrow{OC}$. This rotation
is taken as anti-clockwise. It is natural to take operating by j
once as rotating the vector through 90° anti-clockwise. This
fixes the line on which to represent imaginary numbers as one
perpendicular to the 'real number line'. These lines are called
the 'real axis' and 'imaginary axis' respectively and are shown
in fig. 75. Such a diagram is called an Argand diagram after its
inventor, Robert Argand (1806).

As with graphs, the axes are usually lettered OX and OY.

The point A, or vector $\overrightarrow{OA}$, represents the real number 4.

The point B, or vector $\overrightarrow{OB}$, represents the imaginary number
$j3$.

The point P, or vector $\overrightarrow{OP}$, is taken to represent the complex
number $4+j3$. In this way any complex number can be repre-
sented as a vector.

On the diagram, multiplication of any number by j rotates its
vector through 90° anti-clockwise, and multiplication by $-j$,
through 90° clockwise.

Thus $j(4+j3) = -3+j4$; vector $\overrightarrow{OP}$ is rotated through 90° into the position $\overrightarrow{OP_1}$.

In electrical theory, alternating currents, e.m.f.'s, etc., are represented by rotating vectors which can be treated analytically as complex numbers.

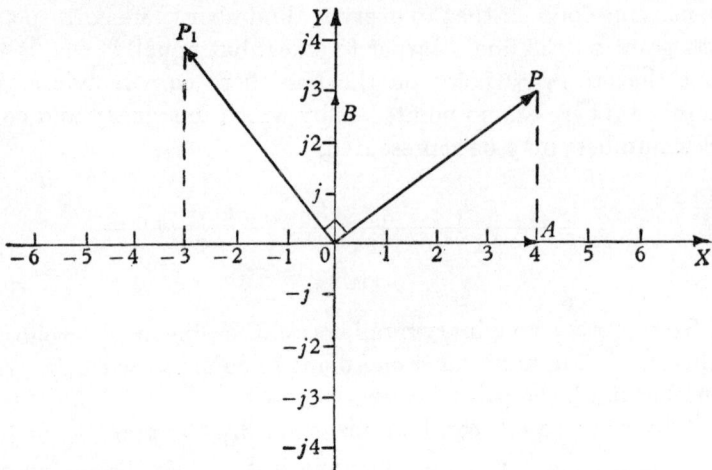

Fig. 75

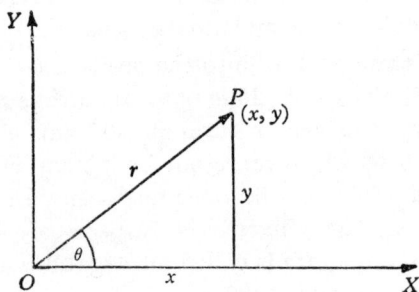

Fig. 76

11·22. Polar form of a complex number

z is the letter normally used to denote a complex number.

Let $z = x + jy$ be a complex number represented by the vector $\overrightarrow{OP}$. In addition to the co-ordinates of P fixing the position of $\overrightarrow{OP}$ it can also be fixed by its length $OP = r$, and the angle θ it makes with $\overrightarrow{OX}$.

r is called the *modulus* of the complex number.

θ is called the *argument* or *amplitude*.

Abbreviations for these are written mod z or $|z|$; arg z or amp z. 'Arg' will be chosen to avoid confusion of 'amp' with amperes.

The complex number itself is written $z = x + jy = r \angle \theta$.

This form, $r \angle \theta$, is called the polar form of the complex number. This is because (r, θ) are the polar co-ordinates of the point P.

The interrelationships between r, θ, x and y are obvious from fig. 76.

$$x = r \cos \theta, \qquad \theta = \tan^{-1} y/x,$$
$$y = r \sin \theta, \qquad r = \sqrt{(x^2 + y^2)}.$$

Note that $r \angle \theta$ is merely an abbreviation for $r(\cos \theta + j \sin \theta)$.

In transforming from the algebraic to the polar form it is wise to use the formula $\theta = \tan^{-1} y/x$ in conjunction with a sketch. The sketch should be used to fix the quadrant in which the vector lies. Alternatively, *both* results $\cos \theta = x/r$, $\sin \theta = y/r$ should be noted in fixing θ.

EXAMPLES

(1) $5 \angle 30° = 5(\cos 30° + j \sin 30°) = \dfrac{5 \sqrt{3}}{2} + j\frac{5}{2}$.

(2) $3 + j4 = r \angle \theta$, where $r = \sqrt{(3^2 + 4^2)} = 5$, $\cos \theta = \frac{3}{5}$ and $\sin \theta = \frac{4}{5}$.

Thus θ is in the first quadrant and is $53° 8'$.

$$\underline{3 + j4 = 5 \angle 53° 8'.}$$

11·221. Principal values of the argument

The position of the vector $\overrightarrow{OP}$ is unaltered if any multiple of $360°$ (2π) is added to or subtracted from θ. The argument of a complex number is therefore not single-valued.

The principal value of the argument is defined to be the value between $-\pi$ and π. Unless the contrary be stated, arg z will be taken to mean the principal value. In cases where no confusion will arise arg z will occasionally be used to imply the general value, $2n\pi$ + principal value.

11·222. Multiplication and division in polar form

$$r_1 \angle \theta_1 \times r_2 \angle \theta_2 = r_1 r_2 (\cos \theta_1 + j \sin \theta_1) \times (\cos \theta_2 + j \sin \theta_2)$$
$$= r_1 r_2 [(\cos \theta_1 \cos \theta_2 - \sin \theta_1 \sin \theta_2)$$
$$+ j(\sin \theta_1 \cos \theta_2 + \cos \theta_1 \sin \theta_2)]$$
$$= r_1 r_2 [\cos (\theta_1 + \theta_2) + j \sin (\theta_1 + \theta_2)]$$
$$= \underline{r_1 r_2 \angle (\theta_1 + \theta_2).}$$

This gives the rule: *multiply the moduli and add the arguments.*
As a particular case $[r \angle \theta]^n$, n a positive integer, is $r^n \angle n\theta$.

$$\frac{r_1 \angle \theta_1}{r_2 \angle \theta_2} = \frac{r_1 (\cos \theta_1 + j \sin \theta_1)}{r_2 (\cos \theta_2 + j \sin \theta_2)}$$

$$= \frac{r_1 (\cos \theta_1 + j \sin \theta_1)(\cos \theta_2 - j \sin \theta_2)}{r_2 \qquad \cos^2 \theta_2 + \sin^2 \theta_2}$$

$$= \frac{r_1}{r_2} [(\cos \theta_1 \cos \theta_2 + \sin \theta_1 \sin \theta_2)$$
$$+ j(\sin \theta_1 \cos \theta_2 - \cos \theta_1 \sin \theta_2)]$$

$$= \frac{r_1}{r_2} [\cos (\theta_1 - \theta_2) + j \sin (\theta_1 - \theta_2)]$$

$$= \frac{r_1}{r_2} \angle (\theta_1 - \theta_2)$$

This gives the rule: *divide the moduli and subtract the arguments.*
Note that if m is a positive integer:

$$[r \angle \theta]^{-m} = \frac{1}{[r \angle \theta]^m} = \frac{1 \angle 0}{r^m \angle m\theta} = \frac{1}{-r^m} \angle (0 - m\theta).$$
$$= r^{-m} \angle (-m\theta).$$

Thus $[r \angle \theta]^n = r^n \angle n\theta$ is true when n is a negative integer.

It will have been noticed that in multiplication and division the argument obeys the same algebraic rule as for powers and indices. The reason for this will become apparent later.

11·23. Addition, subtraction, multiplication and division on the Argand diagram

(a) *Addition.* The vector law of addition is satisfied.
Parallelogram law: $\qquad \overrightarrow{OA} + \overrightarrow{OC} = \overrightarrow{OB}.$

Triangle law: $\qquad \overrightarrow{OA} + \overrightarrow{AB} = \overrightarrow{OB}.$
Fig 77 shows

$$\overrightarrow{OA} = 3 + j, \quad \overrightarrow{AB} = 2 + j2 = \overrightarrow{OC}, \quad \overrightarrow{OB} = 5 + j3.$$

(b) *Subtraction.* The vector law of subtraction is satisfied:

$$\overrightarrow{OA} - \overrightarrow{OB} = \overrightarrow{OA} + \overrightarrow{OB_1} = \overrightarrow{OC}.$$

Fig. 78 shows

$$\overrightarrow{OA} = 5 + j, \quad \overrightarrow{OB} = 1 + j3, \quad \overrightarrow{OB_1} = -(1 + j3), \quad \overrightarrow{OC} = 4 - j2.$$

Alternatively, using fig. 77 and the triangle law

$$\vec{OA} + \vec{AB} = \vec{OB}.$$

Therefore $$\vec{OA} = \vec{OB} - \vec{AB},$$

$$3 + j = (5 + j3) - (2 + j2).$$

(c) *Multiplication.* For this purpose use is made of the rule for multiplication of complex numbers in their polar form.

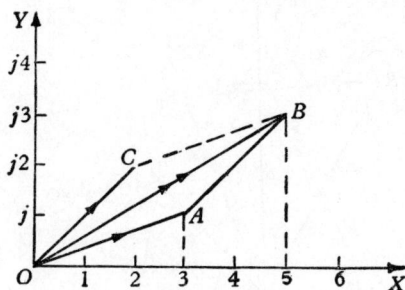

Fig. 77

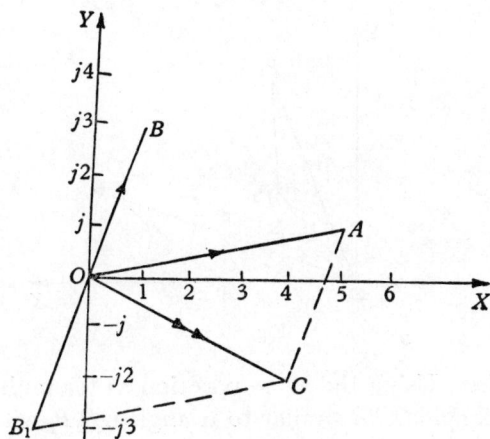

Fig. 78

In fig. 79, let

$$\vec{OP_1} = z_1 = r_1 \angle \theta_1, \quad \vec{OP_2} = z_2 = r_2 \angle \theta_2.$$

Then $OP_1 = r_1$, $X\hat{O}P_1 = \theta_1$ and $OP_2 = r_2$, $X\hat{O}P_2 = \theta_2$.

Let $\vec{OA} = 1 \angle 0$, that is, the real number 1.

Construct triangle OP_1R similar to triangle OAP_2, with OP_1, OA corresponding sides.

Then
$$\frac{OR}{OP_2} = \frac{OP_1}{OA} \quad \text{or} \quad \frac{OR}{r_2} = \frac{r_1}{1}, \quad \text{giving } OR = r_1r_2.$$

Also $\quad X\hat{O}R = P_1\hat{O}R + X\hat{O}P_1 = X\hat{O}P_2 + X\hat{O}P_1 = \theta_1 + \theta_2.$

If $\overrightarrow{OR} = z$, then $|z| = r_1r_2$ and $\arg z = \theta_1 + \theta_2.$

Thus
$$z = z_1z_2.$$

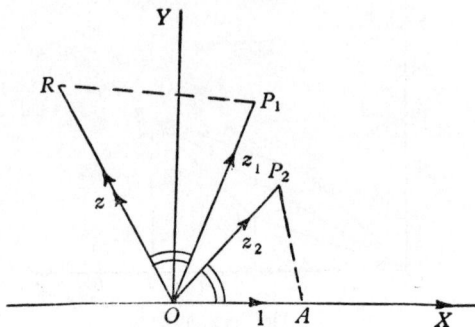

Fig. 79

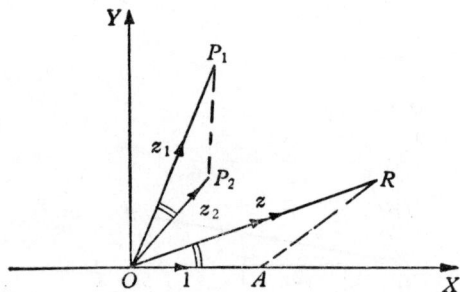

Fig. 80

(d) *Division.* Using the same notation as for multiplication, construct triangle OAR similar to triangle OP_2P_1 with OA, OP_2 corresponding sides (fig. 80).

Then
$$\frac{OR}{OP_1} = \frac{OA}{OP_2} \quad \text{or} \quad \frac{OR}{r_1} = \frac{1}{r_2}, \quad \text{giving} \quad OR = |z| = \frac{r_1}{r_2}.$$

Also $\quad X\hat{O}R = P_2\hat{O}P_1 = X\hat{O}P_1 - X\hat{O}P_2 = \theta_1 - \theta_2.$

Thus
$$\arg z = \theta_1 - \theta_2.$$

Therefore
$$z = z_1/z_2.$$

EXAMPLES

(1) Express $\dfrac{(3-j4)(1+j)}{2+j3}$ in polar form.

$$z = \frac{(3-j4)(1+j)}{2+j3} = \frac{(7-j)}{(2+j3)}\frac{(2-j3)}{(2-j3)} = \frac{11-j23}{13}.$$

$\therefore$ If $z = r \angle \theta$, $r\cos\theta = \frac{11}{13}$, $r\sin\theta = -\frac{23}{13}$.

$$r^2 = \frac{11^2 + 23^2}{13^2} = \frac{650}{169}, \quad r \simeq 1{\cdot}96.$$

θ is negative acute (principal value) and $\tan\theta = -\frac{23}{11}$. Thus

$$\theta \simeq -64° \, 27',$$

$$\underline{z = 1{\cdot}96 \angle -64° \, 27'.}$$

It may be noted here that the notation $\diagdown 64° \, 27'$ is often used to denote the negative angle $\angle -64° \, 27'$.

Alternative method. Using the rules of multiplication and division:

$$|z| = \frac{|3-j4||1+j|}{|2+j3|} = \frac{5\sqrt{2}}{\sqrt{13}} \simeq 1{\cdot}96.$$

$$\arg z = \arg(3-j4) + \arg(1+j) - \arg(2+j3)$$

$$= -53° \, 8' + 45° - 56° \, 19'$$

$$= -64° \, 27'.$$

And again: $\qquad \underline{z = 1{\cdot}96 \angle -64° \, 27'.}$

This alternative method is not quite as short as it seems, all the details of calculation having been omitted.

The reader is advised to choose one of the above methods and keep to it.

(2) If the ratio $\dfrac{z-j}{z-1}$ is purely imaginary, show that the point denoting z on the Argand diagram always lies on a circle, centre $\frac{1}{2}(1+j)$ and radius $\frac{1}{2}\sqrt{2}$. [L.U.]

Let $z = x + jy$, that is, z is represented by the point with co-ordinates (x, y) on the Argand diagram. Then

$$\frac{z-j}{z-1} = \frac{x+j(y-1)}{(x-1)+jy} = \frac{\{x+j(y-1)\}\{(x-1)-jy\}}{(x-1)^2+y^2}.$$

The real part of this expression is $\dfrac{x(x-1)+y(y-1)}{(x-1)^2+y^2}.$

For the given ratio to be purely imaginary this real part must be zero.

Thus
$$x(x-1)+y(y-1)=0,$$
$$x^2-x+y^2-y=0,$$
$$(x-\tfrac{1}{2})^2+(y-\tfrac{1}{2})^2=\tfrac{1}{2}.$$

This represents a circle centre $(\tfrac{1}{2},\tfrac{1}{2})$, radius $\dfrac{1}{\sqrt{2}}=\dfrac{\sqrt{2}}{2}$. (See Appendix B.)

Therefore the point z is always on the circle, centre $\tfrac{1}{2}(1+j)$ and radius $\tfrac{1}{2}\sqrt{2}$.

EXERCISE 41

Express in polar form:

1. $-3.$ **2.** $j5.$ **3.** $-j2.$

4. $2-j3.$ **5.** $\dfrac{5+j2}{5-j2}.$ **6.** $\dfrac{2+j3}{4-j}.$

7. $\dfrac{(2+j)^2}{1-j}.$ **8.** $\dfrac{(6+j)(3-j2)}{(1+j3)}.$

On the Argand diagram state the locus of:

9. $|z|=3.$ **10.** $|z-2|=4.$ **11.** $\arg z=\tfrac{1}{4}\pi.$

Find the equation of the locus of the point $z=x+jy$ if:

12. $\arg\left(\dfrac{z-2}{z+2}\right)=\tfrac{1}{4}\pi.$ **13.** $|z+j|^2-|z-j|^2=4.$

14. $2|z+1|=3|z-2|.$

Express in the form $a+jb$:

15. $2\angle 30^\circ.$ **16.** $5\angle -120^\circ.$ **17.** $1\cdot2\angle 101^\circ.$

18. (a) Find the moduli and arguments of $1+j,\ -1+j,\ 1-j.$

(b) The points $A,\ B,\ C,\ D$ in an Argand diagram are $9+j,\ 4+j13,\ -8+j8,\ -3-j4.$ Prove that $ABCD$ is a square.

[L.U.]

ANSWERS

1. $3\angle\pi.$ **2.** $5\angle\tfrac{1}{2}\pi.$ **3.** $2\angle-\tfrac{1}{2}\pi.$

4. $3\cdot606\angle-56^\circ\,19'.$ **5.** $1\angle 43^\circ\,36'.$

6. $0\cdot875\angle 70^\circ\,21'.$ **7.** $3\cdot535\angle 98^\circ\,8'.$

8. $6 \cdot 936 \angle -95° 48'$.   **9.** Circle, centre O, radius 3.

10. Circle, centre $2 + 0j$, radius 4.

11. Straight line through O making an angle $\frac{1}{4}\pi$ with OX.

12. $x^2 + y^2 - 4y = 4$. Circle centre $(0, 2)$, radius $2\sqrt{2}$.

13. The line $y = 1$.

14. $5x^2 + 5y^2 - 44x + 32 = 0$. A circle.

15. $\sqrt{3} + j$.   **16.** $-2 \cdot 5 - j2 \cdot 5 \sqrt{3}$.

17. $-0 \cdot 229 + j1 \cdot 178$.

18. (a) $\sqrt{2} \angle 45°$; $\sqrt{2} \angle 135°$; $\sqrt{2} \angle -45°$.

(b) Prove all sides same length, and any one angle a right angle.

11·3. Exponential form of a complex number

11·31. Convergence of a power series with complex variable

Let $S = z_1 + z_2 + z_3 + \ldots + z_n + \ldots$ be a series of complex terms, where $z_n = x_n + jy_n$.

Let $\qquad X = x_1 + x_2 + \ldots + x_n + \ldots$

and $\qquad Y = y_1 + y_2 + \ldots + y_n + \ldots$.

Then $\qquad S = X + jY$.

The series S is defined to be convergent if the series X and Y are convergent.

If X and Y are both absolutely convergent, then S is absolutely convergent. $\hfill (1)$

As $z_n = x_n + jy_n$, then $|x_n|$ and $|y_n|$ are both less than, or equal to, $|z_n|$ [as $|z_n| = \sqrt{(|x_n|^2 + |y_n|^2)}$].

Thus, from the comparison test for convergency, if $\Sigma|z_n|$ is convergent then both the series $\Sigma|x_n|$ and $\Sigma|y_n|$ are convergent. That is, if $\Sigma|z_n|$ is convergent Σx_n and Σy_n are absolutely convergent, and therefore Σz_n is absolutely convergent (from (1)).

It has thus been proved that if the series of the *moduli* of the complex terms is convergent, the series of the complex terms is absolutely convergent and therefore convergent. Hence a series of complex terms can be tested for convergency using the ratio test on the 'modulus' series.

11·32. Convergence of series for e^z, z complex

e^z, z complex, is defined by the series

$$e^z = 1 + z + \frac{z^2}{2!} + \frac{z^3}{3!} + \dots + \frac{z^n}{n!} + \dots. \tag{1}$$

Testing the 'modulus' series:

$$\left| \frac{u_{n+1}}{u_n} \right| = \left| \frac{z^n}{n!} \frac{(n-1)!}{z^{n-1}} \right| = \frac{|z|}{n}.$$

If $z = r \angle \theta$, then $|z| = r$.

$$\therefore \lim_{n \to \infty} \left| \frac{u_{n+1}}{u_n} \right| = \lim_{n \to \infty} \frac{r}{n} = 0 \quad \text{for all finite } r.$$

Thus the series for e^z is absolutely convergent, and therefore convergent, for all z.

11·33. Exponential form of a complex number

Putting $z = j\theta$ in equation (1) of § 11·32

$$e^{j\theta} = 1 + j\theta - \frac{\theta^2}{2!} - j\frac{\theta^3}{3!} + \frac{\theta^4}{4!} + \dots$$

$$= \left(1 - \frac{\theta^2}{2!} + \frac{\theta^4}{4!} - \frac{\theta^6}{6!} + \dots \right) + j\left(\theta - \frac{\theta^3}{3!} + \frac{\theta^5}{5!} - \dots \right)$$

$$= \cos\theta + j\sin\theta. \tag{1}$$

Writing $-\theta$ for θ:

$$e^{-j\theta} = \cos(-\theta) + j\sin(-\theta) = \cos\theta - j\sin\theta. \tag{2}$$

An *identity* for the polar form of a complex number has been obtained which can replace the *symbol* $r \angle \theta$:

$$z = x + jy = r(\cos\theta + j\sin\theta) \equiv r\, e^{j\theta} \quad \text{(from (1))}.$$

The rules for multiplication and division obtained previously, are now seen to be simply the actual rules for powers and indices.

Thus $\quad r_1 \angle \theta_1 \times r_2 \angle \theta_2 \equiv r_1 e^{j\theta_1} \times r_2 e^{j\theta_2}$

$$= r_1 r_2 \, e^{j(\theta_1 + \theta_2)} = r_1 r_2 \angle (\theta_1 + \theta_2)$$

and $\quad \dfrac{r_1 \angle \theta_1}{r_2 \angle \theta_2} \equiv \dfrac{r_1 e^{j\theta_1}}{r_2 e^{j\theta_2}} = \dfrac{r_1}{r_2} e^{j(\theta_1 - \theta_2)} = \dfrac{r_1}{r_2} \angle (\theta_1 - \theta_2).$

It will be found useful henceforth to use the identity $r\, e^{j\theta}$ rather than the symbol $r \angle \theta$.

11·4. De Moivre's theorem. Roots of a complex number

De Moivre's theorem states that one value of

$$(\cos\theta + j\sin\theta)^n \quad \text{is} \quad \cos n\theta + j\sin n\theta. \qquad (1)$$

The phrase 'one value of' is especially important when n is a fraction.

The original proof of the theorem will not be given. It is sufficient to see that the result follows immediately on using the exponential form of a complex number.

$$(\cos\theta + j\sin\theta)^n \equiv (e^{j\theta})^n = e^{jn\theta} \equiv (\cos n\theta + j\sin n\theta).$$

Symbolically: $[r\angle\theta]^n = r^n \angle n\theta$ for n any integer or fraction.
Equation (1) of § 11·33 is not in its fullest form, which is

$$e^{j\theta} = \cos\theta + j\sin\theta = \cos(\theta + 2k\pi) + j\sin(\theta + 2k\pi), \qquad (2)$$

k any integer. This result follows from the fact that the addition (or subtraction) of multiples of 2π to any angle does not alter the value of any trigonometric ratio of the angle. Thus

$$(e^{j\theta})^n = e^{jn\theta} = \cos n(\theta + 2k\pi) + j\sin n(\theta + 2k\pi). \qquad (3)$$

Suppose it is required to find the cube roots of any complex number, say $x + jy$. It is first put into its full polar form

$$x + jy = r\{\cos(\theta + 2k\pi) + j\sin(\theta + 2k\pi)\}.$$

Using equation (3) above

$$(x + jy)^{\frac{1}{3}} = r^{\frac{1}{3}} \left\{ \cos\frac{(\theta + 2k\pi)}{3} + j\sin\frac{(\theta + 2k\pi)}{3} \right\}.$$

As k can be *any* integer it may appear that there are innumerable cube roots of any complex number. On closer examination this is not so:

$$\sqrt[3]{(x + jy)} = \sqrt[3]{r} \left\{ \cos\left(\frac{\theta}{3} + \frac{2k\pi}{3}\right) + j\sin\left(\frac{\theta}{3} + \frac{2k\pi}{3}\right) \right\}.$$

Whenever k is a multiple of 3, multiples of 2π will simply be added to (or subtracted from if k is negative) the angle $\frac{1}{3}\theta$. Thus there will only be three different cube roots. They may be obtained by giving to k any three consecutive integral values. It is usual to take them starting at zero. Thus $k = 0, 1, 2$ will give the three different roots.

Similarly $(x+jy)^{1/n}$, n a positive integer, will have exactly n different roots.

Numerical examples will make this clearer.

EXAMPLES

(1) Find the three values of $\sqrt[3]{(3+j4)}$ in their $a+jb$ form.

$$3+j4 = 5(\cos 53° 8' + j \sin 53° 8')$$

$$= 5\{\cos (53° 8' + k\,360°) + j \sin (53° 8' + k\,360°)\}.$$

$$\sqrt[3]{(3+j4)} = \sqrt[3]{5}\left\{\cos\left(\frac{53° 8' + k\,360°}{3}\right) + j \sin\left(\frac{53° 8' + k\,360°}{3}\right)\right\}.$$

Taking $k = 0, 1, 2$, the three roots are:

$1\cdot71(\cos 17° 43' + j \sin 17° 43')$, $1\cdot71(\cos 137° 43' + j \sin 137° 43')$

and $1\cdot71(\cos 257° 43' + j \sin 257° 43')$.

In $a+jb$ form these are

$$1\cdot629 + j\,0\cdot520, \quad -1\cdot265 + j\,1\cdot151 \quad \text{and} \quad -0\cdot365 - j\,1\cdot670.$$

Note that taking $k=3$ would give the angle as $\dfrac{53° 8'}{3} + 360°$, which would just repeat the first value given by $k=0$.

A shorter method of working makes use of symbols:

$$3 + j4 = 5\angle(53° 8' + k\,360°).$$

$$\therefore \ \sqrt[3]{(3+j4)} = \sqrt[3]{5}\angle(17° 43' + k\,120°).$$

The three roots are thus:

$$1\cdot71\angle 17° 43', \quad 1\cdot71\angle 137° 43' \quad \text{and} \quad 1\cdot71\angle 257° 43'.$$

The representation of the three roots on an Argand diagram is instructive (fig. 81).

The vectors on the Argand diagram representing the three roots are each of length $\sqrt[3]{5} = 1\cdot71$, and the angular intervals between them are simply $\dfrac{360°}{3} = 120°$.

The argument of the first root is $\dfrac{\arg(3+j4)}{3}$.

This gives an easy geometrical construction for finding, in general, the nth roots of a complex number. The stages are:

(i) Convert to polar form $r\angle\theta$.

(ii) Draw a circle radius $\sqrt[n]{r}$.

(iii) Mark off the point whose angular distance from OX is θ/n.

(iv) Mark off further points at angular intervals $360°/n$ from the first point.

Then the vectors drawn to these points represent the n roots of the original complex number.

(2) Find the n values of $(-3)^{1/n}$ in polar form.

$$-3 = 3(-1) = 3.1 \angle (\pi + 2k\pi).$$

$$\therefore \; (-3)^{1/n} = \sqrt[n]{3} \; \left| \frac{\pi}{n} + \frac{2k\pi}{n} \right. \qquad (k = 0, 1, 2, ..., (n-1)).$$

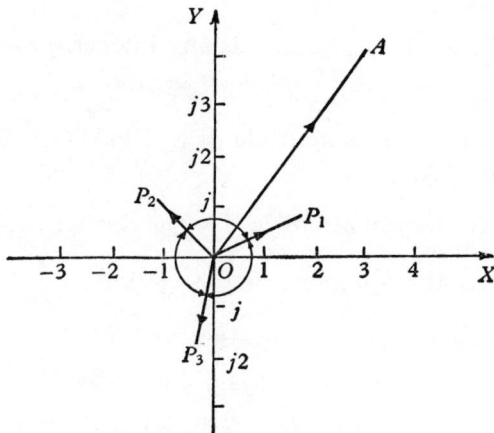

Fig. 81

(3) Put $(1-j)^8$ in the form $a+jb$.

$$(1-j) = \sqrt{2} \left(\frac{1}{\sqrt{2}} - j\frac{1}{\sqrt{2}} \right) = \sqrt{2} \left[\cos\left(-\tfrac{1}{4}\pi\right) + j\sin\left(-\tfrac{1}{4}\pi\right) \right]$$

$$= \sqrt{2} \angle -\tfrac{1}{4}\pi.$$

$$\therefore \; (1-j)^8 = (\sqrt{2})^8 \angle -\tfrac{8}{4}\pi = 2^4 \angle -2\pi$$

$$= 16 \angle 0 = \underline{\underline{16}}.$$

EXERCISE 42

1. (i) Simplify $(2+j3)(3-j2)$ and express the result in polar form.

(ii) Hence find the three cube roots of $(2+j3)(3-j2)$ in their polar form.

2. Find, in polar form, the six values of $(5+j2)^{\frac{1}{6}}$.

3. (i) Express $\dfrac{(1+j)(j-2)}{(1+j3)}$ in its polar form.

(ii) Find the three values of its cube roots in their $a+jb$ form.

(iii) Exhibit the original number and its three cube roots on an Argand diagram.

4. Find the three cube roots of $-7+j24$ in their $a+jb$ form.
[L.U.]

5. If $z=-1+j\sqrt{3}$ and n is any integer, prove that

$$2^{2n}+2^n.z^n+z^{2n}$$

is zero if n is not a multiple of 3. Find its value when n is a multiple of 3. [L.U.]

6. Solve the equation $x^7-1=0$, giving all 7 roots.

7. Solve the equation $x^4+x^2+1=0$.

8. Put $(3-j4)e^{px}+(3+j4)e^{qx}$, where

$$p=-0\cdot5+j4, \quad q=-0\cdot5-j4,$$

in the form $e^{\alpha x}(A\cos\beta x+B\sin\beta x)$ and give the values of A, B, α, β. [Sec. A]

9. (i) Draw the vectors representing the complex numbers $z=\sqrt{3}\,e^{j\pi/3}$ and $z=2\,e^{j7\pi/6}$, and also the vector representing their sum.

(ii) Express their sum in the forms $x+jy$ and $r\,e^{j\theta}$. [Sec. A]

Answers

1. (i) $13\angle22°37'$.

(ii) $2\cdot35\angle7°32'$, $2\cdot35\angle127°32'$, $2\cdot35\angle247°32'$.

2. $1\cdot32\angle(3°\,38'+k\,60°)$ $(k=0,1,2,3,4,5)$.

3. (i) $1\angle126°\,52'$;

(ii) $0\cdot74+j0\cdot673$, $-0\cdot953+j0\cdot304$, $0\cdot213-j0\cdot977$.

4. $2\cdot383+j1\cdot694$, $-2\cdot66+j1\cdot216$, $0\cdot276-j2\cdot912$.

5. 3.2^{2n}. *Hint.* First prove that $z = 2\,e^{j\frac{2}{3}\pi}$. Then $z^n = 2^n\,e^{j\frac{2}{3}n\pi}$, etc. 2^{2n} will be found a common factor of the expression.

6. $1\Big/\dfrac{2k\pi}{7}$ $(k = 0, 1, 2, ..., 6)$.

7. Treat as quadratic in x^2.

$$\tfrac{1}{2} + j\frac{\sqrt{3}}{2},\ -\tfrac{1}{2} + j\frac{\sqrt{3}}{2},\ -\tfrac{1}{2} - j\frac{\sqrt{3}}{2},\ \tfrac{1}{2} - j\frac{\sqrt{3}}{2};$$

or $1\angle 60°$, $1\angle 120°$, $1\angle 240°$, $1\angle 300°$.

8. $A = 6$, $B = 8$, $\alpha = -0.5$, $\beta = 4$.

9. (ii) $-\dfrac{\sqrt{3}}{2} + j\tfrac{1}{2}$, $e^{j\frac{5}{6}\pi}$.

11·5. Application to certain integrals

Integrals of the type $\displaystyle\int e^{ax} \cos bx\, dx$ and $\displaystyle\int e^{ax} \sin bx\, dx$ can be found by expressing $\cos bx$ and $\sin bx$ as the real and imaginary parts of $\cos bx + j \sin bx$, that is of e^{jbx}.

Let
$$C = \int e^{ax} \cos bx\, dx \quad \text{and} \quad S = \int e^{ax} \sin bx\, dx.$$

Then
$$C + jS = \int e^{ax} (\cos bx + j \sin bx)\, dx = \int e^{ax} e^{jbx}\, dx$$

$$= \int e^{(a+jb)x}\, dx = \frac{1}{(a+jb)} e^{(a+jb)x}.$$

Thus
$$C + jS = \frac{(a - jb)}{(a^2 + b^2)} e^{(a+jb)x}$$

$$= \frac{e^{ax}}{(a^2 + b^2)} (a - jb)(\cos bx + j \sin bx).$$

Equating real and imaginary parts:

$$C = \frac{e^{ax}}{(a^2 + b^2)} (a \cos bx + b \sin bx)$$

$$S = \frac{e^{ax}}{(a^2 + b^2)} (a \sin bx - b \cos bx)$$

(arbitrary constants have been omitted).

EXERCISE 43

1. Find $\int e^{-2x} \cos x \, dx$. State another integral whose value can be deduced.

2. Evaluate $\int_0^{\frac{1}{2}\pi} e^x \sin 2x \, dx$.

ANSWERS

1. $e^{-2x} (\sin x - 2 \cos x)/5 + C$;

$$\int e^{-2x} \sin x \, dx = -\frac{e^{-2x}}{5} (2 \sin x + \cos x) + C.$$

2. $\frac{2}{5} (e^{\frac{1}{2}\pi} + 1) \approx 2 \cdot 32$.

CHAPTER 12

PARTIAL DIFFERENTIATION

12·1. Introduction. Functions of more than one variable

Some of the simplest functions met with in practice are functions, not of one, but of two or more variables.

For example, the area of a rectangle is a function of two variables, its length and breadth.

In the derivation of formulae in this chapter only functions of two variables will be taken. It should be noted that all the formulae can be extended immediately to functions of any finite number of variables.

12·2. Partial and total derivatives

Taking the formula $A = lb$ for the area of a rectangle, the area can alter its value in three ways: (i) the length (l) may change and the breadth (b) keep constant; (ii) b may change and l remain constant; (iii) both l and b may change.

The word 'partial' applies to changes such as those in (i) and (ii) above; that is, when only one of the variables changes. The word 'total' is used to describe changes in which all the variables change.

The rate of change of a function when all the variables but one are kept constant is called the *partial derivative* with respect to that variable. The rate of change if all the variables are changing is called the *total derivative*. The symbol 'd' or 'f'' in functional notation, is reserved for the latter. Thus a new symbol is required for the former.

If $z = f(x, y)$

$\partial z/\partial x$, $\partial f/\partial x$ or f_x are all symbols used to denote the first partial derivative of the function when x varies but y is treated as a constant.

dz/dx, df/dx are symbols reserved for the first derivative with respect to x when both x *and* y can vary.

Total derivatives will be dealt with later. At present it is noted that to obtain *partial* derivatives introduces nothing new except new symbols.

Thus if

$$z = x^2 - 3xy + y^2, \quad \partial z/\partial x = 2x - 3y \quad \text{and} \quad \partial z/\partial y = -3x + 2y.$$

For $\partial z/\partial x$ all that is necessary is to think of y as being a constant, and so on.

Before giving a more formal definition of a partial derivative, the case of the rectangle will be worked from first principles.

In fig. 82, breadth, b, being kept constant, let the length, l, be increased by a small amount δl. Let δA be the change in area.

Then

$$A = l \times b$$

and

$$A + \delta A = (l + \delta l)\, b.$$

$$\therefore \ \delta A = b\, \delta l.$$

$$\therefore \ \delta A/\delta l = b.$$

$$\frac{\partial A}{\partial l} = \lim_{\delta l \to 0} \left(\frac{\delta A}{\delta l} \right)_{b\,\text{const.}} = \underline{b}.$$

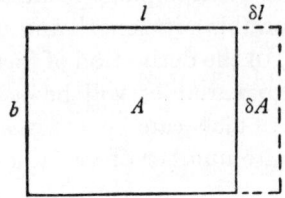

Fig. 82

The symbolism $(\ )_{b\,\text{const.}}$, showing that b has been kept constant, is often omitted if no ambiguity can arise.

12·21. Definition of partial derivatives

If z is a function of x and y, $z = f(x, y)$, then the partial derivatives $\partial z/\partial x$, $\partial z/\partial y$ are defined as

$$\frac{\partial z}{\partial x} = \lim_{\delta x \to 0} \left(\frac{\delta z}{\delta x} \right)_{y\,\text{const.}} \quad \text{or} \quad \lim_{\delta x \to 0} \left\{ \frac{f(x + \delta x, y) - f(x, y)}{\delta x} \right\},$$

$$\frac{\partial z}{\partial y} = \lim_{\delta y \to 0} \left(\frac{\delta z}{\delta y} \right)_{x\,\text{const.}} \quad \text{or} \quad \lim_{\delta y \to 0} \left\{ \frac{f(x, y + \delta y) - f(x, y)}{\delta y} \right\}.$$

As stated previously no new methods are required and partial derivatives can be written down on sight. All the well-known rules for products, quotients, function-of-a-function apply.

For example, if $z = f(y/x)$, then

$$\frac{\partial z}{\partial x} = \frac{df(y/x)}{d(y/x)} \frac{\partial(y/x)}{\partial x}$$

$$= \{f'(y/x)\}\,(-y/x^2).$$

Note the use here of $f'(y/x)$ to denote the awkward expression

$$\frac{d}{d(y/x)} f(y/x).$$

EXAMPLES

(1) If $z = x^3 - 2x^2y + 3xy^2 - y^4$, find $\partial z/\partial x$ and $\partial z/\partial y$.

$$\partial z/\partial x = 3x^2 - 2y(2x) + 3y^2(1) - 0$$
$$= \underline{3x^2 - 4xy + 3y^2}.$$

$$\partial z/\partial y = 0 - 2x^2(1) + 3x(2y) - 4y^3$$
$$= \underline{-2x^2 + 6xy - 4y^3}.$$

(2) Find $\partial z/\partial x$ and $\partial z/\partial y$ in the following cases:

(i) e^{3x-2y};   (ii) $\sin(4x - 3y) + \sqrt{(x^2 + y^2)}$.

(i) $\dfrac{\partial z}{\partial x} = e^{3x-2y} \dfrac{\partial}{\partial x}(3x - 2y) = \underline{3\,e^{3x-2y}}$

$\dfrac{\partial z}{\partial y} = \underline{-2\,e^{3x-2y}}.$

(ii) $\dfrac{\partial z}{\partial x} = 4\cos(4x - 3y) + \dfrac{1}{2\,\sqrt{(x^2 + y^2)}}\,2x$

$\qquad = \underline{4\cos(4x - 3y) + \dfrac{x}{\sqrt{(x^2 + y^2)}}}.$

$\dfrac{\partial z}{\partial y} = -3\cos(4x - 3y) + \dfrac{1}{2\,\sqrt{(x^2 + y^2)}}\,2y$

$\qquad = \underline{-3\cos(4x - 3y) + \dfrac{y}{\sqrt{(x^2 + y^2)}}}.$

(3) If $z = \sin^{-1}\left(\dfrac{x - y}{x + y}\right)$, prove that $x\dfrac{\partial z}{\partial x} + y\dfrac{\partial z}{\partial y} = 0$.

$$z = \sin^{-1}\left(\frac{x - y}{x + y}\right)$$

$$\therefore \quad \frac{x - y}{x + y} = \sin z. \tag{1}$$

Differentiating with respect to x:

$$\frac{(x + y)\,1 - (x - y)\,1}{(x + y)^2} = \cos z\,\frac{\partial z}{\partial x}.$$

$$\therefore \quad \frac{2y\sec z}{(x + y)^2} = \frac{\partial z}{\partial x}$$

and $\qquad\qquad x\dfrac{\partial z}{\partial x} = \dfrac{2xy}{(x + y)^2}\sec z. \tag{2}$

Differentiating (1) with respect to y:

$$\frac{(x+y)(-1)-(x-y)1}{(x+y)^2}=\cos z\frac{\partial z}{\partial y}.$$

$$\therefore\quad -\frac{2x}{(x+y)^2}\sec z=\frac{\partial z}{\partial y}$$

and

$$y\frac{\partial z}{\partial y}=-\frac{2xy}{(x+y)^2}\sec z. \tag{3}$$

Adding (2) and (3): $x\dfrac{\partial z}{\partial x}+y\dfrac{\partial z}{\partial y}=0.$

Note the use of the function-of-a-function rule:

$$\frac{\partial}{\partial x}(\sin z)=\frac{d}{dz}(\sin z)\frac{\partial z}{\partial x}=\cos z\frac{\partial z}{\partial x}.$$

(4) If $z=x^n f(y/x)$, prove $x\dfrac{\partial z}{\partial x}+y\dfrac{\partial z}{\partial y}=nz.$

$$\frac{\partial z}{\partial x}=nx^{n-1}f(y/x)+x^n f'(y/x)\frac{\partial(y/x)}{\partial x}$$

$$=nx^{n-1}f(y/x)+x^n f'(y/x)(-y/x^2)$$

$$=nx^{n-1}f(y/x)-x^{n-2}yf'(y/x).$$

$$\therefore\ x\frac{\partial z}{\partial x}=nx^n f(y/x)-x^{n-1}yf'(y/x). \tag{1}$$

$$\frac{\partial z}{\partial y}=x^n f'(y/x)\frac{\partial(y/x)}{\partial y}=x^n f'(y/x)\frac{1}{x}.$$

$$\therefore\ y\frac{\partial z}{\partial y}=x^{n-1}yf'(y/x). \tag{2}$$

Adding (1) and (2):

$$x\frac{\partial z}{\partial x}+y\frac{\partial z}{\partial y}=nx^n f(y/x)$$

$$=nz.$$

EXERCISE 44

Find $\partial z/\partial x$ and $\partial z/\partial y$ in the following cases:

1. $z=x^2+2xy+y^2.$ 2. $z=\sqrt{(x^2-y^2)}.$

3. $z=ax^2+2hxy+by^2+2gx+2fy+c.$

4. $z=\tan(x^2+y^2).$ 5. $z=\sinh(x+y).$

6. $z = \tan^{-1}(xy)$. **7.** $z = \cosh^{-1}(x/y)$.

8. If any one of x, y, z is expressed in terms of the other two by the formula

$$2xyz - 5x + 2y - 3z = 1,$$

prove that

$$\left(\frac{\partial y}{\partial z}\right)_x \left(\frac{\partial z}{\partial x}\right)_y \left(\frac{\partial x}{\partial y}\right)_z = -1. \qquad \text{[L.U.]}$$

9. If $z = x^3 - 3x^2y - 2y^3$, prove that $x\dfrac{\partial z}{\partial x} + y\dfrac{\partial z}{\partial y} = 3z$.

10. If $V = x^2 + y^2 + z^2$, prove that $xV_x + yV_y + zV_z = 2V$.

11. If $z = x^{n-1}y$, prove that $x\dfrac{\partial z}{\partial x} + y\dfrac{\partial z}{\partial y} = nz$.

12. If $f(y/x)$ is any differentiable function of y/x, find the value of $x\dfrac{\partial u}{\partial x} + y\dfrac{\partial u}{\partial y}$ when

$$u = f(y/x) + \sqrt{(x^2 + y^2)}. \qquad \text{[L.U.]}$$

Answers

1. $2(x+y)$ in both cases.

2. $f_x = \dfrac{x}{\sqrt{(x^2 - y^2)}}$; $f_y = -\dfrac{y}{\sqrt{(x^2 - y^2)}}$.

3. $2(ax + hy + g)$; $2(hx + by + f)$.

4. $2x\sec^2(x^2 + y^2)$; $2y\sec^2(x^2 + y^2)$.

5. $f_x = f_y = \cosh(x+y)$. **6.** $\dfrac{y}{1 + x^2y^2}$; $\dfrac{x}{1 + x^2y^2}$.

7. $\dfrac{1}{\sqrt{(x^2 - y^2)}}$; $-\dfrac{x}{y\sqrt{(x^2 - y^2)}}$.

8. *Hint.* Differentiate the whole expression first with respect to x, then y, then z. Hence find $\left(\dfrac{\partial z}{\partial x}\right)_y$, etc.

12. $\sqrt{(x^2 + y^2)}$.

Hint. In exs. 4, 5 and 6, f_y may be deduced from f_x by symmetry. Merely interchange x and y.

12·3. Higher derivatives

In general, $\partial z/\partial x$ and $\partial z/\partial y$ will be functions of x and y and can be differentiated again partially with respect to x and/or y. In this manner four second partial derivatives may be obtained:

$$\frac{\partial}{\partial x}\left(\frac{\partial z}{\partial x}\right)=\frac{\partial^2 z}{\partial x^2} \quad \text{or} \quad f_{xx},$$

$$\frac{\partial}{\partial y}\left(\frac{\partial z}{\partial y}\right)=\frac{\partial^2 z}{\partial y^2} \quad \text{or} \quad f_{yy},$$

$$\frac{\partial}{\partial y}\left(\frac{\partial z}{\partial x}\right)=\frac{\partial^2 z}{\partial y\,\partial x} \quad \text{or} \quad f_{yx},$$

$$\frac{\partial}{\partial x}\left(\frac{\partial z}{\partial y}\right)=\frac{\partial^2 z}{\partial x\,\partial y} \quad \text{or} \quad f_{xy}.$$

In this symbolism the variable on the *right* signifies which differentiation was carried out first. In some books this is reversed. Thus, $\partial^2 z/\partial x\,\partial y$ would mean differentiation with respect to x first.

In this book the first notation given above will be followed.

If the function and its derivatives are continuous functions of x and y, $\partial^2 z/\partial y\,\partial x$ and $\partial^2 z/\partial x\,\partial y$ can be proved to be identical. Unless stated to the contrary the reader can assume

$$\partial^2 z/\partial y\,\partial x = \partial^2 z/\partial x\,\partial y$$

for work of this standard.

Higher derivatives than the second can be formed but are not of great practical use. They will not be dealt with here.

No new rules are needed. The following example should be sufficient to indicate the procedure in finding second derivatives.

EXAMPLE

If $z = \tan^{-1}(x/y)$, prove that

$$x^2\frac{\partial^2 z}{\partial x^2}+2xy\frac{\partial^2 z}{\partial x\,\partial y}+y^2\frac{\partial^2 z}{\partial y^2}=0. \qquad \text{[L.U.]}$$

$$\frac{\partial z}{\partial x}=\frac{1}{1+x^2/y^2}\frac{1}{y}=\frac{y}{x^2+y^2}.$$

$$\frac{\partial z}{\partial y}=\frac{1}{1+x^2/y^2}\left(-\frac{x}{y^2}\right)=-\frac{x}{x^2+y^2}.$$

$$\therefore \quad x\frac{\partial z}{\partial x}+y\frac{\partial z}{\partial y}=0. \qquad (1)$$

Differentiating this result with respect to x and then with respect to y:

$$\frac{\partial z}{\partial x} + x\frac{\partial^2 z}{\partial x^2} + y\frac{\partial^2 z}{\partial x\,\partial y} = 0. \tag{2}$$

$$x\frac{\partial^2 z}{\partial y\,\partial x} + \frac{\partial z}{\partial y} + y\frac{\partial^2 z}{\partial y^2} = 0. \tag{3}$$

Multiplying (2) by x and (3) by y and adding:

$$\left(x\frac{\partial z}{\partial x} + y\frac{\partial z}{\partial y} \right) + \left(x^2\frac{\partial^2 z}{\partial x^2} + y^2\frac{\partial^2 z}{\partial y^2} \right) + 2xy\frac{\partial^2 z}{\partial x\,\partial y} = 0.$$

Thus, using (1),

$$x^2\frac{\partial^2 z}{\partial x^2} + 2xy\frac{\partial^2 z}{\partial x\,\partial y} + y^2\frac{\partial^2 z}{\partial y^2} = 0.$$

Note. In this type of question, *if* a simple relationship can be found between $\partial z/\partial x$ and $\partial z/\partial y$ the method given above usually gives quicker results than finding separate expressions for

$$\partial^2 z/\partial x^2, \quad \partial^2 z/\partial y^2 \quad \text{and} \quad \partial^2 z/\partial x\,\partial y$$

Exercise 45

1. If $z = A\sin(x+at) + B\cos(x-at)$, where A, B, a are constants, prove that $\dfrac{\partial^2 z}{\partial t^2} = a^2\dfrac{\partial^2 z}{\partial x^2}$ and $\dfrac{\partial^2 z}{\partial x\,\partial t} = \dfrac{\partial^2 z}{\partial t\,\partial x}$.

2. If $z = \log_e(x^2 + y^2)$, prove that $\partial^2 z/\partial x^2 + \partial^2 z/\partial y^2 = 0$.

3. If $r^2 = x^2 + y^2$, prove that $\left(\dfrac{\partial^2 r}{\partial x^2}\right)\left(\dfrac{\partial^2 r}{\partial y^2}\right) = \left(\dfrac{\partial^2 r}{\partial x\,\partial y}\right)^2$.

4. Show that if $z = f(x - ct)$, then $\dfrac{\partial^2 z}{\partial x^2} = \dfrac{1}{c^2}\dfrac{\partial^2 z}{\partial t^2}$.

12·4. Total increment

Let $z = f(x, y)$.

Due to small increments δx in x and δy in y, let the change in z be δz.

Then $\qquad z + \delta z = f(x + \delta x, y + \delta y)$,

but $\qquad z = f(x, y)$.

$$\therefore \; \delta z = f(x + \delta x, y + \delta y) - f(x, y). \tag{1}$$

δz is called the *total* increment or change due to changes in both x and y.

In order to treat the right-hand side of (1) in stages in which only one of x and y varies, $f(x, y + \delta y)$ is added and subtracted.

Thus

$$\delta z = [f(x + \delta x, y + \delta y) - f(x, y + \delta y)] + [f(x, y + \delta y) - f(x, y)].$$

In the first square bracket only x changes, and in the second, only y changes.

Hence, by the mean-value theorem (see § 1·9) the expression in the first square brackets may be written

$$\delta x f_x(x + \theta_1 \delta x, y + \delta y),$$

where $0 < \theta_1 < 1$ and f_x denotes the partial derivative with respect to x.

Similarly, the expression in the second square brackets may be written
$$\delta y f_y(x, y + \theta_2 \delta y), \quad \text{where} \quad 0 < \theta_2 < 1.$$

Thus $\quad \delta z = \delta x f_x(x + \theta_1 \delta x, y + \delta y) + \delta y f_y(x, y + \theta_2 \delta y).$

If z and its derivatives are continuous $f_x(x + \theta_1 \delta x, y + \delta y)$ must tend to the limit $f_x(x, y)$ as δx and $\delta y \to 0$.

It may therefore be written $f_x(x, y) + \epsilon_1$, where $\epsilon_1 \to 0$ as δx and $\delta y \to 0$.

Similarly, $f_y(x, y + \theta_2 \delta y)$ may be written $f_y(x, y) + \epsilon_2$ where $\epsilon_2 \to 0$ as δx and $\delta y \to 0$.

$$\therefore \quad \delta z = \delta x \{ f_x(x, y) + \epsilon_1 \} + \delta y \{ f_y(x, y) + \epsilon_2 \}$$
$$= \delta x f_x(x, y) + \delta y f_y(x, y) + \delta x \epsilon_1 + \delta y \epsilon_2. \tag{2}$$

Hence, if δx and δy are *small* and therefore ϵ_1 and ϵ_2 small

$$\delta z \simeq \delta x f_x(x, y) + \delta y f_y(x, y). \tag{3}$$

In different notation:

$$\delta z \simeq \frac{\partial z}{\partial x} \delta x + \frac{\partial z}{\partial y} \delta y. \tag{4}$$

This is a useful formula for finding small changes or errors in practical cases. It may be extended to functions of any number of variables. Thus, if $u = f(x, y, z)$

$$\delta u \simeq \frac{\partial u}{\partial x} \delta x + \frac{\partial u}{\partial y} \delta y + \frac{\partial u}{\partial z} \delta z. \tag{5}$$

EXAMPLES

(1) The volume of a certain gas at 320 K, pressure 80 N/m², is 0·18 m³. Find the approximate change in volume if the

pressure be increased to 81 N/m² and the temperature raised to
322 K.

$$pV = kT,$$

$$V = \frac{kT}{p}.$$

$$\therefore \ \delta V \simeq \frac{\partial V}{\partial p} \delta p + \frac{\partial V}{\partial T} \delta T$$

$$= -\frac{kT}{p^2} \delta p + \frac{k}{p} \delta T$$

$$= -\frac{kT}{p} \frac{1}{p} \delta p + \frac{kT}{p} \frac{\delta T}{T}.$$

$$\therefore \ \delta V \simeq -V \frac{1}{p} \delta p + V \frac{1}{T} \delta T. \tag{1}$$

Now $p = 80$, $\delta p = 1$, $V = 0.18$, $T = 320$, $\delta T = 2$.

Thus
$$\delta V \simeq -\frac{0.18}{80} \times 1 + \frac{0.18}{320} \times 2$$

$$\simeq -0.0011 \text{ m}^3.$$

The change in volume is a *decrease* of approximately 0.0011 m³.

(2) The period of a simple pendulum is given by $T = 2\pi \sqrt{(l/g)}$.
If the length, l, is measured 1 % too large and the period T 2 %
too small, find the approximate percentage error in the value of
g calculated from the given formula.

When dealing with proportionate, or percentage, errors or
changes, it is best to take logs to the base e before differentiating.

$$g = 4\pi^2 \frac{l}{T^2}.$$

$$\therefore \ \log_e g = \log_e (4\pi^2) + \log_e l - 2 \log_e T.$$

But
$$\delta(\log_e g) \simeq \frac{\partial}{\partial l}(\log_e g) \, \delta l + \frac{\partial}{\partial T}(\log_e g) \, \delta T$$

$$= \frac{1}{l} \delta l - \frac{2}{T} \delta T.$$

Now
$$\delta(\log_e g) \simeq \frac{d}{dg}(\log_e g) \, \delta g = \frac{1}{g} \delta g.$$

$$\therefore \ \frac{1}{g} \delta g \simeq \frac{1}{l} \delta l - \frac{2}{T} \delta T. \tag{1}$$

The proportionate error has thus been obtained. For percentage errors:

$$\left(\frac{\delta g}{g}\right)100 \fallingdotseq \left(\frac{\delta l}{l}\right)100 - 2\left(\frac{\delta T}{T}100\right).$$

Or, percentage error in $g \fallingdotseq$ percentage error in $l - 2 \times$ percentage error in T.

Putting in the values:

$$\text{Percentage error in } g \fallingdotseq 1 - 2(-2) = 5.$$

g is approximately 5 % too large.

Note. When practice has been gained the full working given above, may be curtailed considerably.

12·5. Total derivative. Total differential

Referring to 12·4, equation (2): if x and y, and therefore z, are continuous functions of another variable t, let δx, δy, δz be the changes in x, y, z due to a small increment δt in t. Dividing the above equation by δt:

$$\frac{\delta z}{\delta t} = \frac{\delta x}{\delta t}f_x(x,y) + \frac{\delta y}{\delta t}f_y(x,y) + \frac{\delta x}{\delta t}\epsilon_1 + \frac{\delta y}{\delta t}\epsilon_2$$

or

$$\frac{\delta z}{\delta t} = \frac{\delta x}{\delta t}\frac{\partial z}{\partial x} + \frac{\delta y}{\delta t}\frac{\partial z}{\partial y} + \frac{\delta x}{\delta t}\epsilon_1 + \frac{\delta y}{\delta t}\epsilon_2.$$

Now as $\delta t \to 0$, and therefore δx, δy approach zero,

$$\frac{\delta z}{\delta t} \to \frac{dz}{dt}, \quad \frac{\delta x}{\delta t} \to \frac{dx}{dt}, \quad \frac{\delta y}{\delta t} \to \frac{dy}{dt} \quad \text{and} \quad \epsilon_1 \text{ and } \epsilon_2 \to 0.$$

Thus

$$\frac{dz}{dt} = \frac{\partial z}{\partial x}\frac{dx}{dt} + \frac{\partial z}{\partial y}\frac{dy}{dt}. \tag{1}$$

$$dz = \frac{\partial z}{\partial x}dx + \frac{\partial z}{\partial y}dy \text{ is called the } total \ differential \text{ of } z. \tag{2}$$

dz/dt is the total rate of change of z with respect to t, or the *total derivative*, when both x and y vary with t.

Note that this formula is *exact* and not an approximation as that in §12·4, equation (4). It can be extended to any finite number of variables.

Example

In a triode valve the current flowing may be taken as given by the formula $i = A(V_a + \mu V_g)^{\frac{3}{2}}$, where A, μ are constants and V_a, V_g are the anode and grid voltages. If the rate of change of

V_g is $\frac{1}{10}$ volt/sec. and that of V_a is $\frac{1}{2}$ volt/sec., find the rate of change of current at the instant when $V_a = 250$ volts, $V_g = 5$ volts.

$$\frac{di}{dt} = \frac{\partial i}{\partial V_a}\frac{dV_a}{dt} + \frac{\partial i}{\partial V_g}\frac{dV_g}{dt}.$$

But $\quad i = A(V_a + \mu V_g)^{\frac{3}{2}} \quad$ and $\quad \frac{dV_a}{dt} = \frac{1}{2}, \quad \frac{dV_g}{dt} = \frac{1}{10}.$

$$\therefore \frac{di}{dt} = \frac{1}{2}A\,\frac{3}{2}(V_a + \mu V_g)^{\frac{1}{2}} + \frac{1}{10}A\frac{3}{2}(V_a + \mu V_g)^{\frac{1}{2}}\,\mu.$$

$$= \frac{3}{2}A\,\sqrt{(V_a + \mu V_g)}\,(\tfrac{1}{2} + \tfrac{1}{10}\mu).$$

When $V_a = 250$, $V_g = 5$, this gives

$$\frac{di}{dt} = \frac{3}{4}A\,\sqrt{(250 + 5\mu)}\,(1 + \tfrac{1}{5}\mu).$$

EXERCISE 46

1. The area of a triangle is calculated from the formula $\triangle = \frac{1}{2}bc\sin A$. b, c and A are measured correct to 1%. If A is measured as $45°$, prove that the percentage error in the area cannot be more than about $2\cdot8\%$. (*Note.* Remember that 'δA' must be in radians.)

2. The current flowing in a circuit is given by $i = E/Z$, where E is the voltage and Z the impedance. Find the rate of change of the current when $E = 80$ volts, $Z = 100\ \Omega$ and E is increasing at 1 volt/sec., whilst Z is increasing at $2\ \Omega$/sec. Is the current increasing or decreasing?

3. The side c of a triangle is calculated from the formula $c^2 = a^2 + b^2 - 2ab\cos C$. Find the relation between δc and $\delta a, \delta b, \delta C$.

4. The impedance, Z, of a series circuit is given by

$$Z = \{R^2 + (\omega L - 1/\omega C)^2\}^{\frac{1}{2}},$$

where the symbols have their usual meanings. If ω and R are kept constant and L and C increased by 1% find the approximate percentage change in Z.

5. The discharge through a rectangular notch is given by $\frac{2}{3}C\sqrt{(2g)}LH^{\frac{3}{2}}$, where C is a constant, L is the length of the notch and H the height of the water surface above the sill. Find the approximate increase in discharge when $H = 1\cdot4$ m, $L = 90$ mm, $C = 0\cdot3$, $g = 9\cdot81$ m/s^2 and L is increased by 5 mm and H by 150 mm. (In the formula all lengths are in metres.)

6. For belts on grooved pulleys the power transmitted is given by $P = \dfrac{(T_1 - T_2)\, 2\pi R n}{60}$. If the difference in tensions, $(T_1 - T_2)$, is increased by 1 %, the radius of the smallest pulley, R, increased by 2 % and the number of rev/min, n, is decreased by 3 %, find the approximate percentage change in transmitted power.

7. If the radius of a right circular cone increases at 20 mm./min and the height at 30 mm/min, find the rate at which the volume is increasing when the radius is 60 mm and the height 180 mm.

8. The height, h, of an object is calculated from the angles of elevation A, B at the ends of a horizontal base-line, of length c, in a vertical plane containing the object and on the same side of it. If δh is the greatest possible error in h due to errors δA, δB, δc in the measured quantities, show that

$$\frac{\delta h}{h} = \frac{|\delta c|}{c} + \frac{h}{c}\{\operatorname{cosec}^2 A\, |\delta A| + \operatorname{cosec}^2 B\, |\delta B|\}.$$

[L.U.]

Answers

2. $-0{\cdot}006$ amp./sec.; decreasing.

3. $\delta c = \dfrac{1}{c}[(a - b\cos C)\,\delta a + (b - a\cos C)\,\delta b + ab\sin C\,\delta c].$

4. $\dfrac{(\omega^2 L^2 - 1/\omega^2 C^2)}{[R^2 + (\omega L - 1/\omega C)^2]}.$

5. About $0{\cdot}03$ m^3/s.

6. No change.   **7.** $0{\cdot}18$ mm^3/min.

8. *Hint.* The formula for the calculation of c is

$$c = h(\cot A - \cot B).$$

CHAPTER 13

ORDINARY DIFFERENTIAL EQUATIONS

13·1. Introduction

13·11. Formation. Order and degree of an equation

The formation of differential equations has already been discussed in § 1·5. This discussion is now recapitulated and extended.

If y is a function of x containing three arbitrary constants A, B, C, as long as y and its derivatives are differentiable, three equations can be obtained by differentiating, successively, three times. Together with the original equation, $y=f(x)$, this gives four equations. Thus, theoretically, the three arbitrary constants can be eliminated from these four equations. This would leave a relationship containing, in general, y, dy/dx, d^2y/dx^2, d^3y/dx^3 and functions of x. Such a relationship is called an ordinary differential equation; 'ordinary' because only derivatives with respect to a single variable, x, are involved.

It is obvious that in like manner *partial* differential equations can be obtained from functions of more than one variable. These are beyond the scope of the present volume.

Summarizing: a differential equation is a relationship between the derivatives of a function, which may also contain the function itself and other functions of x.

Order. The highest derivative occurring gives the *order* of the differential equation.

Thus

$$\frac{d^3y}{dx^3} - 2\frac{dy}{dx} + y = \sin x \text{ is said to be of the third order.}$$

Degree. The highest power of the highest derivative occurring gives the *degree* of the equation. Thus

$$\left(\frac{d^3y}{dx^3}\right)^2 + \frac{d^3y}{dx^3} - 2\left(\frac{dy}{dx}\right)^3 = 4x$$

is said to be of the second degree.

Differential equations can be broadly classified by means of their order and degree.

13·12. Solutions

$dy/dx = f(x)$ is a simple differential equation. Its solution is
$y = \int f(x)\, dx + C$, where C is arbitrary. Unless special information
is given there are innumerable special solutions, obtained by
giving to C any numerical value.

In general, the solution of a differential equation involves
finding y as a function of x, or in the form $f(x, y) = 0$.

If the highest derivative occurring is d^4y/dx^4, whatever process
is used in solution, it must be equivalent to integrating *four*
times. This would involve four arbitrary constants in the
solution.

It appears, then, that the *full solution to a given differential
equation will contain as many arbitrary constants as the order
of the equation.* Such a solution is called a *General Solution* (G.S.)
or *Complete Primitive* (C.P.).

If a solution satisfying a given equation is found which has
less than the required number of arbitrary constants for a general
solution, it is called a *Particular Integral* (P.I.).

A particular integral can always be obtained from the general
solution by giving special values to some or all of the arbitrary
constants.

EXAMPLE

$$d^2y/dx^2 = 3.$$
$$\therefore \ dy/dx = 3x + A,$$
and
$$y = \tfrac{3}{2}x^2 + Ax + B.$$

This is the general solution.

Particular integrals, or solutions, would be

$$y = \tfrac{3}{2}x^2, \quad y = \tfrac{3}{2}x^2 - x, \quad y = \tfrac{3}{2}x^2 + 5x - 7, \quad \text{etc.}$$

Quite often particular solutions are required which must be
made to fit given initial conditions.

EXAMPLE

Solve $d^2y/dx^2 = 3$, given $dy/dx = 1$ when $x = 0$ and $y = 0$ when
$x = 2$.

From above
$$dy/dx = 3x + A$$
when $x = 0$, $dy/dx = 1$. $\therefore \ 1 = 0 + A$, $A = 1$.
$$\therefore \ dy/dx = 3x + 1.$$

Thus $\qquad\qquad y = \tfrac{3}{2}x^2 + x + B,$

when $x = 2$, $y = 0$.

$$\therefore \quad 0 = 6 + 2 + B, \quad B = -8.$$

The particular solution required is therefore

$$\underline{y = \tfrac{3}{2}x^2 + x - 8.}$$

13·13. Differential equations in engineering

Many relationships in engineering, and science, are more easily expressed in the form of a differential equation than in any other form.

Some of the harder types cannot be accurately solved; only approximate numerical solutions can be found. Computers are extensively used for this purpose.

The rest of this chapter will deal with methods of solution of the simpler types of differential equations. Only types which have well-known applications to engineering will be considered.

13·2. First-order equations

13·21. *y* absent or *x* absent

That is $\qquad dy/dx = f(x) \quad$ or $\quad dy/dx = F(y).$

These can be solved by direct integration:

$$y = \int f(x)\,dx \quad \text{or} \quad x = \int \frac{1}{F(y)}\,dy.$$

EXAMPLES

(1) $$x^3 \frac{dy}{dx} = 1 + x.$$

$$\therefore \quad \frac{dy}{dx} = \frac{1}{x^3} + \frac{1}{x^2}.$$

$$y = \int \left(\frac{1}{x^3} + \frac{1}{x^2} \right) dx = \underline{-\frac{1}{2x^2} - \frac{1}{x} + C.}$$

(2) $$\frac{1}{y^3} \frac{dy}{dx} = a; \; a \text{ constant.}$$

$$\therefore \quad x = \int \frac{1}{ay^3}\,dy = \underline{-\frac{1}{2ay^2} + C.}$$

13·211. $\dfrac{d^n y}{dx^n} = f(x)$

It is convenient to consider here the type $d^n y/dx^n = f(x)$, as this type can also be solved by direct integration, step by step. Thus, if

$$\frac{d^4 y}{dx^4} = \frac{w}{EI},$$

$$EI \frac{d^3 y}{dx^3} = wx + A,$$

$$EI \frac{d^2 y}{dx^2} = \tfrac{1}{2} wx^2 + Ax + B,$$

$$EI \frac{dy}{dx} = \tfrac{1}{6} wx^3 + \tfrac{1}{2} Ax^2 + Bx + C$$

and $\qquad EI y = \tfrac{1}{24} wx^4 + \tfrac{1}{6} Ax^3 + \tfrac{1}{2} Bx^2 + Cx + D.$

EXERCISE 47

Solve:

1. $dy/dx - 3y + 2 = 0.$ **2.** $x^2 \dfrac{dy}{dx} - x + 2 = 0.$

3. $5y \dfrac{dy}{dx} - 3y^2 + 4 = 0.$ **4.** $EI \dfrac{d^2 y}{dx^2} = x^2 - 1.$

5. Find a function of x which has the values 1 and 2 when $x = 0$ and 1 respectively, and such that its rate of change with respect to x is proportional to the cube of the value of the function.

ANSWERS

1. $x = \tfrac{1}{3} \log_e C(3y - 2).$ **2.** $y = \log_e Cx + \dfrac{2}{x}.$

3. $\tfrac{5}{6} \log_e C(3y^2 - 4) = x.$ **4.** $EI y = \tfrac{1}{12} x^4 - \tfrac{1}{2} x^2 + Ax + B.$

5. $y^2 = \dfrac{4}{(4 - 3x)} \left(\text{the D.E. is } \dfrac{dy}{dx} = ky^3 \right).$

13·22. Variables separable

Equations of this type can be arranged in the form

$$f(x)\, dx = F(y)\, dy;$$

that is, the variables x and y can be separated.

The equation will seldom be found in the simple form given above. To test for this type solve for dy/dx, then cross-multiplication should tranfer all the y's to the dy and all the x's to the dx.

EXAMPLES

(1) Solve $x^3 \dfrac{dy}{dx} = (1+x)(1-y^2)$ given $y = 0$ when $x = -1$.

[L.U.]

Cross-multiplying: $\dfrac{dy}{1-y^2} = \dfrac{(1+x)}{x^3} dx.$

$$\therefore \int \frac{1}{(1-y^2)} dy = \int \left(\frac{1}{x^3} + \frac{1}{x^2}\right) dx,$$

$$\tfrac{1}{2} \log_e C\left(\frac{1+y}{1-y}\right) = -\frac{1}{2x^2} - \frac{1}{x},$$

$$\log_e C\left(\frac{1+y}{1-y}\right) = -\frac{1}{x^2} - \frac{2}{x},$$

when $x = -1$, $y = 0$.

$$\therefore \log_e C = -1 + 2 = 1.$$

$$\therefore C = e.$$

Thus $\log_e\left(\dfrac{1+y}{1-y}\right) = -\dfrac{1}{x^2} - \dfrac{2}{x} - 1 = -\dfrac{(1+x)^2}{x^2},$

$$\frac{1+y}{1-y} = \exp\left[-\left(\frac{1+x}{x}\right)^2\right], \quad [Note. \ \exp(x) \equiv e^x.]$$

$$y = \frac{\exp\left[-\left(\dfrac{1+x}{x}\right)^2\right] - 1}{\exp\left[-\left(\dfrac{1+x}{x}\right)^2\right] + 1}.$$

Although not essential, the solution is usually solved for y in terms of x, or x in terms of y, when this is straightforward.

(2) Solve $dy \operatorname{cosec} x = e^y dx.$

Cross-multiplying: $e^{-y} dy = \sin x\, dx.$

$$\therefore \int e^{-y} dy = \int \sin x\, dx$$

$$-e^{-y} = -\cos x + A \quad (A \text{ arbitrary}),$$
$$e^{-y} = \cos x + C \quad (C \text{ arbitrary}).$$
$$-y = \log_e (C + \cos x),$$

$$y = \log_e \frac{1}{(C + \cos x)}.$$

EXERCISE 48

Solve the following equations:

1. $xy\dfrac{dy}{dx} = 1 + y^2.$ **2.** $xy\dfrac{dy}{dx} = 1 + x.$

3. $(1 + x^2)\dfrac{dy}{dx} = (1 + y^2).$ [L.U.]

4. $(1 + x^2)\dfrac{dy}{dx} = x(1 + y).$ [L.U.]

5. $y(4 - x^2)\dfrac{dy}{dx} = 1 + y^2.$ [L.U.]

6. $dy/dx = \tan y \cot x$, given $x = \tfrac{1}{4}\pi$ when $y = \tfrac{1}{4}\pi$.

ANSWERS

1. $y^2 = Ax^2 - 1.$ **2.** $y^2 = 2x + 2\log_e Ax.$

3. $\tan^{-1} y = \tan^{-1} x + A$, giving $y = \dfrac{x + c}{1 - xc}$, where $c = \tan A$.

4. $y = A\sqrt{(1 + x^2)} - 1.$ **5.** $(1 + y^2)^2 = A\left(\dfrac{2 + x}{2 - x}\right).$

6. $y = x.$

13·23. Exact equations

Many first-order equations can be seen to be exact differentials if rearranged and carefully inspected.

Thus

$$x\frac{dy}{dx} + y = 0 \quad \text{gives} \quad x\,dy + y\,dx = 0.$$

The left-hand side is easily seen to be an exact differential, viz. $d(xy)$.

The solution is therefore $xy = C$.

Similarly in $x\cos y\,dy + \sin y\,dx = 0$ the left-hand side is $d(x\sin y)$. The solution is thus, $x\sin y = C$.

Many equations which are not exact can be made so by multiplying through by a suitable factor, called an *integrating factor*.

For example, in $y\,dx - x\,dy = 0$, divide through by $-x^2$: then

$$\frac{-y}{x^2}\,dx + \frac{1}{x}\,dy = 0$$

$$\therefore\ yd(1/x) + \frac{1}{x}\,dy = 0$$

$$\therefore\ d(y/x) = 0$$

$$\therefore\ y/x = C \quad \text{or} \quad \underline{y = Cx}.$$

Here $-1/x^2$ is the integrating factor.

It is obviously useful to have a test for exactness of an equation.

Let $P\,dx + Q\,dy = 0$, where P and Q are functions of x and y.

For the equation to be exact the left-hand side must be the differential of a certain function, say z, of x and y, i.e.

$$d(z) = P\,dx + Q\,dy.$$

Comparing this with $dz = \dfrac{\partial z}{\partial x}\,dx + \dfrac{\partial z}{\partial y}\,dy$ (see §12·5, (2)) it is required that

$$\partial z/\partial x = P \quad \text{and} \quad \partial z/\partial y = Q. \tag{1}$$

Partially differentiating the first with respect to y and the second with respect to x:

$$\frac{\partial^2 z}{\partial x\,\partial y} = \frac{\partial P}{\partial y} = \frac{\partial Q}{\partial x}.$$

A necessary condition for exactness is thus

$$\partial P/\partial y = \partial Q/\partial x. \tag{2}$$

That is, the equation *cannot* be exact if $\partial P/\partial y \ne \partial Q/\partial x$. It has *not* been proved that the equation *must* be exact if equation (2) above is satisfied. This, however, can be proved, although the proof is beyond the scope of this book.

Thus, the condition $\partial P/\partial y = \partial Q/\partial x$ is both a necessary *and* sufficient condition for an equation to be exact.

EXAMPLES

(1)
$$3x\,dy + x^2\,dx - y^2\,dy + 3y\,dx = 0.$$
$$(3x - y^2)\,dy + (x^2 + 3y)\,dx = 0.$$

Here
$$P \equiv x^2 + 3y \quad \text{and} \quad Q \equiv 3x - y^2.$$
$$\partial P/\partial y = 3, \quad \partial Q/\partial x = 3.$$

The equation is exact.

Rearranging:
$$3(x\,dy + y\,dx) + x^2\,dx - y^2\,dy = 0,$$
$$3d(xy) + \tfrac{1}{3}d(x^3) - \tfrac{1}{3}d(y^3) = 0.$$
$$\therefore \quad \underline{3xy + \tfrac{1}{3}x^3 - \tfrac{1}{3}y^3 = C.}$$

(2)
$$y + \frac{x}{2}\frac{dy}{dx} = x^2.$$

$$2y\,dx + x\,dy = 2x^2\,dx.$$

The right-hand side could be integrated on sight.
For the left-hand side,

$$\partial P/\partial y = 2, \quad \partial Q/\partial x = 1.$$

The equation is therefore not exact as it stands.

Multiplying by the integrating factor x, however, makes it exact.
$$2xy\,dx + x^2\,dy = 2x^3\,dx,$$
$$d(x^2y) = \tfrac{1}{2}d(x^4),$$
$$\underline{x^2y = \tfrac{1}{2}x^4 + C.}$$

Usually, if an equation is exact, the solution is obvious on sight. If not, the following example demonstrates how the solution can be deduced:

(3)
$$\cos x\,dy + 1\,dx = y\sin x\,dx.$$

Rearranging:
$$(1 - y\sin x)\,dx + \cos x\,dy = 0,$$
$$\frac{\partial(1 - y\sin x)}{\partial y} = -\sin x = \frac{\partial(\cos x)}{\partial x}.$$

Therefore the equation is exact.
If $z = f(x, y) = 0$ is a solution
$$\partial z/\partial x = 1 - y\sin x \quad \text{and} \quad \partial z/\partial y = \cos x \quad (\text{see } \S\,13{\cdot}23,\,(1)).$$

Integrating the first, treating y as a constant, and the second, treating x as a constant:

$$z = x + y\cos x + \text{a function of } y, \tag{1}$$
and
$$z = y\cos x + \text{a function of } x. \tag{2}$$

On inspection of (1) and (2) it is easily seen that $z = x + y\cos x$ should satisfy. This may be verified by testing in the original equation.

Thus the solution is
$$\underline{x + y\cos x = C.}$$

EXERCISE 49

Solve:

1. $x\,dy + y\,dx - x\,dx = 0.$ **2.** $-y\,dx + x\,dy + x^3\,dx = 0.$

3. $y\sec^2 x + \tan x\dfrac{dy}{dx} = 0.$

Show that the following equations are exact and solve them:

4. $\cos x\,\cos y\,dx - \sin x\,\sin y\,dy = 0.$

5. $3x^2y^2\,dx + 2x^3y\,dy - 2x\,dx = 0.$

6. $e^x\sin y\,dx + e^x\cos y\,dy = 0.$

7. $2xy - y^2 + y + (x^2 - 2xy + x)\dfrac{dy}{dx} = 0.$

8. Show that the equation $x + y + (y - x)\dfrac{dy}{dx} = 0$ is not exact, but that it is made exact by dividing by $(x^2 + y^2)$. Hence solve it.

ANSWERS

1. $xy - \tfrac{1}{2}x^2 = C.$ **2.** $y/x + \tfrac{1}{2}x^2 = C.$

3. $y\tan x = C.$ **4.** $\sin x\,\cos y = C.$

5. $x^3y^2 - x^2 = C.$ **6.** $e^x\sin y = C.$

7. $xy(x - y + 1) = C.$

8. $\log_e C(x^2 + y^2) = 2\tan^{-1}(y/x).$

13·24. First-order linear equations. Integrating factor

A differential equation of the general form

$$P_n\frac{d^ny}{dx^n} + P_{n-1}\frac{d^{n-1}y}{dx^{n-1}} + \ldots + P_2\frac{d^2y}{dx^2} + P_1\frac{dy}{dx} + Py = Q,$$

where the P's and Q are functions of x, is called a *linear* equation.

A first-order linear equation is thus one which can be reduced to the form $dy/dx + Py = Q$, P and Q being functions of x.

This equation can be made exact by multiplying by an integrating factor $\exp\left(\displaystyle\int P\,dx\right).$

Before demonstrating this it should be noted that

$$\frac{d}{dx}\left(\int P\,dx\right) = P.$$

Hence $\qquad \dfrac{d}{dx}\left[\exp\left(\int P\,dx\right)\right]=P\exp\left(\int P\,dx\right).$

Multiplying the first-order linear equation by $\exp\left(\int P\,dx\right)$

$$\frac{dy}{dx}\exp\left(\int P\,dx\right)+P\exp\left(\int P\,dx\right)y=Q\exp\left(\int P\,dx\right).$$

But

$$\frac{d}{dx}\left[y\exp\left(\int P\,dx\right)\right]=\frac{dy}{dx}\exp\left(\int P\,dx\right)+P\exp\left(\int P\,dx\right)y.$$

Thus $\qquad \dfrac{d}{dx}\left[y\exp\left(\int P\,dx\right)\right]=Q\exp\left(\int P\,dx\right).$

The right-hand side contains functions of x only.

$$\therefore\ y\exp\left(\int P\,dx\right)=\text{integral of } Q\exp\left(\int P\,dx\right) \text{ with respect to } x.$$

$$(1)$$

The best method is to work out $\exp\left(\int P\,dx\right)$ first, in practical cases. In this connexion note that $e^{\log_e f(x)}$ is just $f(x)$.

Examples

(1) Solve $\qquad\qquad dy/dx+3y=2x.$

Integrating factor (I.F.) is $\exp\left(\int 3\,dx\right)=e^{3x}.$

Using formula (1) above:

$$y\,e^{3x}=\int 2x\,e^{3x}\,dx$$

$$=\tfrac{2}{3}x\,e^{3x}-\frac{2}{3}\int 1\,e^{3x}\,dx\quad\text{(by parts)}$$

$$y\,e^{3x}=\tfrac{2}{3}x\,e^{3x}-\tfrac{2}{9}e^{3x}+C.$$

$$\therefore\ \underline{y=\tfrac{2}{3}(x-\tfrac{1}{3})+C\,e^{-3x}.}$$

(2) $\qquad\qquad\qquad x\dfrac{dy}{dx}+y=x^2+2x.$

$$\frac{dy}{dx}+\frac{1}{x}y=x+2.$$

I.F. is
$$\exp\left(\int \frac{1}{x}\,dx\right) = e^{\log_e x} = x.$$

$$\therefore\ yx = \int (x+2)\,x\,dx = \tfrac{1}{3}x^3 + x^2 + C.$$

$$\therefore\ y = \tfrac{1}{3}x^2 + x + \frac{C}{x}.$$

EXERCISE 50

Solve:

1. $dy/dx + 2y = 3\,e^x.$ **2.** $dy/dx + y\cot x = \cos x.$

3. $\dfrac{dy}{dx} + \dfrac{1}{x}y = 2x + \dfrac{1}{x}.$

4. $R\dfrac{dq}{dt} + \dfrac{q}{C} = V,$ given $q = 0$ at $t = 0.$

5. $\sec x\,dy/dx = \sin x - y.$ **6.** $(x^2 - 1)\dfrac{dy}{dx} = 2xy + x.$

ANSWERS

1. $y = e^x + C\,e^{-2x}.$

2. $y = \dfrac{C}{\sin x} - \dfrac{\cos 2x}{4\sin x}$ or $\tfrac{1}{2}\sin x + \dfrac{A}{\sin x}.$

3. $y = \tfrac{2}{3}x^2 + 1 + C/x.$ **4.** $q = CV(1 - e^{-t/CR}).$

5. $y = \sin x - 1 + C\,e^{-\sin x}.$ $\left[\textit{Note.}\ \int \cos x\,e^{\sin x}\,dx = e^{\sin x}.\right]$

6. $y = C(x^2 - 1) - \tfrac{1}{2}.$

13·3. Second-order equations

13·31. $\dfrac{d^2y}{dx^2} = f(y)$

Writing
$$p \equiv dy/dx,$$

$$\frac{d^2y}{dx^2} = \frac{d}{dx}\left(\frac{dy}{dx}\right) = \frac{d}{dy}\left(\frac{dy}{dx}\right)\frac{dy}{dx} = \frac{dp}{dy}p.$$

This type therefore reduces to

$$p\frac{dp}{dy} = f(y)$$

or
$$p\,dp = f(y)\,dy.$$

Thus
$$\tfrac{1}{2}p^2 = \int f(y)\,dy.$$

This gives p as a function of y.

A further integration will then give x in terms of y.

13·311. Simple harmonic motion

An important example of the above type is the simple harmonic motion equation
$$d^2x/dt^2 = -n^2x.$$

Let
$$p = dx/dt,$$

then
$$p\frac{dp}{dx} = -n^2x,$$

$$\frac{p^2}{2} = -\int n^2x\,dx = -\frac{n^2x^2}{2} + C.$$

$$\therefore\ p^2 = 2C - n^2x^2. \tag{1}$$

This equation shows that for p to be real, $2C$ must be positive. It is convenient to write it in the form a^2n^2 (a arbitrary).

Thus
$$p^2 = n^2(a^2 - x^2) = (dx/dt)^2. \tag{2}$$

Normally, x is distance and t the time.

Equation (2) then gives the velocity v, as
$$v^2 = n^2(a^2 - x^2),$$

$$v = \pm\,n\,\sqrt{(a^2 - x^2)}. \tag{3}$$

When $x = \pm a$, the velocity is zero. Distance a is the furthest the body can move from the centre $x = 0$. 'a' is called the *amplitude* of the motion.

From (3):
$$\frac{dx}{dt} = \pm\,n\,\sqrt{(a^2 - x^2)}.$$

$$\therefore\ \pm\int\frac{1}{\sqrt{(a^2 - x^2)}}\,dx = \int n\,dt$$

$$\pm\sin^{-1}\frac{x}{a} = nt + \alpha \quad (\alpha\ \text{arbitrary}).$$

Hence
$$x = \pm\,a\sin(nt + \alpha).$$

As both a and α are arbitrary the ambiguity of sign can be absorbed by changing the sign of a or by replacing α by $\alpha + \pi$.

Hence the solution can be written

$$x = a \sin(nt + \alpha). \tag{4}$$

Equation (4) is

$$x = (a \cos \alpha) \sin nt + (a \sin \alpha) \cos nt.$$

$$\therefore \underline{x = A \cos nt + B \sin nt \quad (A, B \text{ arbitrary}).} \tag{5}$$

The reader will have met s.h.m. before, but may not have had the formulae derived by the above method.

Equation (4) shows the motion to be periodic and of period $2\pi/n$.

13·32. Linear equations with constant coefficients

These are equations which can be reduced to

$$\frac{d^2y}{dx^2} + a\frac{dy}{dx} + by = f(x); \quad a, b \text{ constants.}$$

Higher order equations of this type can be dealt with in the same way.

For first year Higher National Certificate work only equations of the form $\dfrac{d^2y}{dx^2} + a\dfrac{dy}{dx} + by = 0$ will be considered. Cases in which a function of x occurs on the right-hand side are left for second year work.

13·321. General solutions

Suppose that, by any means, $y = u$ and $y = v$, where u, v are functions of x, are found to satisfy the equation. Then

$$y = Au + Bv, \quad A, B \text{ arbitrary,}$$

is the general solution. (Note that it contains the required two arbitrary constants for a second order equation.) For

$$\frac{d^2u}{dx^2} + a\frac{du}{dx} + bu = 0 \quad (u \text{ satisfies}),$$

$$\frac{d^2v}{dx^2} + a\frac{dv}{dx} + bv = 0 \quad (v \text{ satisfies}).$$

Multiplying the first by A and the second by B and adding:

$$\left(A\frac{d^2u}{dx^2} + B\frac{d^2v}{dx^2} \right) + a\left(A\frac{du}{dx} + B\frac{dv}{dx} \right) + b(Au + Bv) = 0.$$

$$\therefore y = Au + Bv \text{ satisfies the equation.} \tag{1}$$

13·322. Method of solution

(a) Consider the equation

$$\frac{d^2y}{dx^2} - 3\frac{dy}{dx} + 2y = 0.$$

It can be written

$$\left(\frac{d^2y}{dx^2} - \frac{dy}{dx}\right) = 2\left(\frac{dy}{dx} - y\right)$$

or

$$\frac{d}{dx}\left(\frac{dy}{dx} - y\right) = 2\left(\frac{dy}{dx} - y\right).$$

Letting

$$z = \left(\frac{dy}{dx} - y\right), \tag{1}$$

$$\frac{dz}{dx} = 2z \quad \text{or} \quad \frac{1}{z}dz = 2dx,$$

giving

$$2x = \log Cz \quad \text{or} \quad z = A\,e^{2x}. \tag{2}$$

From (1) and (2), the original equation reduces to

$$\frac{dy}{dx} - y = A\,e^{2x}. \tag{3}$$

Again, the original equation may be written

$$\frac{d^2y}{dx^2} - 2\frac{dy}{dx} = \frac{dy}{dx} - 2y$$

or

$$\frac{d}{dx}\left(\frac{dy}{dx} - 2y\right) = \left(\frac{dy}{dx} - 2y\right).$$

Letting

$$z = \frac{dy}{dx} - 2y \quad \text{gives} \quad \frac{dz}{dx} = z,$$

which has a solution

$$z = C\,e^x.$$

Thus the original equation reduces to

$$\frac{dy}{dx} - 2y = C\,e^x. \tag{4}$$

Eliminating dy/dx from (3) and (4) by subtraction

$$y = A\,e^{2x} - C\,e^x,$$

or, as A, C are arbitrary

$$\underline{y = A\,e^{2x} + B\,e^x.} \tag{5}$$

This is the general solution.

The method of solution may seem intricate but a rule is soon found.

Consider
$$\frac{d^2y}{dx^2} + a\frac{dy}{dx} + by = 0. \tag{6}$$

Let the equation $m^2 + am + b = 0$ have two *real different* roots α, β.

Note how this quadratic is easily formed from the original equation: m^2 replaces d^2y/dx^2, m replaces dy/dx and 1 replaces y.

In this case the quadratic is $(m - \alpha)(m - \beta) = 0$, or

$$m^2 - (\alpha + \beta)m + \alpha\beta = 0. \tag{7}$$

Thus, the original differential equation is of the form

$$\frac{d^2y}{dx^2} - (\alpha + \beta)\frac{dy}{dx} + \alpha\beta y = 0. \tag{8}$$

This may be written in either of the forms

$$\frac{d}{dx}\left(\frac{dy}{dx} - \alpha y\right) = \beta\left(\frac{dy}{dx} - \alpha y\right), \tag{9}$$

or
$$\frac{d}{dx}\left(\frac{dy}{dx} - \beta y\right) = \alpha\left(\frac{dy}{dx} - \beta y\right). \tag{10}$$

As in the numerical example above, (9) gives
$$dy/dx - \alpha y = D\,e^{\beta x},$$
and (10) gives
$$dy/dx - \beta y = C\,e^{\alpha x}.$$

Eliminating dy/dx gives

$$\underline{y = A\,e^{\alpha x} + B\,e^{\beta x} \quad (A, B \text{ arbitrary}).} \tag{11}$$

The following rule now applies and solutions may be written down very quickly:

Form the quadratic $m^2 + am + b = 0$. Solve it. If it has two real, different, roots α, β, the solution to the differential equation is

$$y = A\,e^{\alpha x} + B\,e^{\beta x}.$$

The quadratic $m^2 + am + b = 0$ is called the *auxiliary* equation (A.E.).

EXAMPLE

Solve
$$\frac{d^2y}{dx^2} + 5\frac{dy}{dx} + 6y = 0.$$

A.E. is
$$m^2 + 5m + 6 = 0,$$
$$(m + 3)(m + 2) = 0.$$

The roots are $m = -3$, $m = -2$,

$$\therefore \quad \text{the solution is } \underline{y = A\,e^{-3x} + B\,e^{-2x}}.$$

(b) If the auxiliary equation happens to have equal roots, say both of them k, then the original equation must be of the form

$$\frac{d^2y}{dx^2} - 2k\frac{dy}{dx} + k^2y = 0,$$

giving the A.E. $(m - k)^2 = 0$.

The equation can be written

$$\frac{d^2y}{dx^2} - k\frac{dy}{dx} = k\frac{dy}{dx} - k^2y,$$

$$\frac{d}{dx}\left(\frac{dy}{dx} - ky\right) = k\left(\frac{dy}{dx} - ky\right).$$

Putting $z = dy/dx - ky,$

$$dz/dx = kz.$$

$$\therefore \quad z = A\,e^{kx}.$$

Thus $dy/dx - ky = A\,e^{kx},$

$$e^{-kx}\frac{dy}{dx} + (-k\,e^{-kx})\,y = A.$$

The left-hand side is $\dfrac{d}{dx}(y\,e^{-kx}).$

Thus, integrating, $y\,e^{-kx} = Ax + B.$

$$\underline{y = (Ax + B)\,e^{kx}}. \tag{12}$$

This gives the rule in the case where the A.E. has equal roots, k.

EXAMPLE

$$\frac{d^2y}{dx^2} - 4\frac{dy}{dx} + 4y = 0.$$

A.E. is $m^2 - 4m + 4 = 0,$

$$(m - 2)^2 = 0.$$

$$\therefore \quad m = 2.$$

General solution: $\underline{y = (Ax + B)\,e^{2x}}.$

(c) If the A.E., $m^2 + am + b = 0$, has complex roots, then

$$a^2 - 4b < 0.$$

The roots are

$$-\frac{a \pm \sqrt{(a^2 - 4b)}}{2} = -\frac{a}{2} \pm \frac{j}{2}\sqrt{(4b - a^2)},$$

where $\sqrt{(4b - a^2)}$ is real as $a^2 - 4b < 0$.

Thus the roots are of the form $p \pm jq$ (conjugates).

The method of (a) then gives as a general solution:

$$y = A\,e^{(p+jq)x} + B\,e^{(p-jq)x}$$
$$= e^{px}\,(A\,e^{jqx} + B\,e^{-jqx}),$$

i.e.
$$y = e^{px}\,[A(\cos qx + j\sin qx) + B(\cos qx - j\sin qx)]$$
$$= e^{px}\,[(A + B)\cos qx + j(A - B)\sin qx].$$

As A, B are arbitrary, this reduces to

$$y = e^{px}\,(C\cos qx + D\sin qx) \quad (C, D \text{ arbitrary})$$

or
$$y = R\,e^{px}\sin(qx + \alpha) \quad (R, \alpha \text{ arbitrary}).$$

EXAMPLE

Solve $\qquad\qquad d^2y/dx^2 + dy/dx + y = 0.$

A.E. is $\qquad\qquad m^2 + m + 1 = 0,$

Roots are $\qquad -\tfrac{1}{2} \pm \dfrac{\sqrt{(1-4)}}{2} = -\tfrac{1}{2} \pm j\dfrac{\sqrt{3}}{2}.$

Solution: $\qquad y = e^{-\frac{1}{2}x}\left(C\cos\dfrac{\sqrt{3}}{2}x + D\sin\dfrac{\sqrt{3}}{2}x\right).$

Summary

To solve a linear differential equation (second order) with constant coefficients:

(i) Form the auxiliary equation and solve it.

(ii) (a) If its roots are real and unequal, say α, β, then
$$y = A\,e^{\alpha x} + B\,e^{\beta x}.$$

$\quad$ (b) If its roots are equal, say k, k, then
$$y = (Ax + B)\,e^{kx}.$$

$\quad$ (c) If its roots are complex, say $p \pm jq$, then
$$y = e^{px}\,(C\cos qx + D\sin qx).$$

13·323. Alternative method of solution

To solve $\dfrac{d^2y}{dx^2} + a\dfrac{dy}{dx} + by = 0$ many books adopt what may be called the 'trial' method. y is tried equal to $C\,e^{mx}$, C arbitrary.

This gives $C\,e^{mx}(m^2+am+b)=0$, and the auxiliary equation, $m^2+am+b=0$, is again obtained. If m_1, m_2 are its roots, then $y=A\,e^{m_1 x}+B\,e^{m_2 x}$ is the general solution, A, B, arbitrary (see §13·321, equation (1)).

Two objections have been made to this method:

(i) The case of equal roots is not covered and has to be dealt with separately using a new method.

(ii) 'Trying' $y=C\,e^{mx}$ in the first place leads one to suspect that the answer was known before the trial. Otherwise, why e^{mx}?

EXAMPLES

Solve the following equations:

(1) $3\dfrac{d^2y}{dx^2}+\dfrac{dy}{dx}-10y=0$.

(2) $\dfrac{d^2y}{dx^2}+6\dfrac{dy}{dx}+9y=0$.

(3) $4\dfrac{d^2y}{dx^2}-4\dfrac{dy}{dx}+5y=0$, given $y=1$, $\dfrac{dy}{dx}=0$ when $x=0$.

(1) A.E. is
$$3m^2+m-10=0,$$
$$(3m-5)\,(m+2)=0.$$

G.S. is
$$y=A\,e^{\frac{5}{3}x}+B\,e^{-2x}.$$

(2) A.E. is
$$m^2+6m+9=0,$$
$$(m+3)^2=0.$$

G.S. is
$$y=(Ax+B)\,e^{-3x}.$$

(3) A.E. is
$$4m^2-4m+5=0,$$
$$m=\frac{4\pm j8}{8}=\tfrac{1}{2}\pm j.$$

G.S. is
$$y=e^{\frac{1}{2}x}\,(C\cos x+D\sin x).$$

When $x=0$, $y=1$, $\qquad\qquad\therefore\ 1=C.$ $\hfill(1)$

$$\therefore\ y=e^{\frac{1}{2}x}\,(\cos x+D\sin x),$$
$$dy/dx=e^{\frac{1}{2}x}\,(D\cos x-\sin x)+\tfrac{1}{2}e^{\frac{1}{2}x}\,(\cos x+D\sin x).$$

When $x=0$, $\qquad\qquad dy/dx=0$,
$$\therefore\ 0=D+\tfrac{1}{2},\quad D=-\tfrac{1}{2}. \hfill(2)$$

The particular solution required is
$$y=e^{\frac{1}{2}x}\,(\cos x-\tfrac{1}{2}\sin x).$$

EXERCISE 51

Solve:

1. $\dfrac{d^2y}{dx^2} - 7\dfrac{dy}{dx} + 12y = 0.$ **2.** $\dfrac{d^2y}{dx^2} + 2\dfrac{dy}{dx} + y = 0.$

3. $\dfrac{d^2y}{dx^2} + 2\dfrac{dy}{dx} + 5y = 0.$ **4.** $\dfrac{d^2y}{dx^2} + \dfrac{1}{y^3} = 0.$

5. $2\dfrac{d^2y}{dx^2} - 3\dfrac{dy}{dx} + 4y = 0.$ **6.** $\dfrac{d^2y}{dx^2} - 10\dfrac{dy}{dx} + 16y = 0.$

7. $\dfrac{d^2y}{dx^2} - 10\dfrac{dy}{dx} + 25y = 0.$ **8.** $\dfrac{d^2y}{dx^2} + 6\dfrac{dy}{dx} + 10y = 0.$

9. $\dfrac{d^2y}{dx^2} + 3\dfrac{dy}{dx} - 10y = 0$, given $y = 2$, $\dfrac{dy}{dx} = 1$ when $x = 0$.

10. $\dfrac{d^2x}{dt^2} + 4\dfrac{dx}{dt} + 5x = 0$, given $x = 0$, $\dfrac{dx}{dt} = 3$ when $t = 0$.

ANSWERS

1. $y = A\,e^{3x} + B\,e^{4x}.$ **2.** $y = (Ax + B)\,e^{-x}.$

3. $y = e^{-x}(A\cos 2x + B\sin 2x).$

4. $Ay^2 = (B + x)^2 - A^2.$

5. $y = e^{\frac{3}{4}x}\left(A\cos\dfrac{\surd(23)}{4}\,x + B\sin\dfrac{\surd(23)}{4}\,x\right).$

6. $y = A\,e^{2x} + B\,e^{8x}.$ **7.** $y = (Ax + B)\,e^{5x}.$

8. $y = e^{-3x}(A\cos x + B\sin x).$ **9.** $y = \frac{1}{7}(11\,e^{2x} + 3\,e^{-5x}).$

10. $x = 3\,e^{-2t}\sin t.$

MISCELLANEOUS EXERCISES ON CHAPTER 13

1. If $dx/dt = -x$ and $dy/dt = x - 2y$ and if $x = a$ and $y = b$ when $t = 0$, obtain the relations between (i) x and t, (ii) y and t. Hence deduce that $y = x + (b - a)(x/a)^2$.   [Sec. A]

2. (i) Solve $dy/dt + y = 2t\,e^{-t}$. If $y = -3$ when $t = 0$, show that y has a maximum value when $t = 3$ and a minimum when $t = -1$.

(ii) Solve

$$\dfrac{d^2y}{dx^2} + 4\dfrac{dy}{dx} + 8y = 0.$$ [Sec. A]

3. If $d^2y/dx^2 = x\sin x$, find y, given $dy/dx = 0$ when $x = \frac{1}{2}\pi$, and $y = 1$ when $x = 0$. [Sec. A]

4. Solve the equation $2\dfrac{d^2y}{dt^2} + 3\dfrac{dy}{dt} + y = 0$, given that $y = 1$, $dy/dt = -2$ when $t = 0$. Show that y is zero for only one positive value of t and find the least value of y. Sketch the graph for positive values of t. [Sec. A]

5. Find the value of 'a' given that $x^3\dfrac{dy}{dx} = a - x$ and that $y = 0$ when $x = 2$ and when $x = 6$. [L.U.]

6. Solve $(1+x)\dfrac{dy}{dx} + (1+2x)y = x^2(1+x)^3$. [L.U.]

7. (i) Solve $di/dt + 3i = 10\sin t$, given $i = 0$ when $t = 0$.
[Sec. A]

(ii) Solve the equation $d^2x/dt^2 + dx/dt + 64\cdot25x = 0$. [Sec. A]

8. Solve: (i) $dy/dx + 0\cdot5y = e^x$, given $y = 0$ when $x = 0$.
[Sec. A]

(ii) $d^2y/dx^2 + 4y = 0$, given $y = 0$, $dy/dx = 2$ when $x = 0$.

9. (i) Solve $dv/dt + kv = 0$ given that $v = V$ when $t = 0$. If $v = 0\cdot4V$ when $t = 3$, find k.

(ii) Solve $\dfrac{d^2y}{dx^2} + 1\cdot6\dfrac{dy}{dx} - 0\cdot36y = 0$, given that $y = 0$, $dy/dx = 2$ when $x = 0$. [Sec. A]

10. The differential equation for a long thin rod clamped vertically at its lower end and carrying a weight W at its upper end is $d^2y/dx^2 = n^2(a-y)$, where $n^2 = W/EI$.

Solve this equation subject to the condition that $y = 0$ at $x = 0$, $y = a$ at $x = l$. (*Hint.* Let $z = a - y$.)

Find also the smallest value of l for which $dy/dx = 0$ at $x = 0$.

ANSWERS

1. (i) $x = ae^{-t}$;   (ii) $y = ae^{-t} + (b-a)e^{-2t}$.

2. (i) $y = (t^2 + A)e^{-t}$;   (ii) $y = e^{-2x}(A\cos 2x + B\sin 2x)$.

3. $y = -x\sin x - 2\cos x - x + 3$.

4. $y = 3e^{-t} - 2e^{-\frac{1}{2}t}$; $y = 0$ when $t = 2\log_e\frac{3}{2}$; minimum y is $-\frac{1}{3}$.

5. $a = 3$.

6. I.F. is $\dfrac{e^{2x}}{(1+x)}$; $y = \dfrac{(1+x)(2x^2 - 2x + 1)}{4} + C(1+x)e^{-2x}$.

7. (i) $(3 \sin t - \cos t + e^{-3t})$;   (ii) $x = e^{-\frac{1}{2}t}(A \cos 8t + B \sin 8t)$.

8. (i) $y = \frac{2}{3}(e^x - e^{-\frac{1}{2}x})$;   (ii) $y = \sin 2x$.

9. (i) $v = V e^{-kt}$; $k = \frac{1}{3} \log_e \frac{5}{2}$;   (ii) $y = e^{0.2x} - e^{-1.8x}$.

10. $y = a\left[1 - \dfrac{\sin n(l-x)}{\sin nl}\right]$; $\dfrac{\pi}{2n}$.

CHAPTER 14

SOME APPLICATIONS OF DIFFERENTIAL EQUATIONS TO MECHANICAL AND ELECTRICAL ENGINEERING

14·1. Motion in a straight line

Let P be a point moving on a line $X'OX$ (fig. 83).

Let its distance from O at any time t be x.

Then its velocity is given by $v = dx/dt = \dot{x}$. (Dots are often used to signify differentiation with respect to time.)

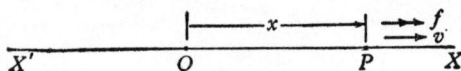

Fig. 83

Its acceleration is given by

$$\frac{d^2x}{dt^2} = \frac{dv}{dt} = \frac{dv}{dx} \times \frac{dx}{dt} = v\frac{dv}{dx} = \frac{1}{2}\frac{d}{dx}(v^2).$$

Three forms for the acceleration are: $\dfrac{d^2x}{dt^2}, \dfrac{dv}{dt}$ and $v\dfrac{dv}{dx}$. Which of these three is used in any particular problem will depend on which two out of the three quantities x, v, t it is desired to connect.

For example, if it is required to find a relationship between velocity and distance $v\dfrac{dv}{dx}$ would be used for the acceleration.

It will be assumed that the reader is familiar with the simpler laws of mechanics:

Accelerating force $=$ (Mass) $\times$ (Acceleration).

Power $\qquad = $ force $\times$ velocity $=$ rate of doing work.

Accelerating force $=$ total force acting $-$ resistive force.

14·11. Terminal velocity

If a body, subject to a constant propelling force, moves against a resistance which increases as the velocity increases, a time will come when the velocity is large enough for the resistance to equal the propelling force. From this point onwards there will be no accelerating force and the body will move with constant velocity. This velocity is called the *terminal velocity*.

For example, suppose that the resistance is proportional to v^2. Then the resistance is given by $R = kv^2$ (k constant).

Let the constant propelling force be P. As the velocity grows there will be a value V say when

$$P = R = kV^2.$$

After this the velocity remains constant at the terminal velocity V.

Thus $$V = \sqrt{(P/k)}.$$

EXAMPLES

(1) A small body falls from rest in a medium whose resistance per unit mass of the body is k times the square of the velocity. Show that the distance it must fall to acquire a velocity u is

$$\frac{1}{2k} \log_e \frac{g}{(g - ku^2)},$$

and that the velocity approaches a limiting value $\sqrt{(g/k)}$. [L.U.]

Resistance to motion $= M k v^2$ (absolute units), where $M =$ mass of body.

$$\therefore \text{ Accelerating force} = Mg - M kv^2 = M(g - kv^2).$$

Thus $$M(g - kv^2) = Mv \frac{dv}{dx}.$$

$\left(v \dfrac{dv}{dx} \text{ is used as it is desired to connect the velocity and distance.} \right)$

$$\therefore \ g - kv^2 = v \frac{dv}{dx} \quad \text{(separate variables)},$$

$$\frac{v\, dv}{g - kv^2} = dx.$$

$$-\frac{2kv\, dv}{g - kv^2} = -2k\, dx.$$

Integrating $$\log_e C(g - kv^2) = -2kx. \tag{1}$$

The body falls from rest. Therefore when $x = 0$, $v = 0$.

From (1): $\qquad\qquad\qquad \log_e Cg = 0,$

$$\therefore\ C = 1/g.$$

Thus $\qquad\qquad\qquad \log_e \frac{(g - kv^2)}{g} = -2kx,$

$$x = -\frac{1}{2k} \log_e \frac{(g - kv^2)}{g},$$

$$x = \frac{1}{2k} \log_e \frac{g}{(g - kv^2)}. \qquad (2)$$

$\therefore$ Distance to acquire a velocity u is $\underline{\dfrac{1}{2k} \log \left(\dfrac{g}{g - ku^2} \right)}.$

As $x \to \infty$, $\dfrac{g}{g - kv^2} \to \infty$, i.e.

$$g \to kv^2, \quad v \to \sqrt{(g/k)}.$$

Thus, *however far the body falls its velocity will never exceed the limiting value* $\sqrt{(g/k)}$.

Alternatively:

Let V = terminal velocity.

$\therefore$ When $v = V$, the accelerating force is zero.

Thus $\qquad\qquad\qquad Mg - MkV^2 = 0,$

$$V^2 = g/k,$$

$$\underline{V = \sqrt{(g/k)}.}$$

(2) A train is drawn by an engine which exerts a constant pull at all speeds, and the total resistance to motion varies as the square of the speed. The combined mass of engine and train is 300 Mg, the maximum speed on the level is 120 km/hr, and the power then developed is 1·2 MW. Write down the equation of motion on the level, and find the distance travelled from rest in acquiring a speed of 90 km/hr.

$$120 \text{ km/hr} = \frac{120 \times 10^3}{3600} = \frac{100}{3} \text{ m/s.}$$

$$90 \text{ km/hr} = 25 \text{ m/s.}$$

Rate of working at 120 km/hr is $1·2 \times 10^6$ W. If P newtons is the constant pull of the engine

$$P \times \tfrac{100}{3} = 1·2 \times 10^6,$$

giving $\qquad\qquad\qquad P = 36 \times 10^3 \text{ N.} \qquad (1)$

Resistance, R newtons, at speed v metres per second, is kV^2 (k constant). At maximum speed of $\frac{100}{3}$ m/s, pull = resistance.

Thus
$$k(\tfrac{100}{3})^2 = 36 \times 10^3,$$

giving
$$k = 32 \cdot 4.$$

In general,
$$R = 32 \cdot 4v^2 \text{ N.} \tag{2}$$

The equation of motion is

$$P - R = 300 \times 10^3 \times \text{(acceleration)},$$

i.e.
$$36 \times 10^3 - 32 \cdot 4v^2 = 300 \times 10^3 \times v\frac{dv}{dx}.$$

Thus
$$\int_0^{25} \frac{300 \times 10^3 \, v \, dv}{(36 \times 10^3 - 32 \cdot 4v^2)} = \int_0^X dx,$$

as $v = 0$ when $x = 0$. (X = distance gone when speed is 25 m/s.)
This gives

$$\frac{-300 \times 10^3}{64 \cdot 8} \left[\log_e (36 \,.\, 10^3 - 32 \cdot 4v^2)\right]_0^{25} = X,$$

$$\frac{-10^5}{21 \cdot 6} \left[\log_e \frac{(36 \,.\, 10^3 - 32 \cdot 4 \,.\, 25^2)}{36 \,.\, 10^3}\right] = X,$$

$$\frac{10^5}{21 \cdot 6} \left[\log_e \frac{10^3}{(10^3 - 0 \cdot 9 \times 625)}\right] = X,$$

$$\frac{10^5}{21 \cdot 6} \log_e \left(\frac{1000}{437 \cdot 5}\right) = X,$$

giving
$$X = \frac{10^5}{21 \cdot 6} \log_e 2 \cdot 286,$$

$$\underline{X \simeq 3828 \text{ m.}}$$

EXERCISE 52

1. In example (1) above, if the body is projected upwards with the limiting velocity $\sqrt{(g/k)}$, show that its speed on again reaching the point of projection will be $\sqrt{(g/2k)}$.

2. The effective power of a ship of mass 10,000 Mg is 4,500 kW, and its full speed is 9 m./s. Assuming that the resistance to motion varies as the square of the speed and that the power is constant, find the distance travelled from rest in attaining a speed of 6 m./s.

3. A train of mass 320 Mg travels along the level at a uniform speed of 32 m/s against a resistance of 63 N/Mg. It then climbs an incline of 1 in 160. Assuming that the power and resistances remain constant, show that when the speed is v metres per second the retardation is $\dfrac{7(2v-35)}{100v}$ m/s^2. Find the time taken for the speed to fall from 32 to 24 m/s.

4. A body is allowed to fall from rest under gravity in a liquid whose resistance to the motion is k times the velocity. Show that the velocity tends to a limiting value as the distance increases. Find also an expression for the distance fallen after the first t sec.

[L.U.]

ANSWERS

2. About 190 m.   **3.** About 114 s.

4. Terminal velocity $= Mg/k$,

$$x = \frac{Mgt}{k} - \frac{M^2g}{k^2}\left(1 - e^{-kt}/M\right),$$

where $M =$ mass of body, and resistance $= kv$.

14·2. Simple harmonic motion and damped harmonic motion

14·21. Simple harmonic motion

Simple harmonic motion is motion in a straight line under a force which is always directed towards a fixed point and is proportional to the distance of the body from that point.

Problems on S.H.M. will have already been met with in an Ordinary National Certificate course.

Let the force per unit mass at distance x from O be n^2x, i.e. acceleration at distance x is $-n^2x$. The $-$ sign occurs as the direction of the acceleration is along XO not OX (see fig. 84).

The equation of motion is

$$\ddot{x} = -n^2x. \tag{1}$$

This has been solved by one method (see § 13·311). It may also be solved as a linear differential equation (with constant coefficients)
$$d^2x/dt^2 + n^2x = 0.$$

A.E. is
$$m^2 + n^2 = 0,$$

$$m = \pm jn.$$

∴ G.S. is $\quad x = e^{o.t}(C\cos nt + D\sin nt)$

or $\qquad \underline{x = C\cos nt + D\sin nt}\quad$ (C, D arbitrary),

$\qquad \underline{x = a\sin(nt+\alpha)}\qquad$ (a, α arbitrary).

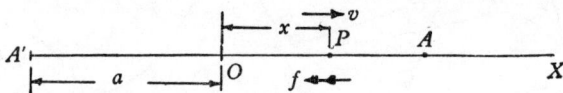

Fig. 84

14·22. Damped oscillations. Logarithmic decrement

If the body referred to in § 14·21 above has an additional resistance to motion which is proportional to the velocity, say $2k\dot{x}$ per unit mass, the equation of motion becomes

$$\ddot{x} = -n^2x - 2k\dot{x},$$

i.e.
$$\frac{d^2x}{dt^2} + 2k\frac{dx}{dt} + n^2x = 0. \tag{1}$$

Solving in the normal manner:
A.E. is
$$m^2 + 2km + n^2 = 0,$$
$$m = -k \pm \sqrt{(k^2 - n^2)}. \tag{2}$$

Three possibilities exist:
(i) $k > n$. $\sqrt{(k^2 - n^2)}$ is real and m has two real values.
Both are negative as $\sqrt{(k^2 - n^2)} < k$.
Let the values be $-\lambda_1$ and $-\lambda_2$.
Then
$$\underline{x = A\,e^{-\lambda_1 t} + B\,e^{-\lambda_2 t}}. \tag{3}$$

(ii) $k = n$. $\sqrt{(k^2 - n^2)} = 0$ and m has two equal values, each $-k$. Then

$$x = (A + Bt) e^{-kt}. \tag{4}$$

(iii) $k < n$. $\sqrt{(k^2 - n^2)}$ is imaginary and equals $j \sqrt{(n^2 - k^2)}$. The values are $-k + j \sqrt{(n^2 - k^2)}$ and $-k - j \sqrt{(n^2 - k^2)}$. If $\omega^2 = n^2 - k^2$, the general solution is

$$\left. \begin{array}{l} x = e^{-kt} (A \cos \omega t + B \sin \omega t), \\[2mm] x = R e^{-kt} \sin (\omega t + \alpha). \end{array} \right\} \tag{5}$$

or

Equations (3) and (4) show the motion to be non-oscillatory in character when $k > n$ or $k = n$, i.e. when the 'damping' factor k is comparatively large.

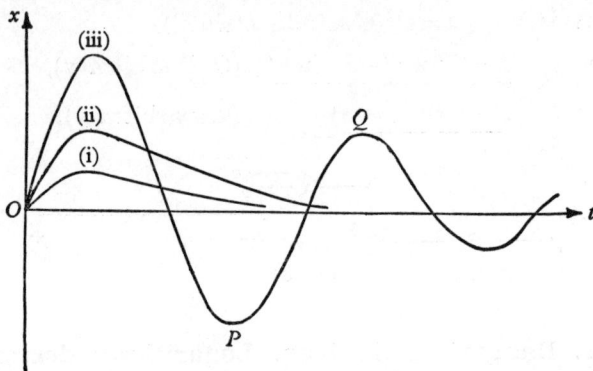

Fig. 85

When $k > n$, the motion is said to be *heavily damped*. When $k = n$, the motion is said to be *critically damped*.

'Dead-beat' measuring instruments are constructed so that the motion of the needle is heavily damped. In this way any tendency for the needle to oscillate about the correct deflexion is avoided.

Equation (5) shows that if $k < n$ the motion *is* oscillatory, but due to the factor e^{-kt} the amplitude of the oscillations dies away.

As the damping of the oscillations is due to the term $2k\dot{x}$ in the differential equation, this term is called the damping term. The greater k is, the greater the damping.

Fig. 85 shows the three typical cases (i), (ii), (iii).

No generality is lost, for graphical purposes, in sketching all the curves through the origin; its only meaning being that when $t = 0$, x has been taken also to be zero.

Equation (5) then becomes $x = R e^{-kt} \sin \omega t$.

Logarithmic decrement. Taking the oscillatory case,

$$x = R e^{-kt} \sin \omega t:$$

$$dx/dt = R e^{-kt}(\omega \cos \omega t - k \sin \omega t).$$

$$\therefore \; dx/dt = 0 \quad \text{when} \quad \tan \omega t = \omega/k.$$

$\therefore \; \omega t = \theta + n\pi$, where $\tan \theta = \omega/k$ and n is any integer, for maximum and minimum x.

Two successive values are $\omega t = \theta$ and $\omega t = \theta + \pi$.

$\therefore$ Two successive extreme positions, such as P and Q in fig. 85, are given by

$$R e^{-k\theta/\omega} \sin \theta \quad \text{and} \quad R \exp(-k\theta/\omega - k\pi/\omega).\sin(\theta + \pi).$$

Their ratio is

$$\frac{\exp(-k\theta/\omega - k\pi/\omega)}{\exp(-k\theta/\omega)} \quad \text{(numerically)}$$

$$= e^{-k\pi/\omega},$$

i.e. *The ratio of the amplitudes at any two successive extreme positions on opposite sides of $x = 0$ is*

$$e^{-k\pi/\omega}. \tag{6}$$

$k\pi/\omega$ is called the *logarithmic decrement.*

Note. If the ratio is called r

$$r = e^{-k\pi/\omega}, \quad \log_e r = -k\pi/\omega.$$

Damped oscillations are important also in electricity (see §14·42).

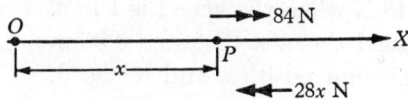

Fig. 86

EXAMPLES

(1) A body, mass 7 kg, moves under a force always directed towards a fixed point O and of magnitude $28x$ newtons, where x metres is the distance from O. It is also acted upon by a constant force 84 N in the direction of x increasing (fig. 86). If it is released from O with zero velocity, find its position at any time t seconds afterwards.

The equation of motion is

$$7\ddot{x} = 84 - 28x,$$

$$\ddot{x} + 4x = 12. \qquad (1)$$

This equation can be transformed into a s.h.m. equation as follows:

$$\ddot{x} + 4(x - 3) = 0,$$

Let $\qquad z = x - 3.$

Then $\qquad \dot{z} = \dot{x}$ and $\ddot{z} = \ddot{x}.$

Thus (1) becomes $\qquad \ddot{z} + 4z = 0.$

$$\therefore \; z = A \cos 2t + B \sin 2t.$$

$$\therefore \; x = 3 + A \cos 2t + B \sin 2t.$$

When $t = 0$, $x = 0$,

$$\therefore \; 0 = 3 + A, \quad A = -3.$$

$$\therefore \; x = 3 - 3 \cos 2t + B \sin 2t,$$

$$\dot{x} = 6 \sin 2t + 2B \cos 2t.$$

When $t = 0$, $\dot{x} = 0$ (body sets off from rest).

$$\therefore \; 0 = 2B, \quad B = 0.$$

Thus $\qquad x = 3 - 3 \cos 2t$

or $\qquad x = 3(1 - \cos 2t)$

$$= 6 \sin^2 t.$$

(2) A body, mass 3 kg, is hung from the end of a spring of stiffness 300 N/m (fig. 87). The resistance to motion is proportional to the velocity and is 48 N when the speed is 1 m/s. The body is pulled down a distance 0·05 m below the equilibrium position and released. If the displacement from the equilibrium position is x metres after t sec., find the equation of motion and solve it. State the period of the oscillations and find the ratio of successive maximum displacements.

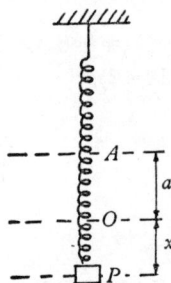

Fig. 87

In the equilibrium position, O, let the extension be a metres.

Tension in spring $= a \times 300$ N.

Thus $\qquad 3 \times 9\cdot81 = 300a,$

giving $\qquad a = 0\cdot0981$ m $= AO.$

The body oscillates about O as centre. Note that as $a = 0.0981$ m and the body is only pulled down a further 0.05 m before being released, the spring will never become slack.

At extension x metres from O, net returning force is $x \times 300$ N.

Speed is $\dot{x}$ metres per second.

$\therefore$ Resistance $= \dot{x}48$ N.

Acceleration is $\ddot{x}$ metres/s².

Equation of motion is

$$3\ddot{x} = -300x - 48\dot{x},$$

giving
$$\ddot{x} + 16\dot{x} + 100x = 0. \tag{1}$$

A.E. is
$$m^2 + 16m + 100 = 0,$$

$$m = -8 \pm j6.$$

$$\therefore \quad x = e^{-8t}(A\cos 6t + B\sin 6t). \tag{2}$$

When $x = 0.05$ m, $\quad t = 0$.

$$\therefore \quad 0.05 = A.$$

Thus
$$x = e^{-8t}(0.05\cos 6t + B\sin 6t),$$

$$\dot{x} = e^{-8t}(6B\cos 6t - 0.3\sin 6t) - 8e^{-8t}(0.05\cos 6t + B\sin 6t).$$

When $t = 0$, $\dot{x} = 0$. $\quad \therefore \quad 0 = 6B - 0.4,$

$$B = 0.2/3.$$

Thus
$$\underline{x = e^{-8t}(0.05\cos 6t + 0.2/3\sin 6t).}$$

Period of oscillations is $\frac{2}{6}\pi$ sec. $\simeq 1.05$ sec.

Ratio of two successive maximum displacements is $e^{-k\pi/\omega}$, where $k = 8$, $\omega = 6$ (see §14·22, (6)), i.e.

$$e^{-8\pi/6} = e^{-4\pi/3} \simeq \underline{0.015}.$$

Thus the oscillations in this case are damped out very quickly.

EXERCISE 53

1. An engine has a crankshaft 0.1 m long and the line of stroke passes through the axis of the crankshaft. If the speed of the crank is uniform and the connecting rod long compared with the crank, show that the motion of the piston is approximately S.H. Taking the speed of the crank as 300 rev./min., find the acceleration of the piston at the ends of the stroke and the velocity of the piston at $\frac{1}{4}$ and $\frac{1}{2}$ stroke.

2. The complete period of a small oscillation of a simple pendulum is 2 sec. and the angular retardation due to air resistance is $0\cdot04 \times$ (angular velocity of pendulum). The bob is held at rest so that the string makes a small angle $\alpha = 1°$ with the downward vertical and let go. Show that after 10 complete oscillations the string will make an angle of about $40'$ with the vertical.   [L.U.]

3. A body, mass 4 kg, is oscillating harmonically in such a way that the force acting on it is $16x$ newtons when its distance from the centre of oscillation is x metres. If a damping resistance which is proportional to the velocity $(\dot{x})$ is to be introduced so as just to make the motion non-oscillating, find the value of the damping factor in units of newtons per metre per second.

ANSWERS

1. $98\cdot7$ m/s^2; $2\cdot72$ m/s; $3\cdot14$ m/s (approx.).

2. Note that if the motion has no air resistance $\ddot{\theta} + \pi^2\theta = \theta$.

3. 16 N/m/s.

14·3. Bending of beams

14·31. Shear force and bending moment

The beams considered here will be assumed to be of small cross-section and to be straight and horizontal when undeflected. The x-axis will be taken along the beam and the y-axis vertically downwards. Thus, deflexions and forces are positive when in a downward direction. All the loads on the beam will be assumed to be vertical.

In general, loads are of two types: (i) Concentrated loads, W, applied at definite points (reactions at supports fall into this category). (ii) Distributed loads, w per unit length. w is often constant, for example a uniform beam of weight w per unit length; but in a general case w may be a function of x, the distance from one end of the beam.

Shear force (S). At a point x the shear force, S, is the resultant of all the forces on the beam to the right of the point, measured positive if in the direction OY.

The shear force is thus discontinuous at a concentrated load.

In fig. 88, let the beam be uniform, of weight w per unit length.

$$0 < x \leqslant a, \qquad\qquad S = W + w(l - x), \qquad\qquad (1)$$

$$a < x \leqslant l, \qquad\qquad S = w(l - x). \qquad\qquad (2)$$

At the point $x = a$ the shear force jumps by an amount W, the point load at A. Fig. 89 shows the shear-force graph. Both lines have the same slope, $-w$, as is also seen from equations (1) and (2).

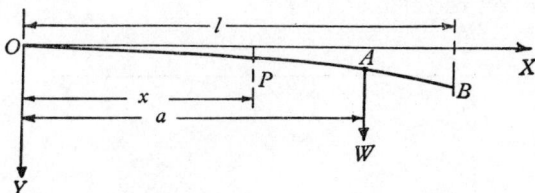

Fig. 88

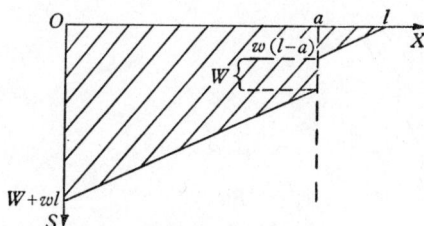

Fig. 89

Bending moment (M). The bending moment, M, at a point x of the beam is the sum of the moments about the point (in the direction from OX to OY taken positive) of all the forces to the right of the point (i.e. 'hogging' moments are positive). All the forces on the beam, to the right of the point, are then equivalent to a force S and a couple M applied at the point.

Referring to fig. 88 again:

$$0 < x \leqslant a: \qquad M = W(a - x) + \frac{w(l - x)^2}{2}. \qquad\qquad (3)$$

$$a < x \leqslant l. \qquad M = \frac{w(l - x)^2}{2}. \qquad\qquad (4)$$

The bending moment diagram is shown in fig. 90. It will be noted that although there is a discontinuity at the point $x = a$ on the shear-force diagram, the bending moment diagram is continuous at that point; though its *slope* is discontinuous there.

Relations between S, M and w. If the beam has a distributed load w in a region, important relations between S, M and w can be obtained.

In fig. 91, for convenience of lettering, the portion of the beam considered is shown separated.

Consider the equilibrium of an element, length δx, of the beam between x, where the shearing force and bending moment are S and M, and the point $x + \delta x$ where they are $S + \delta S$ and $M + \delta M$.

The load on the element is $w\,\delta x$.

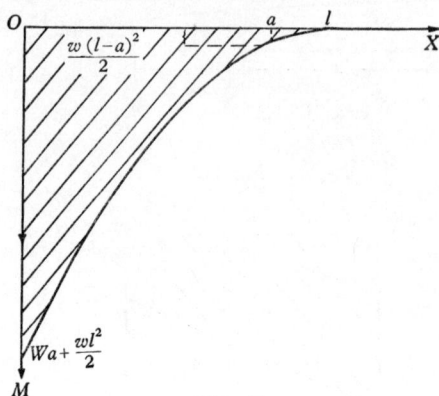

Fig. 90

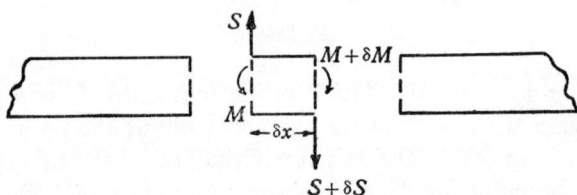

Fig. 91

The conditions for equilibrium give:

Forces.
$$\delta S + w\,\delta x = 0.$$

Torques. $\quad \delta M + S\,\delta x - (w\,\delta x)\,c\,\delta x = 0 \quad (0 < c < 1).$

Thus
$$\delta S/\delta x = -w$$

and
$$\delta M/\delta x = -S + wc\,\delta x.$$

These give, as $\delta x \to 0$: $\quad dS/dx = -w,$ \hfill (5)

$$dM/dx = -S. \hfill (6)$$

Differentiating (6) $\quad d^2M/dx^2 = -dS/dx = w.$ \hfill (7)

(5) and (6) give the slopes of the graphs of S and M against x. This may be easily verified in the example already given, using equations (1) and (2), and (3) and (4).

It is important to realize that these relations only hold in a region free from concentrated loads.

Shear-force and bending-moment diagrams form part of an Ordinary National Certificate course. No examples will be given here, but the relations derived are required for the work which follows.

14·32. Differential equation for the deflexion of a beam

Let P be a point distant x from one end of a beam, at which the deflexion is y (fig. 92).

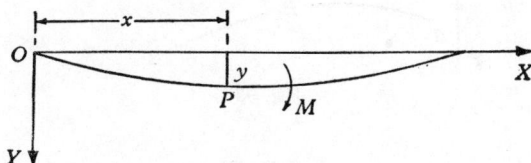

Fig. 92

There is a standard relationship, proved in most books on strength of materials or elasticity, between the bending moment M, the radius of curvature R, the second moment of area of cross-section I at P, and the elasticity E of the material of the beam. This relationship is

$$M/I = E/R. \tag{1}$$

It is true for thin beams whose deflexions are small and the bending of such a nature that all plane sections remain plane sections.

Now

$$\frac{1}{R} = \frac{d^2y}{dx^2} \bigg/ \left[1 + \left(\frac{dy}{dx}\right)^2\right]^{\frac{3}{2}}.$$

If the bending is small and the beam thin, dy/dx is small. Thus

$$\frac{1}{R} \simeq \frac{d^2y}{dx^2}.$$

In this case (1) reduces to

$$M = EI \frac{d^2y}{dx^2}. \tag{2}$$

EI is called the *flexural rigidity* of the beam.

14·33. Applications. Sign of bending moment

By equating $EI\dfrac{d^2y}{dx^2}$ to the value of M a differential equation is obtained. When solved, this gives the deflexion at any point.

Care must be taken in determining the sign of the bending moment.

With axes as shown in fig. 93, $EI\dfrac{d^2y}{dx^2}$ at P is equal to:

Clockwise (hogging) moment of forces to the right of P

or Anti-clockwise (hogging) moment of forces to the left of P.

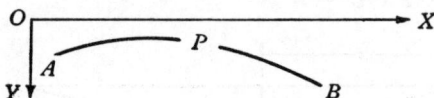

Fig. 93

If the y-axis were reversed in direction, 'sagging' moments would be taken.

In §14·31 it was shown that $d^2M/dx^2 = w$, for a distributed load. Thus

$$\frac{d^2}{dx^2}\left(EI\frac{d^2y}{dx^2}\right) = w. \tag{1}$$

For a beam of constant cross-section I is constant. In this case (1) above simplifies to

$$EI\frac{d^4y}{dx^4} = w. \tag{2}$$

This is an alternative starting point to using $M = EI\dfrac{d^2y}{dx^2}$, and is useful in cases when it is difficult to find M by normal means.

14·34. Boundary conditions

When integrating either $M = EI\dfrac{d^2y}{dx^2}$ or $EI\dfrac{d^4y}{dx^4} = w$ to obtain y, arbitrary constants arise. The value of these may be found if the boundary conditions are known. The most common are:

(a) *At a free end.*

$M = 0$ and therefore, $d^2y/dx^2 = 0$.

Also, unless there is a concentrated load at the end, $S = 0$ and therefore $dM/dx = 0$. i.e.

$$\frac{d}{dx}\left(EI\frac{d^2y}{dx^2}\right) = 0,$$

so that if I is constant, $d^3y/dx^3 = 0$.

(b) *At a clamped end* (horizontal).

Here, y and dy/dx are zero.

(c) *At a freely hinged end.*

Here, $y = 0$ and, as $M = 0$, $d^2y/dx^2 = 0$.

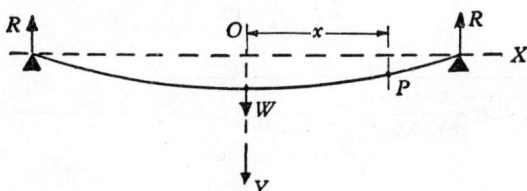

Fig. 94

EXAMPLES

(1) A light beam of uniform cross-section and length l with a point load W at the middle is simply supported at both ends (fig. 94).

The beam is obviously symmetrical about its centre.

Thus, the y-axis is taken through the centre.

By symmetry, both reactions are given by $R = \frac{1}{2}W$.

At P:
$$M = -R(\tfrac{1}{2}l - x) = -\tfrac{1}{2}W(\tfrac{1}{2}l - x).$$

$$\therefore\ EI\frac{d^2y}{dx^2} = -\frac{W}{2}\left(\frac{l}{2} - x\right).$$

Integrating
$$EI\frac{dy}{dx} = \frac{W}{4}\left(\frac{l}{2} - x\right)^2 + A.$$

$$[Note.\ \int(\tfrac{1}{2}l - x)\,dx = -\tfrac{1}{2}(\tfrac{1}{2}l - x)^2.]$$

When $x = 0$, $dy/dx = 0$, $\therefore\ 0 = \tfrac{1}{4}W\,\tfrac{1}{4}l^2 + A.$

Thus
$$EI\frac{dy}{dx} = \frac{W}{4}\left(\frac{l}{2} - x\right)^2 - \frac{Wl^2}{16}.$$

$$\therefore\ EIy = -\tfrac{1}{12}W(\tfrac{1}{2}l - x)^3 - \tfrac{1}{16}Wl^2x + B.$$

When $x = \tfrac{1}{2}l,\ y = 0,$ $\therefore\ 0 = -\tfrac{1}{32}Wl^3 + B.$

Thus $\underline{EIy = -\tfrac{1}{12}W(\tfrac{1}{2}l - x)^3 - \tfrac{1}{16}Wl^2x + \tfrac{1}{32}Wl^3.}$

At O, $x=0$, $y=\dfrac{W}{EI}\dfrac{l^3}{48}.$

This is the maximum deflexion.

(2) A light uniform beam, length l, carries a uniform load w per unit length, and is clamped horizontally at the same level at both ends $x=0$ and $x=l$ (fig. 95).

The bending moment at any point P is difficult to find in this case, as torques of unknown magnitude are introduced by the clamps. It is therefore best to start from the equation

$$EI\frac{d^4y}{dx^4}=w.$$

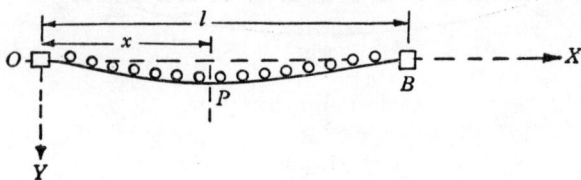

Fig. 95

The boundary conditions (see § 14·34) are

$$x=0, \quad y=dy/dx=0,$$
$$x=l, \quad y=dy/dx=0.$$

Integrating $EI\dfrac{d^4y}{dx^4}=w,$

$$EI\frac{d^3y}{dx^3}=wx+A,$$

$$EI\frac{d^2y}{dx^2}=\frac{wx^2}{2}+Ax+C, \tag{1}$$

$$EI\frac{dy}{dx}=\frac{wx^3}{6}+\frac{Ax^2}{2}+Cx+D.$$

When $x=0$, $dy/dx=0$, $\therefore$ $D=0$.

Thus, $EI\dfrac{dy}{dx}=\dfrac{wx^3}{6}+\dfrac{Ax^2}{2}+Cx.$ $\tag{2}$

$$\therefore\ EIy=\tfrac{1}{24}wx^4+\tfrac{1}{6}Ax^3+\tfrac{1}{2}Cx^2+F.$$

When $x=0$, $y=0$, $\therefore$ $F=0$

$$\therefore\ EIy=\tfrac{1}{24}wx^4+\tfrac{1}{6}Ax^3+\tfrac{1}{2}Cx^2. \tag{3}$$

When $x = l$, $dy/dx = 0$.

$\therefore$ From (2):
$$0 = \tfrac{1}{6}wl^3 + \tfrac{1}{2}Al^2 + Cl,$$
$$0 = \tfrac{1}{6}wl^2 + \tfrac{1}{2}Al + C. \tag{4}$$

When $x = l$, $y = 0$.

$\therefore$ From (3):
$$0 = \tfrac{1}{24}wl^4 + \tfrac{1}{6}Al^3 + \tfrac{1}{2}Cl^2,$$
$$0 = \tfrac{1}{12}wl^2 + \tfrac{1}{3}Al + C. \tag{5}$$

Solving equations (4) and (5):
$$A = -\tfrac{1}{2}wl, \quad C = \tfrac{1}{12}wl^2.$$

Substitution into equation (3) gives
$$EIy = \tfrac{1}{24}wx^4 - \tfrac{1}{12}wlx^3 + \tfrac{1}{24}wl^2x^2$$

or
$$\underline{EIy = \tfrac{1}{24}wx^2(l-x)^2}. \tag{6}$$

Any required information can now be found.
For example, from (1):
Bending moment at any point is given by
$$M = EI\frac{d^2y}{dx^2} = \frac{wx^2}{2} - \frac{wlx}{2} + \frac{wl^2}{12}. \tag{7}$$

Maximum deflexion at $x = \tfrac{1}{2}l$ is
$$y = \frac{w}{EI}\frac{l^4}{384}.$$

Shearing force is given by
$$S = -dM/dx = -wx + \tfrac{1}{2}wl. \tag{8}$$

From (7), the torques introduced by the clamps are given by:
At $x = 0$,
$$M = \tfrac{1}{12}wl^2.$$
At $x = l$,
$$M = \tfrac{1}{12}wl^2.$$

From (8), the shearing forces at O and B are given by
At $x = 0$, $S = \tfrac{1}{2}wl$ and at $x = l$, $S = -\tfrac{1}{2}wl$.

(3) A uniform beam, length l, is clamped horizontally at one end and freely supported at the other end at the same level (fig. 96).

Let $w = $ weight per unit length of the beam.

The reaction at the support, R, and the torque, C, introduced by the clamp, are unknown. The reaction at O is $wl - R$ ($wl = $ weight of beam).

The equation $EI\dfrac{d^4y}{dx^4}=w$ could again be used; but as an alternative $EI\dfrac{d^2y}{dx^2}=M$ will be used, even though the expression for M will contain unknowns.

At P, bending moment M is given by

$$M=\frac{w(l-x)^2}{2}-R(l-x).$$

Thus

$$EI\frac{d^2y}{dx^2}=\frac{w(l-x)^2}{2}-R(l-x). \tag{1}$$

$$\therefore\ EI\frac{dy}{dx}=-\frac{w(l-x)^3}{6}+\frac{R(l-x)^2}{2}+B \quad (B \text{ arbitrary}).$$

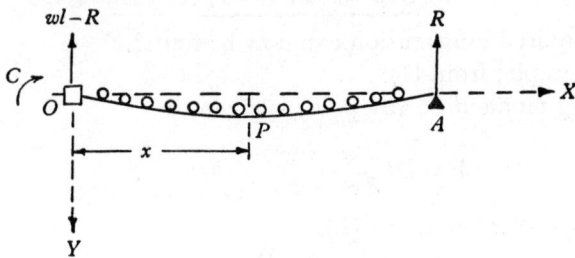

Fig. 96

When $x=0$, $dy/dx=0$,

$$\therefore\ 0=-\tfrac{1}{6}wl^3+\tfrac{1}{2}Rl^2+B.$$

Thus

$$EI\frac{dy}{dx}=-\frac{w(l-x)^3}{6}+\frac{R}{2}(l-x)^2+\frac{wl^3}{6}-\frac{Rl^2}{2}, \tag{2}$$

$$EIy=\tfrac{1}{24}w(l-x)^4-\tfrac{1}{6}R(l-x)^3+\tfrac{1}{6}wl^3x-\tfrac{1}{2}Rl^2x+D. \tag{3}$$

When $x=l$, $y=0$, $\therefore\ 0=\tfrac{1}{6}wl^4-\tfrac{1}{2}Rl^3+D.$ $\tag{4}$

Also when $x=0$, $y=0$.

$$\therefore\ 0=\tfrac{1}{24}wl^4-\tfrac{1}{6}Rl^3+D. \tag{5}$$

From (4) and (5): $D=\tfrac{1}{48}wl^4,\quad R=\tfrac{3}{8}wl.$

Substituting in (3):

$$EIy=\tfrac{1}{24}w(l-x)^4-\tfrac{1}{16}wl(l-x)^3+\tfrac{1}{6}wl^3x-\tfrac{3}{16}wl^3x+\tfrac{1}{48}wl^4,$$

$$\underline{EIy=\tfrac{1}{24}w(l-x)^4-\tfrac{1}{16}wl(l-x)^3-\tfrac{1}{48}wl^3x+\tfrac{1}{48}wl^4.} \tag{6}$$

The reader should verify that starting from $EI\dfrac{d^4y}{dx^4}=w$ would give the same solution.

The reactions at O and A are:

$$R=\tfrac{3}{8}wl, \quad wl-R=\tfrac{5}{8}wl.$$

Substituting for R in equations (1) and (2):

$$EI\frac{d^2y}{dx^2}=M=\tfrac{1}{2}w(l-x)^2-\tfrac{3}{8}wl(l-x).$$

$$EI\frac{dy}{dx}=-\tfrac{1}{6}w(l-x)^3+\tfrac{3}{16}wl(l-x)^2-\tfrac{1}{48}wl^3. \tag{7}$$

Thus at $x=0$, $\qquad M=\tfrac{1}{2}wl^2-\tfrac{3}{8}wl^2=\tfrac{1}{8}wl^2.$

The torque at O is $\qquad C=\tfrac{1}{8}wl^2.$

If it is desired to find the point on the beam where the deflexion is greatest:

From (7), when $dy/dx=0$:

$$-\tfrac{1}{6}w(l-x)^3+\tfrac{3}{16}wl(l-x)^2-\tfrac{1}{48}wl^3=0$$

which gives $\qquad \tfrac{1}{6}wx^3-\tfrac{5}{16}wlx^2+\tfrac{1}{8}wl^2x=0,$

$$\tfrac{1}{48}wx(8x^2-15xl+6l^2)=0.$$

$$x=0 \quad\text{or}\quad x=\left(\frac{15\pm\surd(33)}{16}\right)l.$$

Neglecting the value greater than l:

$$x=0 \quad\text{or}\quad x\simeq0\!\cdot\!58l.$$

At $x=0$, deflexion is zero.

Thus $x\simeq0\!\cdot\!58l$ is the point of maximum deflexion.

The value of the maximum deflexion could then be found from equation (6).

(4) *Non-continuous, asymmetrical loading.* Consider the case of a uniform beam, length l, weight w per unit length, simply supported at its ends and with a point load W distant $\tfrac{2}{3}l$ from one end (fig. 97).

$$R_1+R_2=wl+W. \tag{1}$$

Taking moments round A for whole beam:

$$R_2l=W\,.\tfrac{2}{3}l+wl\,.\tfrac{1}{2}l. \tag{2}$$

From (1) and (2)

$$R_2=\tfrac{1}{2}wl+\tfrac{2}{3}W, \quad R_1=\tfrac{1}{2}wl+\tfrac{1}{3}W. \tag{3}$$

Because of the asymmetrical placing of the point load the two regions $0 < x < \frac{2}{3}l$ and $\frac{2}{3}l < x < l$ must be treated separately.

$0 < x < \frac{2}{3}l$ (*point* P_1): Taking the y-axis upwards

$$EI\frac{d^2y}{dx^2} = M = R_1 x - \tfrac{1}{2}wx^2.$$

From (3):

$$EI\frac{d^2y}{dx^2} = \left(\frac{wl}{2} + \frac{W}{3}\right)x - \frac{wx^2}{2}.$$

$$\therefore\ EI\frac{dy}{dx} = \left(\frac{wl}{2} + \frac{W}{3}\right)\frac{x^2}{2} - \frac{wx^3}{6} + C. \tag{4}$$

$$EIy = (\tfrac{1}{2}wl + \tfrac{1}{3}W)\tfrac{1}{6}x^3 - \tfrac{1}{24}wx^4 + Cx + D.$$

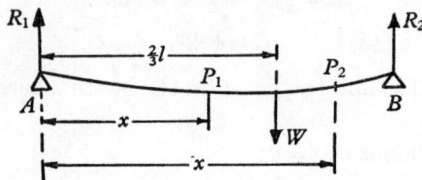

Fig. 97

But when $x = 0$, $y = 0$, therefore $D = 0$ and

$$EIy = (\tfrac{1}{2}wl + \tfrac{1}{3}W)\tfrac{1}{6}x^3 - \tfrac{1}{24}wx^4 + Cx. \tag{5}$$

$\frac{2}{3}l < x < l$ (*point* P_2):

$$M = EI\frac{d^2y}{dx^2} = \left(\frac{wl}{2} + \frac{W}{3}\right)x - \frac{wx^2}{2} - W[x - \tfrac{2}{3}l],$$

$$EI\frac{dy}{dx} = \left(\frac{wl}{2} + \frac{W}{3}\right)\frac{x^2}{2} - \frac{wx^3}{6} - \frac{W}{2}[x - \tfrac{2}{3}l]^2 + C. \tag{6}$$

Note that as the slope of the beam must be the same in both regions when $x = \frac{2}{3}l$, it is legitimate to put the same arbitrary constant C in (6) as in (4).

$$EIy = (\tfrac{1}{2}wl + \tfrac{1}{3}W)\tfrac{1}{6}x^3 - \tfrac{1}{24}wx^4 - \tfrac{1}{6}W[x - \tfrac{2}{3}l]^3 + Cx. \tag{7}$$

No new arbitrary constant has been introduced in (7), as equations (5) and (7) must give the same y value when $x = \frac{2}{3}l$.

Now $y = 0$ when $x = l$.

Thus, from (7),

$$0 = \frac{wl^4}{12} + \frac{Wl^3}{18} - \frac{wl^4}{24} - \frac{Wl^3}{6.27} + Cl,$$

giving
$$C = -\frac{wl^3}{24} - \frac{8Wl^2}{6.27},$$

$$C = -\left(\frac{4Wl^2}{81} + \frac{wl^3}{24}\right). \tag{8}$$

Substituting this value into (5) and (7):

$0 < x < \frac{2}{3}l$:

$$EIy = (\tfrac{1}{2}wl + \tfrac{1}{3}W)\tfrac{1}{6}x^3 - \tfrac{1}{24}wx^4 - (\tfrac{4}{81}Wl^2 + \tfrac{1}{24}wl^3)x, \tag{9}$$

$\frac{2}{3}l < x < l$:

$$EIy = (\tfrac{1}{2}wl + \tfrac{1}{3}W)\tfrac{1}{6}x^3 - \tfrac{1}{24}wx^4 - \tfrac{1}{6}W[x - \tfrac{2}{3}l]^3 - (\tfrac{4}{81}Wl^2 + \tfrac{1}{24}wl^3)x. \tag{10}$$

The student might note that the only new term introduced for the region $\frac{2}{3}l < x < l$, beyond the point load, is the term containing the factor $[x - \frac{2}{3}l]$. This was deliberately put in square brackets and integrated as a whole so that its effect could be easily followed through the whole working.

Macaulay's method for dealing with non-continuous loading makes use of this idea to cut down the working required. The method can be studied in detail in most books on strength of materials, etc.

EXERCISE 54

1. A torque of moment M is applied to the free end of a light horizontal cantilever. Find the maximum deflexion.

2. A torque of moment M is applied at each end of a light beam of length l. Find the maximum slope of the beam.

3. A light uniform beam, length l, carries a uniform load w per unit length. It is freely hinged at both ends. Find a formula for the deflexion (y) at any distance x $(< l)$ from one end.

4. A heavy uniform beam, length l, weight w per unit length, carries a concentrated load W at the centre. It is simply supported at both ends. Find the central deflexion.

5. Find the bending moment at any point of a uniform beam, length $2l$, weight W, which is supported at its mid-point. Also find the depression of the ends and show that it is increased by an amount $\dfrac{W_1}{3}\dfrac{l^3}{EI}$ if a weight W_1 is suspended from each end.

6. A uniform beam of weight W and length $6l$ rests symmetrically on two supports at the same level distant $2l$ apart. Find the height of the centre above the level of the supports.

7. A uniform beam, clamped horizontally at one end, carries a load uniformly distributed between the mid-point and the free end. If this load has the same weight per unit length as the beam, prove the deflexion at the free end is $\frac{89}{31}$ times that at the mid-point.   [L.U.]

8. A uniform beam, length $2l$, weight W, is clamped horizontally at one end and is supported at the mid-point so that the free end is at the same level as the clamp. Find the reactions at the supports and the deflexion and slope at the mid-point. [L.U.]

Answers

1. $Ml^2/2EI$.   **2.** $Ml/2EI$.

3. $y = \dfrac{wx}{24EI}(x^3 - 2lx^2 + l^3)$.   **4.** $\dfrac{1}{EI}(\frac{1}{48}Wl^3 + \frac{5}{384}wl^4)$.

5. $\dfrac{W}{4l}(l-x)^2$ at distance x from centre; $\dfrac{Wl^3}{16EI}$.

6. $\dfrac{19Wl^3}{144EI}$.

8. Reaction at support at mid-point is $\frac{6}{5}W$.

Reaction at clamp is $\frac{1}{5}W$ (downwards).

Deflexion at mid-point is $\dfrac{11l^3}{240EI}$ above the ends.

Slope at mid-point is $\dfrac{Wl^2}{60EI}$ numerically.

14·4. Electrical applications

14·41. Simple standard circuits

The applications given here are not exhaustive. They are a sample of some of those most frequently occurring.

The theory is closely connected with that of mechanical oscillations. Many of the results are similar. Even though not a mechanical engineer, the reader will benefit if he compares the results obtained here with those in § 14·22.

Electric circuits will be regarded as built up of 'elements' of three main types: inductance (L), resistance (R) and capacitance (C). The properties of such circuits will be deduced by considering the constituent elements to be 'lumped'. Thus an inductance coil will be looked on as an inductance L in series with a resistance R:

This is not strictly accurate, as each tiny element of the coil. however small, has its own inductance and resistance. The 'lumping', however, gives quite good approximations for most purposes. One main exception is in the case of iron-cored resistances or inductances.

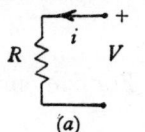

(a)

The following knowledge of electrical theory is assumed; the symbols have their usual meanings:

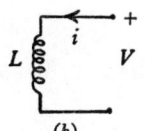

(b)

(a) $V = Ri$ (fig. 98 a).

(b) $V = L\dfrac{di}{dt}$ (fig. 98 b).

(c) $V = q/C$ (fig. 98 c).

(d) $i = dq/dt$.

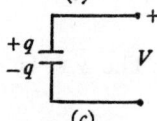

(c)

Fig. 98

(e) The total impressed voltage in a circuit is equal to the sum of the voltage drops over the constituent elements.

(f) In a closed circuit the algebraic sum of the voltages is zero.

(g) The current flowing into a junction is the same as that flowing out.

The simpler electrical circuits give rise to linear differential equations.

14·42. The $L-C-R$ series circuit

This type of circuit (fig. 99) is taken as a general worked example.

From (e) above (§ 14·41):

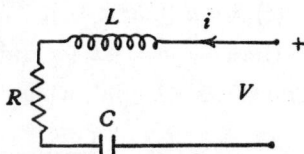

Fig. 99

$$L\frac{di}{dt} + Ri + \frac{q}{C} = V.$$

But $i = dq/dt$. $\quad\therefore\ L\dfrac{d^2q}{dt^2} + R\dfrac{dq}{dt} + \dfrac{q}{C} = V,$ \hfill (1)

or, differentiating again,

$$L\frac{d^3q}{dt^3} + R\frac{d^2q}{dt^2} + \frac{1}{C}\frac{dq}{dt} = \frac{dV}{dt},$$

giving $$L\frac{d^2i}{dt^2} + R\frac{di}{dt} + \frac{1}{C}i = \frac{dV}{dt}.$$ (2)

(1) is solved if the charge at any time is required; (2) is solved if the current is required.

If the impressed voltage V is *zero (closed circuit) or is constant*, $dV/dt = 0$.

Thus (2) becomes

$$L\frac{d^2i}{dt^2} + R\frac{di}{dt} + \frac{1}{C}i = 0.$$ (3)

For convenience in working put

$$2k = \frac{R}{L}, \quad n^2 = \frac{1}{LC}.$$ (4)

Then (3) becomes $$\frac{d^2i}{dt^2} + \frac{R}{L}\frac{di}{dt} + \frac{1}{CL}i = 0,$$

$$\frac{d^2i}{dt^2} + 2k\frac{di}{dt} + n^2i = 0.$$

The A.E. is $$m^2 + 2km + n^2 = 0.$$

$$m = -k \pm \sqrt{(k^2 - n^2)}.$$ (5)

Three typical cases arise:

(a) $k > n$ $\left(\text{i.e. } \dfrac{R}{2L} > \dfrac{1}{\sqrt{(LC)}}, \text{ or } CR^2 > 4L\right)$.

Both values for m are real and both are negative, say $-\lambda_1$, $-\lambda_2$. Then $$i = A e^{-\lambda_1 t} + B e^{-\lambda_2 t}.$$

(b) $k = n$ $(CR^2 = 4L)$.

Both values for m are $-k$. Then

$$i = (A + Bt) e^{-kt}.$$

(c) $k < n$ $(CR^2 < 4L)$.

Here $$m = -k + j\sqrt{(n^2 - k^2)} \quad \text{and} \quad -k - j\sqrt{(n^2 - k^2)}.$$

Call these $-k \pm jn'$, where $n' = \sqrt{(n^2 - k^2)}$. Then

$$i = e^{-kt}(A \cos n't + B \sin n't).$$

In both (a) and (b) the current dies away very rapidly and is non-oscillatory.

Case (a) is said to be heavily damped.

Case (b) is said to be critically damped.

In case (c) the current is oscillatory, but its amplitude is rapidly damped by the factor e^{-kt}.

In all three cases the currents are called *'transient'* as they die out rapidly. Fig. 85 of § 14·22 gives a sketch of the current against time for the three cases.

e^{-kt} is the damping term in each case.

Now k is $R/2L$. Thus the greater the resistance R, the greater the damping.

From (c), for oscillations to be possible, $R^2 < 4L/C$.

Case (b), $R^2 = 4L/C$, is called 'critical' damping, as R is *just* large enough to stop oscillations.

The remarks on the *logarithmic decrement* already made in § 14·22 apply equally well here.

Given initial conditions, the arbitrary constants may be calculated.

Thus in case (c), if when $t = 0$, $i = q = 0$:

$$i = e^{-kt}(A \cos n't + B \sin n't).$$

$$\therefore \quad 0 = A \quad (t = 0,\ i = 0).$$

Thus $$i = B e^{-kt} \sin n't.$$

But $$L \frac{di}{dt} + Ri + \frac{1}{C} q = V,$$

i.e. $$LB e^{-kt}(n' \cos n't - k \sin n't) + RB e^{-kt} \sin n't + \frac{1}{C} q = V.$$

But $q = 0$ at $t = 0$.

$$\therefore \quad LBn' = V$$

$$B = V/n'L \quad \text{(or zero if } V \text{ is zero).}$$

Thus, for V constant, say $V = E$;

$$i = \frac{E}{n'L} e^{-kt} \sin n't$$

or $$i = \frac{E}{L \sqrt{\left(\dfrac{1}{LC} - \dfrac{R^2}{4L^2}\right)}} e^{-Rt/2L} \sin \sqrt{\left(\frac{1}{LC} - \frac{R^2}{4L^2}\right)} t.$$

The case when $V = E \sin \omega t$. That is, when there is a *periodic* impressed voltage.

The solution of the resulting differential equation is second year Higher National Certificate work. In this case, however, the situation when the steady state has been reached can be solved by the application of complex number theory to the calculation of impedances, phase displacements, etc. This is done in lectures on a.c. theory in Higher National Certificate classes, and will not be repeated here.

EXERCISE 55

1. A capacitor capacitance C, is charged to a voltage V and discharged at time $t=0$ into a non-inductive resistance R. Show that the charge on the capacitor at time t is $CV\,e^{-t/RC}$. [This provides a possible method of measuring the value of a large resistance; as if R is large, the charge will not die away too rapidly to be measured.]

2. A battery of voltage V is applied at time $t=0$ to a series circuit of an inductance L and capacitance C. Show that, if the initial charge and current are zero, the current at time t is $(V/nL)\sin nt$, where $n=1/\sqrt{(LC)}$.

3. A voltage $E\cos\omega t$ is applied to a series circuit of inductance L with resistance R. Write down the differential equation for the current i at any time t and solve it. Indicate the transient term. [The D.E. will be linear and of first order.]

4. A capacitor of capacitance C is discharged through a circuit of resistance R and inductance L. Prove that the charge Q at any time t is given by $L\dfrac{d^2Q}{dt^2}+R\dfrac{dQ}{dt}+\dfrac{Q}{C}=0$; hence show that, if R be sufficiently small, the discharge is oscillatory and determine the period of the oscillation. Calculate the frequency if the capacitance is 0·02 microfarad, the inductance 0·0003 henry and the resistance negligible. [L.U.]

5. A leaky capacitor can be taken as a capacitance C in parallel with a resistance $1/G$ (G small). It is charged with an amount Q and then discharged through a resistance R. Show that the charge after time t is given by

$$q=Q\exp\left[-\frac{(1+GR)}{CR}\,t\right].$$

6. If, in the circuit of Q. 5 above, an inductance L is put in series with the resistance R, show that the period of natural oscillations of the circuit is

$$2\pi\left[\frac{1}{LC}-\frac{1}{4}\left(\frac{R}{L}-\frac{G}{C}\right)^2\right]^{-\frac{1}{2}}.$$

ANSWERS

3. $\dfrac{di}{dt} + \dfrac{R}{L}\,i = \dfrac{E}{L}\cos \omega t;$ $i = \dfrac{E}{\sqrt{(\omega^2 L^2 + R^2)}}\cos(\omega t - \alpha) + A\,e^{-Rt/L},$

where $\tan\alpha = \omega L/R,$ $A\,e^{-Rt/L}$ is the transient.

4. Period is $\dfrac{4\pi L}{\sqrt{\left(\dfrac{4L}{C} - R^2\right)}};$ frequency $\dfrac{1}{2\pi\sqrt{(LC)}} \backsimeq 65{,}000.$

CHAPTER 15

HARMONIC ANALYSIS

15·1. Introduction. Periodic functions

Periodic functions arise in many branches of engineering mathematics; for example, in oscillatory electrical circuits, reciprocating engines and in many other cases.

As time is the variable in many cases, in most of the discussion which follows t will be taken as the independent variable and $2T$ as the period. In special cases $2T$ may actually be 2π, and matters are simplified a little.

If
$$f(t + r\,2T) = f(t) \quad (r = \pm 1, \pm 2, \ldots), \tag{1}$$

then $f(t)$ is said to be periodic and of period $2T$. Each time the value of t is changed by a multiple of the period $2T$, the value of the function is repeated. Fig. 100 shows the graph of a typical periodic function.

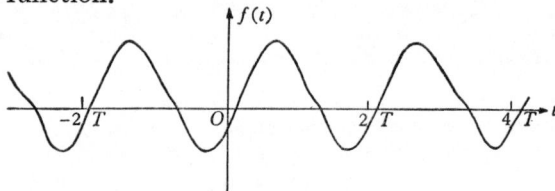

Fig. 100

The simple trigonometric functions repeat their values every time the complete angle is changed by an amount 2π.

Thus
$$\sin 3t = \sin(3t + 2\pi) = \sin 3(t + \tfrac{2}{3}\pi).$$

Here the period is $\tfrac{2}{3}\pi$.

Trigonometric functions can easily be arranged to have period $2T$. For example

$$\sin\frac{\pi}{T}t \quad \text{and} \quad \cos\frac{\pi}{T}t \quad \text{have each a period } 2\pi \Big/ \frac{\pi}{T} = 2T.$$

15·11. Fourier series

As the trigonometric functions are the simplest periodic functions to deal with, it is of obvious interest to find out whether *any* periodic function, $f(t)$, with period $2T$ can be represented in terms of a trigonometric series.

That is, whether $f(t)$ can be expressed in the form

$$f(t) = a_0 + \left(a_1 \cos\frac{\pi}{T}t + a_2 \cos\frac{2\pi}{T}t + ...\right) + \left(b_1 \sin\frac{\pi}{T}t + b_2 \sin\frac{2\pi}{T}t + ...\right)$$

$$= a_0 + \sum_1^\infty a_n \cos\frac{n\pi}{T}t + \sum_1^\infty b_n \sin\frac{n\pi}{T}t, \tag{1}$$

where the a's and b's are constants.

It has been found that this can be done in many cases of practical interest.

The process is called harmonic analysis.

Such a series as (1) is called a Fourier series, after the name of its instigator.

Several important results are now obtained which will be used later.

15·12. Odd and even functions

These have already been discussed in § 9·5.

There it was found that sine functions are odd and cosine functions even.

Thus if an even function is expanded in a Fourier series, it will only contain cosine terms. (Note that a constant term, a_0, may be looked on as a cosine term $a_0 \cos(0t)$.) Similarly, the Fourier series of an odd function will only contain sine terms.

15·13. Special definite integrals

All the definite integrals given below will be required.

$$\int_{-T}^{T} \cos\frac{n\pi}{T}t\, dt = \int_{-T}^{T} \sin\frac{n\pi}{T}t\, dt = 0 \quad (n = 1, 2, ...), \tag{1}$$

$$\int_{-T}^{T} \cos\frac{n\pi}{T}t \sin\frac{m\pi}{T}t\, dt = 0 \quad \text{for all integers } m \text{ and } n, \tag{2}$$

$$\left.\begin{aligned} \int_{-T}^{T} \cos\frac{n\pi}{T}t \cos\frac{m\pi}{T}t\, dt &= 0 \quad \text{if } m \neq n \\ &= T \quad \text{if } m = n \end{aligned}\right\}, \tag{3}$$

$$\left.\begin{aligned} \int_{-T}^{T} \sin\frac{n\pi}{T}t \sin\frac{m\pi}{T}t\, dt &= 0 \quad \text{if } m \neq n \\ &= T \quad \text{if } m = n \end{aligned}\right\}. \tag{4}$$

The special case of (3) when $n = m = 0$ is

$$\int_{-T}^{T} dt = 2T. \tag{5}$$

The proof of the above should be well within the scope of the reader. Case (3) is worked out below; the reader should himself verify the others:

$$m \neq n: \quad \int_{-T}^{T} \cos\frac{n\pi}{T}t \cos\frac{m\pi}{T}t\,dt$$

$$= \frac{1}{2}\int_{-T}^{T}\left\{\cos\frac{(m+n)\pi}{T}t + \cos\frac{(m-n)\pi}{T}t\right\}dt$$

$$= \frac{1}{2}\left[\frac{T}{(m+n)\pi}\sin\frac{(m+n)\pi}{T}t + \frac{T}{(m-n)\pi}\sin\frac{(m-n)\pi}{T}t\right]_{-T}^{T}$$

$$= 0 \quad \text{as } \sin r\pi = 0 \text{ for all integers } r.$$

$m = n$: The integral becomes

$$\int_{-T}^{T}\cos^2\frac{n\pi}{T}t\,dt = \frac{1}{2}\int_{-T}^{T}\left(1 + \cos\frac{2n\pi}{T}t\right)dt$$

$$= \frac{1}{2}\left[t + \frac{T}{2n\pi}\sin\frac{2n\pi}{T}t\right]_{-T}^{T} = \tfrac{1}{2}[(T+0) - (-T+0)].$$

$$= T.$$

All the results would be the same as long as the limits covered a complete period $2T$. For example, instead of from $-T$ to T, 0 to $2T$ could have been taken.

15·14. Calculation of the coefficients in a Fourier series

It will be assumed that an expansion such as 15·11 (2) exists for the function, and can be integrated taking each term separately.

For reasons to appear below it is best to write the series in the form

$$f(t) = \frac{a_0}{2} + \left(a_1\cos\frac{\pi}{T}t + \ldots + a_n\cos\frac{n\pi}{T}t + \ldots\right)$$

$$+ \left(b_1\sin\frac{\pi}{T}t + \ldots + b_n\sin\frac{n\pi}{T}t + \ldots\right). \quad (1)$$

First integrate both sides with respect to t over a period $(-T$ to $T)$. From § 15·13, all the integrals on the right-hand side vanish, except the first.

Thus
$$\int_{-T}^{T} f(t)\,dt = \frac{a_0}{2}\int_{-T}^{T} dt = Ta_0.$$

$$\therefore \ a_0 = \frac{1}{T}\int_{-T}^{T} f(t)\,dt$$

$$= \underline{2 \times \text{mean value of } f(t) \text{ over a period}}. \quad (2)$$

To find a_n, multiply both sides of (1) by $\cos n\pi t/T$ and integrate over a period. All terms on the right-hand side now vanish except

$$a_n \int_{-T}^{T} \cos^2 \frac{n\pi}{T} t \, dt.$$

Thus $\qquad \int_{-T}^{T} f(t) \cos \frac{n\pi}{T} t \, dt = a_n \int_{-T}^{T} \cos^2 \frac{n\pi}{T} t \, dt = T a_n.$

$$\therefore \ a_n = \frac{1}{T} \int_{-T}^{T} f(t) \cos \frac{n\pi}{T} t \, dt$$

$$= 2 \times \text{M.V. of } f(t) \cos \frac{n\pi}{T} t \text{ over a period.} \qquad (3)$$

To find b_n, multiply both sides of (1) by $\sin \dfrac{n\pi}{T} t$ and integrate over a period. All terms on the right-hand side vanish except

$$b_n \int_{-T}^{T} \sin^2 \frac{n\pi}{T} t \, dt.$$

Thus $\qquad \int_{-T}^{T} f(t) \sin \frac{n\pi}{T} t \, dt = b_n \int_{-T}^{T} \sin^2 \frac{n\pi}{T} t \, dt = b_n T.$

$$\therefore \ b_n = \frac{1}{T} \int_{-T}^{T} f(t) \sin \frac{n\pi}{T} t \, dt$$

$$= 2 \times \text{M.V. of } f(t) \sin \frac{n\pi}{T} t \text{ over a period.} \qquad (4)$$

(2), (3) and (4) are similar, in so far as '2 × mean value' occurs on the right-hand side. This is the reason for taking the constant term as $\frac{1}{2}a_0$ in the Fourier series (1).

N.B. If the period is 2π (2), (3) and (4) simplify to

$$a_0 = \frac{1}{\pi} \int_{-\pi}^{\pi} f(t)\, dt; \quad a_n = \frac{1}{\pi} \int_{-\pi}^{\pi} f(t) \cos nt\, dt; \quad b_n = \frac{1}{\pi} \int_{-\pi}^{\pi} f(t) \sin nt\, dt,$$
$$\qquad (5)$$

and the Fourier series is then

$$f(t) = \tfrac{1}{2}a_0 + (a_1 \cos t + \dots + a_n \cos nt + \dots)$$
$$+ (b_1 \sin t + \dots + b_n \sin nt + \dots). \qquad (6)$$

15·15. Conditions for a Fourier series to be valid

The calculations in § 15·14 do not *prove* that $f(t)$ is completely represented by the Fourier series obtained. To prove this the series obtained would have to be proved convergent and its sum equal to $f(t)$.

It can be proved that this is true, provided that $f(t)$ satisfies the condition:

$f(t)$ is continuous from $-T$ to $+T$ with the possible exception of a *finite* number of *finite* discontinuities (see § 1·71).

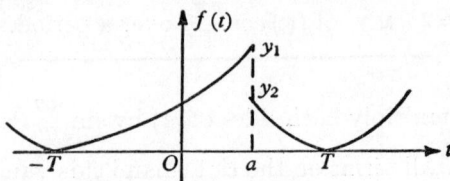

Fig. 101

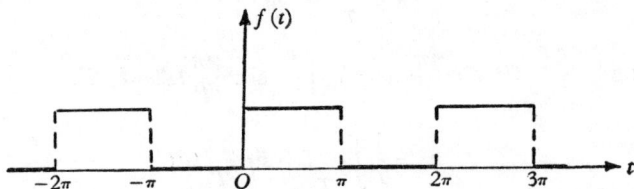

Fig. 102

At a point where $f(t)$ has a finite discontinuity the sum of the Fourier series is found to be the average of the two 'alternative' values of $f(t)$ at that point.

In fig. 101, at the point of discontinuity $t = a$, the sum of the Fourier series for $f(t)$ would give $\frac{1}{2}(y_1 + y_2)$.

EXAMPLES

(1) Find a Fourier series for the $f(t)$ defined as follows:

$$f(t) = 1 \quad \text{when} \quad 0 < t < \pi,$$
$$f(t) = 0 \quad \text{when} \quad \pi < t < 2\pi,$$

and $f(t)$ is periodic outside this range, with period 2π.

A sketch of the function is given in fig. 102. It is an example of what is commonly called a 'square-wave' function.

As the period is 2π the simplified formulae of § 15·14 (6) may be used. Integration is taken over the range 0 to 2π; but since

$f(t)$ is given by two different expressions in the two halves of the range, the integrals are taken in two stages, from 0 to π and π to 2π.

$$a_0 = \frac{1}{\pi}\left[\int_0^\pi 1.dt + \int_\pi^{2\pi} 0.dt\right] = \frac{1}{\pi}\int_0^\pi dt = 1;$$

$$a_n = \frac{1}{\pi}\left[\int_0^\pi 1.\cos nt\,dt + \int_\pi^{2\pi} 0.\cos nt\,dt\right] = \frac{1}{\pi}\int_0^\pi \cos nt\,dt$$

$$= \frac{1}{\pi}\left[\frac{1}{n}\sin nt\right]_0^\pi$$

$$= \frac{1}{n\pi}[\sin n\pi]$$

$$= 0 \quad \text{as} \quad n \neq 0;$$

$$b_n = \frac{1}{\pi}\left[\int_0^\pi 1.\sin nt\,dt + \int_\pi^{2\pi} 0.\sin nt\,dt\right] = \frac{1}{\pi}\int_0^\pi \sin nt\,dt$$

$$= \frac{1}{n\pi}[-\cos nt]_0^\pi = \frac{1}{n\pi}(1-\cos n\pi).$$

n even: $\qquad\qquad b_n = 0 \qquad (\cos n\pi = 1).$

n odd: $\qquad\qquad b_n = \dfrac{2}{n\pi} \qquad (\cos n\pi = -1).$

Putting in the values of a_0, a_n, b_n, in the Fourier series

$$f(t) = \frac{1}{2} + \frac{2}{\pi}(\sin t + \tfrac{1}{3}\sin 3t + \tfrac{1}{5}\sin 5t + \ldots).$$

Note. $t = r\pi$ gives points of discontinuity.
Here $\sin r\pi = 0$, and $f(t) = \frac{1}{2}$.
This is the average value of the two values 0 and 1 for the function at such points.

(2) Find a Fourier series for $f(t)$ of period $2T$ such that:

$$f(t) = 0, \qquad\qquad -T \leqslant t \leqslant 0$$

$$f(t) = \sin \pi t/T, \quad 0 \leqslant t \leqslant T.$$

A sketch of the function is shown in fig. 103. It is often called the half-wave rectified sine wave:

$$a_0 = \frac{1}{T}\left[\int_{-T}^0 0.dt + \int_0^T \sin\frac{\pi t}{T}dt\right] = -\frac{1}{T}\frac{T}{\pi}\left[\cos\frac{\pi t}{T}\right]_0^T = \frac{2}{\pi},$$

$$a_n = \frac{1}{T} \left[\int_{-T}^{0} 0 \cdot \cos \frac{n\pi t}{T} \, dt + \int_{0}^{T} \sin \frac{\pi t}{T} \cos \frac{n\pi t}{T} \right] dt \qquad (1)$$

$$= \frac{1}{2T} \int_{0}^{T} \left\{ \sin \frac{(n+1)\pi t}{T} - \sin \frac{(n-1)\pi t}{T} \right\} dt$$

$$= \frac{1}{2T} \frac{T}{\pi} \left[\frac{1}{(n-1)} \cos \frac{(n-1)\pi t}{T} - \frac{1}{(n+1)} \cos \frac{(n+1)\pi t}{T} \right]_{0}^{T}$$

$$= \frac{1}{2\pi} \left[\frac{\cos(n-1)\pi}{(n-1)} - \frac{1}{(n-1)} - \frac{1}{(n+1)} \cos(n+1)\pi + \frac{1}{(n+1)} \right].$$

If $n = 1$ this result breaks down as $n - 1 = 0$ occurs in the denominator.

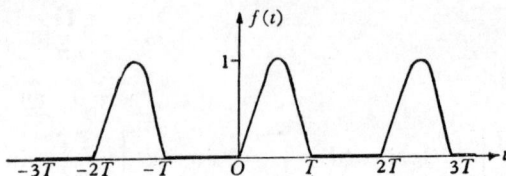

Fig. 103

(i) $n > 1$:

n odd:

$$a_n = \frac{1}{2\pi} \left[\frac{1}{(n-1)} - \frac{1}{(n-1)} - \frac{1}{(n+1)} + \frac{1}{(n+1)} \right] = 0.$$

$$(\cos(n-1)\pi = \cos(n+1)\pi = 1),$$

n even:

$$a_n = \frac{1}{2\pi} \left[-\frac{1}{(n-1)} - \frac{1}{(n-1)} + \frac{1}{(n+1)} + \frac{1}{(n+1)} \right]$$

$$(\cos(n-1)\pi = \cos(n+1)\pi = -1),$$

$$a_n = -\frac{2}{\pi(n^2-1)}.$$

(ii) $n = 1$: Going back to (1) above:

$$a_1 = \frac{1}{T} \int_{0}^{T} \sin \frac{\pi t}{T} \cos \frac{\pi t}{T} \, dt = \frac{1}{2T} \int_{0}^{T} \sin \frac{2\pi t}{T} \, dt$$

$$= 0.$$

Collecting up: $a_n = 0$, n odd; $a_n = -\dfrac{2}{\pi(n^2-1)}$, n even.

$$b_n = \frac{1}{T} \left[\int_{-T}^{0} 0 \sin \frac{n\pi t}{T} \, dt + \int_{0}^{T} \sin \frac{\pi t}{T} \sin \frac{n\pi t}{T} \, dt \right]$$

$$= \frac{1}{2T} \int_{0}^{T} \left\{ \cos \frac{(n-1)\pi t}{T} - \cos \frac{(n+1)\pi t}{T} \right\} dt.$$

If $n = 1$:

$$b_1 = \frac{1}{2T} \int_0^T \left(1 - \cos\frac{2\pi t}{T}\right) dt = \frac{1}{2T}\left[t - \frac{T}{2\pi}\sin\frac{2\pi t}{T}\right]_0^T,$$

$$b_1 = \tfrac{1}{2}.$$

If $n > 1$:

$$b_n = \frac{1}{2T}\left[\frac{T}{\pi(n-1)}\sin\frac{(n-1)\pi t}{T} - \frac{T}{\pi(n+1)}\sin\frac{(n+1)\pi t}{T}\right]_0^T$$

$$= 0 \quad (\text{as } \sin r\pi = 0).$$

Collecting up: $\qquad b_1 = \tfrac{1}{2},$

$$b_n = 0 \quad (n > 1).$$

Thus

$$f(t) = \frac{1}{\pi} + \tfrac{1}{2}\sin\frac{\pi t}{T} - \frac{2}{\pi}\left(\tfrac{1}{3}\cos\frac{2\pi t}{T} + \tfrac{1}{15}\cos\frac{4\pi t}{T} + \ldots\right.$$

$$\left. + \frac{1}{(4r^2-1)}\cos\frac{2r\pi t}{T} + \ldots\right).$$

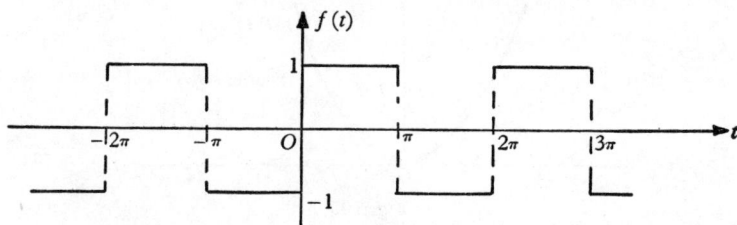

Fig. 104

Exercise 56

1. Find a Fourier series to represent $f(t)$, of period $2T$, which is such that $f(t) = t$ when $-T < t < T$.

Sketch a graph of the function.

(This is called the 'saw-tooth' wave.)

2. Find a Fourier series for the type of square-wave shown in fig. 104.

3. Find a Fourier series to represent $f(t)$, of period 2π, which is such that $f(t) = t$ for $0 < t < \pi$ and $f(t) = -t$ for $-\pi < t < 0$. Sketch a graph of the function.

(This is called a 'triangular' wave.)

ANSWERS

1. $\dfrac{2T}{\pi}\left(\sin\dfrac{\pi t}{T}-\tfrac{1}{2}\sin\dfrac{2\pi t}{T}+\tfrac{1}{3}\sin\dfrac{3\pi t}{T}-\ldots\right).$

2. $\dfrac{4}{\pi}\left[\sin t+\tfrac{1}{3}\sin 3t+\tfrac{1}{5}\sin 5t+\ldots+\dfrac{1}{(2r+1)}\sin(2r+1)t+\ldots\right].$

3. $\dfrac{\pi}{2}-\dfrac{4}{\pi}\left[\dfrac{1}{1^2}\cos t+\dfrac{1}{3^2}\cos 3t+\ldots+\dfrac{1}{(2r+1)^2}\cos(2r+1)t+\ldots\right].$

15·16. Expansion of non-periodic functions over limited ranges

If $f(t)$ is *not* periodic, a Fourier series to represent it over any *limited* range, say from $-T$ to $+T$, may be found.

In the calculation of the coefficients a_0, a_n, b_n, it will be noticed that the periodicity of the *function* was never used (see § 15·14). Thus no new formulae are required.

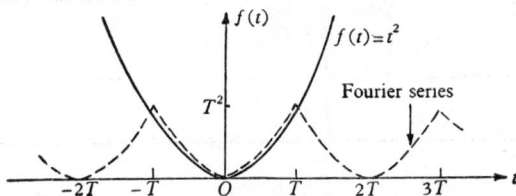

Fig. 105

It is emphasized, however, that the series obtained will *not* represent the given function *outside* the range calculated ($-T$ to $+T$).

Take, for example, $f(t)=t^2$.

A series can be found to represent t^2 within the range $-T$ to $+T$ by the means explained in the previous subsections. Outside this range, the *series* will be periodic, but t^2, of course, is not.

Sketches of the graphs of t^2 and the series are shown in fig. 105. The graph of the series is shown in broken line.

The Fourier series is obtained as follows:

$$f(t)=t^2, \quad -T\leqslant t\leqslant T.$$

$$a_0=\frac{1}{T}\int_{-T}^{T}t^2\,dt=\tfrac{2}{3}T^2.$$

$$a_n = \frac{1}{T} \int_{-T}^{T} t^2 \cos \frac{n\pi t}{T} \, dt$$

$$= \frac{1}{T} \left[\frac{T}{n\pi} t^2 \sin \frac{n\pi t}{T} \, dt \right]_{-T}^{T} - \frac{1}{T} \int_{-T}^{T} 2t \frac{T}{n\pi} \sin \frac{n\pi t}{T} \, dt$$

$$= 0 - \frac{2}{n\pi} \int_{-T}^{T} t \sin \frac{n\pi t}{T} \, dt$$

$$= \frac{2}{n\pi} \frac{T}{n\pi} \left[t \cos \frac{n\pi t}{T} \right]_{-T}^{T} - \frac{2}{n\pi} \frac{T}{n\pi} \int_{-T}^{T} 1 \cos \frac{n\pi t}{T} \, dt$$

$$= \frac{2T}{n^2\pi^2} [T \cos n\pi + T \cos(-n\pi)] - \frac{2T}{n^2\pi^2} \left[\frac{T}{n\pi} \sin \frac{n\pi t}{T} \right]_{-T}^{T}$$

$$= \frac{2T}{n^2\pi^2} 2T \cos n\pi - 0 \quad (\sin r\pi = 0)$$

$$= \frac{4T^2}{n^2\pi^2} \cos n\pi.$$

n even: $\quad a_n = \dfrac{4T^2}{n^2\pi^2} \quad$ or $\quad a_{2r} = \dfrac{4T^2}{4r^2\pi^2}.$

n odd: $\quad a_n = -\dfrac{4T^2}{n^2\pi^2} \quad$ or $\quad a_{2r+1} = -\dfrac{4T^2}{(2r+1)^2 \pi^2}.$

$$b_n = \frac{1}{T} \int_{-T}^{T} t^2 \sin \frac{n\pi t}{T} \, dt$$

$$= -\frac{1}{T} \left[t^2 \frac{T}{n\pi} \cos \frac{n\pi t}{T} \right]_{-T}^{T} + \frac{1}{T} \int_{-T}^{T} 2t \frac{T}{n\pi} \cos \frac{n\pi t}{T} \, dt$$

$$= 0 + \frac{2}{n\pi} \int_{-T}^{T} t \cos \frac{n\pi t}{T} \, dt$$

$$= \frac{2}{n\pi} \left[t \frac{T}{n\pi} \sin \frac{n\pi t}{T} \right]_{-T}^{T} - \frac{2}{n\pi} \int_{-T}^{T} 1 \frac{T}{n\pi} \sin \frac{n\pi t}{T} \, dt$$

$$= 0 - \frac{2T}{n^2\pi^2} \left[-\frac{T}{n\pi} \cos \frac{n\pi t}{T} \right]_{-T}^{T}$$

$$= \frac{2T^2}{n^3\pi^3} [\cos n\pi - \cos(-n\pi)] = 0 \quad \text{for all } n.$$

Thus, for $-T \leqslant t \leqslant T$,

$$t^2 = \frac{T^2}{3} - \frac{4T^2}{\pi^2} \left(\frac{1}{1^2} \cos \frac{\pi t}{T} - \frac{1}{2^2} \cos \frac{2\pi t}{T} + \frac{1}{3^2} \cos \frac{3\pi t}{T} - \dots \right).$$

EXAMPLE

Assuming that $f(x)$ can be expanded as a series in the form

$$a_1 \sin x + a_2 \sin 2x + \dots + a_n \sin nx + \dots$$

for values of x between 0 and π, find a_n.

If $f(x) = x$ from $x = 0$ to $\frac{1}{2}\pi$ and $f(x) = \frac{1}{2}\pi$ from $x = \frac{1}{2}\pi$ to π, evaluate a_n and write down the first four terms of the series.

[Sec. A]

$$f(x) = a_1 \sin x + a_2 \sin 2x + \ldots + a_n \sin nx + \ldots \tag{1}$$

Integrating both sides between the limits 0 and π, after first multiplying by $\sin nx$,

$$\int_0^\pi f(x) \sin nx\, dx = \int_0^\pi a_1 \sin x \sin nx\, dx + \int_0^\pi a_2 \sin 2x \sin nx\, dx + \ldots$$

$$+ \int_0^\pi a_n \sin^2 nx\, dx + \ldots \tag{2}$$

Now
$$\int_0^\pi \sin mx \sin nx\, dx \quad (m \neq n)$$

$$= \frac{1}{2} \int_0^\pi \{\cos(m-n)x - \cos(m+n)x\}\, dx$$

$$= \frac{1}{2} \left[\frac{1}{(m-n)} \sin(m-n)x - \frac{1}{(m+n)} \sin(m+n)x \right]_0^\pi$$

$$= 0 \quad \text{as} \quad \sin r\pi = 0, \quad r \text{ any integer.}$$

Also
$$\int_0^\pi \sin^2 nx\, dx = \frac{1}{2} \int_0^\pi (1 - \cos 2nx)\, dx$$

$$= \frac{1}{2} \left[x - \frac{1}{2n} \sin 2nx \right]_0^\pi = \frac{\pi}{2}.$$

Thus
$$\int_0^\pi f(x) \sin nx\, dx = a_n \frac{\pi}{2}.$$

$$a_n = \frac{2}{\pi} \int_0^\pi f(x) \sin nx\, dx. \tag{3}$$

If
$$f(x) = x, \quad 0 \leqslant x \leqslant \tfrac{1}{2}\pi,$$
$$f(x) = \tfrac{1}{2}\pi, \quad \tfrac{1}{2}\pi \leqslant x \leqslant \pi,$$

then

$$a_n = \frac{2}{\pi} \left[\int_0^{\frac{1}{2}\pi} x \sin nx\, dx + \int_{\frac{1}{2}\pi}^\pi \tfrac{1}{2}\pi \sin nx\, dx \right]$$

$$= \frac{2}{\pi} \left[-\frac{x}{n} \cos nx \right]_0^{\frac{1}{2}\pi} + \frac{2}{\pi} \int_0^{\frac{1}{2}\pi} \frac{1}{n} \cos nx\, dx + \left[-\frac{1}{n} \cos nx \right]_{\frac{1}{2}\pi}^\pi$$

$$= -\frac{2}{\pi n} \left[\frac{\pi}{2} \cos \frac{n\pi}{2} - 0 \right] + \frac{2}{\pi n^2} [\sin nx]_0^{\frac{1}{2}\pi} - \frac{1}{n} [\cos n\pi - \cos \tfrac{1}{2} n\pi]$$

$$= -\frac{1}{n} \cos \frac{n\pi}{2} + \frac{2}{\pi n^2} \sin \frac{n\pi}{2} - \frac{1}{n} \cos n\pi + \frac{1}{n} \cos \frac{n\pi}{2},$$

$$a_n = \frac{2}{\pi n^2} \sin \frac{n\pi}{2} - \frac{1}{n} \cos n\pi. \tag{4}$$

n *even*:
$$a_{2r} = \frac{2}{\pi \cdot 4r^2} \sin r\pi - \frac{1}{2r} \cos 2r\pi$$

$$= 0 - \frac{1}{2r},$$

$$a_{2r} = -\frac{1}{2r}. \tag{5}$$

n *odd*:

$$a_{4r-1} = \frac{2}{\pi(4r-1)^2} \sin(4r-1)\tfrac{1}{2}\pi - \frac{1}{(4r-1)} \cos(4r-1)\pi$$

$$= \frac{2}{\pi(4r-1)^2} \sin(2r\pi - \tfrac{1}{2}\pi) - \frac{1}{(4r-1)} \cos(4r\pi - \pi)$$

$$= \frac{2}{\pi(4r-1)^2} \sin(-\tfrac{1}{2}\pi) - \frac{1}{(4r-1)} \cos(-\pi)$$

$$= -\frac{2}{\pi(4r-1)^2} + \frac{1}{(4r-1)}. \tag{6}$$

$$a_{4r+1} = \frac{2}{\pi(4r+1)^2} \sin(2r\pi + \tfrac{1}{2}\pi) - \frac{1}{(4r+1)} \cos(4r\pi + \pi)$$

$$= \frac{2}{\pi(4r+1)^2} \sin\tfrac{1}{2}\pi - \frac{1}{(4r+1)} \cos\pi$$

$$= \frac{2}{\pi(4r+1)^2} + \frac{1}{(4r+1)}. \tag{7}$$

Putting the first four terms into (1):

$$f(x) = \left(\frac{2}{\pi} + 1\right) \sin x - \tfrac{1}{2} \sin 2x - \left(\frac{2}{9\pi} - \frac{1}{3}\right) \sin 3x - \tfrac{1}{4}\sin 4x + \ldots.$$

EXERCISE 57

1. A function of x is given in the range $-\pi$ to π by:
$$f(x) = -x^2, \quad -\pi < x \leqslant 0,$$
$$f(x) = x^2, \qquad 0 \leqslant x < \pi.$$
Express $f(x)$ as a Fourier series, within this range.

2. Explain why, in Exercise 56, QQ. 1 and 2, you could have assumed the series to have no cosine terms; and in Q. 3 to have no sine terms.

3. $f(t)$ is defined by
$$f(t) = 1, \quad -\tfrac{1}{2}\pi < t < \tfrac{1}{2}\pi,$$
$$f(t) = -1 \quad \text{when} \quad -\pi < t < -\tfrac{1}{2}\pi \text{ and } \tfrac{1}{2}\pi < t < \pi.$$
Sketch the graph of the function within the range $-\pi$ to π.
  Is the function even or odd?

Find a Fourier series for $f(t)$ within this range, explaining why you may omit the calculation of the 'sine' coefficients.

ANSWERS

1. $f(x) = \left(2\pi - \dfrac{8}{\pi}\right) \sin x - \dfrac{2\pi}{2} \sin 2x +$

$$\left(\dfrac{2\pi}{3} - \dfrac{8}{\pi \cdot 3^3}\right) \sin 3x - \dfrac{2\pi}{4} \sin 4x + \dots$$

n odd: Coefficient of $\sin nx$ is $\dfrac{2\pi}{n} - \dfrac{8}{\pi n^3}$.

n even: Coefficient of $\sin nx$ is $-\dfrac{2\pi}{n}$.

2. See § 15·12.

3. Even function

$$f(t) = \frac{4}{\pi} \left[\cos t - \tfrac{1}{3} \cos 3t + \tfrac{1}{5} \cos 5t - \dots + \frac{(-1)^r}{(2r+1)} \cos (2r+1)t + \dots \right].$$

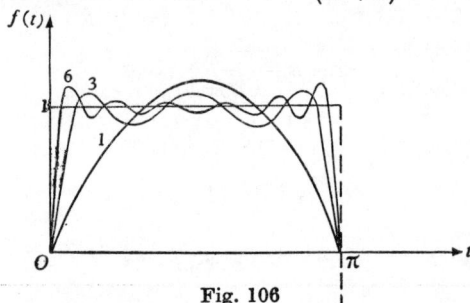

Fig. 106

15·2. Approximations by successive harmonics

The terms in $\sin \pi t / T$, $\cos \pi t / T$, in a Fourier series, are called the 'fundamental' terms.

The rest are called 'harmonics'.

Thus the term in $\sin 3\pi t / T$ is called a *third* harmonic.

It will be noticed, from the exercises already done, that the coefficients of the higher harmonics rapidly become smaller.

In practice a sufficiently good approximation to a given function can be found by taking the first few terms. Rarely is it necessary to take terms beyond the sixth harmonic.

In the case of the square-wave given as Q. 2 in Exercise 56, fig. 106 shows a sketch of the original square-wave and the Fourier series representing it, as far as the first, third and sixth terms.

15·3. Fourier series in engineering problems

15·31. Electric circuit theory

A general periodic voltage is rarely purely sinusoidal in character. The voltage, of period $2\pi/\omega$ say, is expressed as a Fourier series:

$$V = \tfrac{1}{2}a_0 + (a_1 \cos \omega t + \dots + a_n \cos n\omega t + \dots)$$
$$+ (b_1 \sin \omega t + \dots + b_n \sin n\omega t + \dots)$$
$$= \tfrac{1}{2}a_0 + A_1 \sin (\omega t + \alpha_1) + \dots + A_n \sin (n\omega t + \alpha_n) + \dots,$$

where A_n α_n, etc., are known constants.

Only the first three or four harmonics need usually be taken.

Each term $\tfrac{1}{2}a_0$, $A_1 \sin (\omega t + \alpha_1)$, etc., is then taken in turn and treated separately, as far as the electric circuit is concerned.

15·32. Mechanical problems

Fourier series play an important part in the theory of many mechanisms. For example, in the slider-crank mechanism of reciprocating engines, the connecting rod moves in approximate S.H.M. only when the rod is very long. For short rods, the motion is most conveniently expressed by a Fourier series.

Fourier series are also used in advanced examples on the bending of beams.

15·33. 'Wave-form' problems

Types of Fourier series play an important part in the solution of partial differential equations which occur in the study of heat conduction, electric waves in a conductor, and many more 'wave-form' problems.

15·4. Tabulatory method for a Fourier series

When the function is given by a graph or table of values, obviously the method of integration cannot be used to calculate the coefficients.

Approximate mean values are calculated for a series of equidistant ordinates.

17

In this connexion (see § 15·14) note that

$$a_0 = 2 \times \text{M.V. of } f(t) \text{ over the range.}$$

$$a_n = 2 \times \text{M.V. of } f(t) \cos \frac{n\pi t}{T}.$$

$$b_n = 2 \times \text{M.V. of } f(t) \sin \frac{n\pi t}{T}.$$

As explained in § 15·2 values beyond the fifth or sixth harmonics are rarely calculated.

An example should be sufficient to show the method clearly.

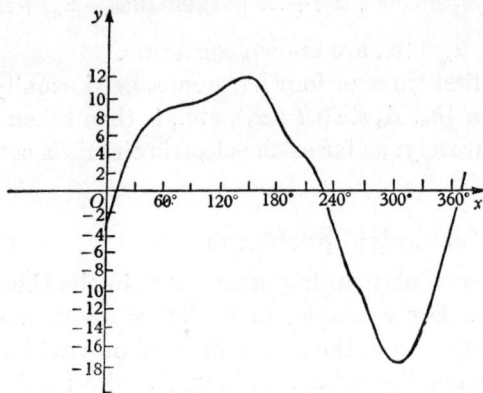

Fig. 107

EXAMPLE

Find a Fourier series, as far as the third harmonic, to represent the periodic function y, given by the values in the following table:

x	0°	30°	60°	90°	120°	150°	180°
y	$-3\cdot5$	$5\cdot2$	$8\cdot5$	9	$10\cdot5$	12	8

x	210°	240°	270°	300°	330°	360°
y	$3\cdot5$	-6	-12	-18	$-14\cdot5$	$-3\cdot5$

A sketch of the function is shown in fig. 107. If a graph only were available a table of values could be made up from it, taking equidistant ordinates.

Period $(2T)$ is 2π and x replaces t in this exercise.

As the period is 2π, $n\pi x/T$ becomes nx, e.g. $a_n = 2$ M.V. of $f(x) \cos nx$ over the range 0 to $2\pi = 2$ M.V. of $y \cos nx$.

A table is now drawn up as shown below:

x	y	$\cos x$	$y \cos x$	$\sin x$	$y \sin x$	$\cos 2x$	$y \cos 2x$	$\sin 2x$	$y \sin 2x$	$\cos 3x$	$y \cos 3x$	$\sin 3x$	$y \sin 3x$
0°	−3·5	1	−3·5	0	0	1	−3·5	0	0	1	−3·5	0	0
30°	5·2	0·866	4·503	0·5	2·6	0·5	2·6	0·866	4·503	0	0	1	5·2
60°	8·5	0·5	4·25	0·866	7·361	−0·5	−4·25	0·866	7·361	−1	−8·5	0	0
90°	9	0	0	1	9	−1	−9	0	0	0	0	−1	−9
120°	10·5	−0·5	−5·25	0·866	9·093	−0·5	−5·25	−0·866	−9·093	1	10·5	0	0
150°	12	−0·866	−10·392	0·5	6	0·5	6	−0·866	−10·392	0	0	1	12
180°	8	−1	−8	0	0	1	8	0	0	−1	−8	0	0
210°	3·5	−0·866	−3·031	−0·5	−1·75	0·5	2·75	0·866	3·031	0	0	−1	−3·5
240°	−6	−0·5	3	−0·866	5·196	−0·5	3	0·866	−5·196	1	−6	0	0
270°	−12	0	0	−1	12	−1	12	0	0	0	0	1	−12
300°	−18	0·5	−9	−0·866	15·588	−0·5	9	−0·866	15·588	−1	18	0	0
330°	−14·5	0·866	−12·557	−0·5	7·25	0·5	−7·25	−0·866	12·577	0	0	−1	14·5
Sum	+2·7		−39·977		72·338		14·1		18·359		2·5		7·2
M.V.	$\frac{2·7}{12}=0·225$		−3·331		6·03		1·175		1·53		0·21		0·6
2M.V.	$a_0 = 0·45$		$a_1 \simeq -6·66$		$b_1 \simeq 12·06$		$a_2 \simeq 2·35$		$b_2 \simeq 3·06$		$a_3 \simeq 0·42$		$b_3 \simeq 1·2$

As far as the third harmonic:

$$y = 0·225 - 6·66 \cos x + 12·06 \sin x + 2·35 \cos 2x + 3·06 \sin 2x + 0·42 \cos 3x + 1·2 \sin 3x.$$

Comments

(a) It will be noticed that the work becomes much easier for the higher harmonics (cf. table of values of $\cos x$ with those of $\cos 3x$). Also, several of the products required occur more than once, thus making the arithmetic appear longer than it is (e.g. $10 \cdot 5 \times 0 \cdot 866$ occurs twice).

(b) If required, the solution can be given as a sum of sines (or cosines) only, e.g. $12 \cdot 06 \sin x - 6 \cdot 66 \cos x$ can be put in the form $R \sin (x - \alpha)$.

(c) If it is obvious from the tables, or a graph, that the function is odd or even, the work is much shortened as cosine terms or sine terms need not be calculated.

EXERCISE 58

1. The value of y, a periodic function of θ, for the twelve values of θ covering a period is as follows:

θ	0°	30°	60°	90°	120°	150°
y	2·34	3·012	3·685	4·149	3·685	2·203

θ	180°	210°	240°	270°	300°	330°
y	0·825	0·573	0·875	1·085	1·189	1·637

Express y as a Fourier series as far as the terms of the second harmonic.

2. The value of a certain e.m.f. (E) at different points in a cycle are as follows:

θ	0°	30°	60°	90°	120°	150°	180°
E	70	866	1293	1400	1307	814	− 70

θ	210°	240°	270°	300°	330°	360°
E	− 866	− 1293	− 1400	− 1307	− 814	70

Sketch a graph of E against θ. Notice that when θ increases by 180° the value of E is numerically unchanged. Deduce that only odd harmonics are present. Find the Fourier series for E as far as the third harmonic.

3. The displacement, y, of a point on a certain portion of a pulley-machine is given by the following table; θ being the angle turned through (in degrees) by the pulley. Find a Fourier series to represent y, as far as the term of the third harmonic.

θ	0°	30°	60°	90°	120°	150°	180°
y	7·2	6·7	5	4·8	4·5	3·0	1·6

θ	210°	240°	270°	300°	330°	360°
y	3·0	4·5	4·8	5	6·7	7·2

ANSWERS

1. $y \simeq 2\cdot105 + 0\cdot55 \cos\theta - 0\cdot512 \cos 2\theta + 1\cdot527 \sin\theta$
$$- 0\cdot0821 \sin 2\theta.$$

2. $E \simeq 1500 \sin\theta + 93 \sin 3\theta + 36 \cos\theta + 28 \cos 3\theta.$

3. $y \simeq 4\cdot73 + 2\cdot08 \cos\theta - 0\cdot1 \cos 2\theta + 0\cdot77 \cos 3\theta.$

APPENDIX A

CURVE TRACING

Various curves have already been met with in the text and drawn in detail.

The quick sketching of curves is largely a matter of experience. The more important points are given below.

A.1. Fundamental points in the case of Cartesian equations

A.11. Symmetry in x and/or y

If no odd powers of x occur in the equation of the curve, then changing the sign of x leaves y values unaltered. Thus the curve must be symmetrical about the y-axis.

Similarly, if no odd powers of y occur, the curve is symmetrical about the x-axis, e.g. $x^2(y^2-2)=x^4+4$ is a curve symmetrical about both axes. In this case the curve need only be plotted in the first quadrant, and the rest filled in from symmetry.

A.12. Asymptotes parallel to the axes

Asymptotes are straight lines which approach the curve at infinity. They may be thought of as tangents whose points of contact are at infinity.

Some asymptotes are obvious at a first inspection of the equation. Thus, if $y=\dfrac{x^2-2}{x-1}$, as $x \to 1$, $y \to \infty$.

The line $x=1$ is therefore an asmptote.

A general rule for finding asymptotes parallel to the axes is:

Transform the equation to a form without fractions. Then, equating to zero the coefficient of the highest power of x gives the asymptotes (if any) parallel to the x-axis. Equating to zero the coefficient of the highest power of y gives the asymptotes (if any) parallel to the y-axis.

For example, if
$$y^2=\frac{x^3(3-2y)}{(x-1)},$$

then
$$(x-1)y^2-x^3(3-2y)=0.$$

The coefficient of the highest power of y is $(x-1)$. Thus $x=1$ is an asymptote parallel to the y-axis.

Similarly, $y=\frac{3}{2}$ is an asymptote parallel to the x-axis.

A. 13. Important standard points on the curve

These include:

(i) Value(s) of y when $x=0$ and vice versa. That is, the points (if any) where the curve cuts the co-ordinate axes.

(ii) Maximum, minimum and points of inflexion. These, if they exist, are found by the normal means.

In this connexion it is useful to remember that around points where d^2y/dx^2 is negative the curve is concave downwards, and where positive, concave upwards.

As dy/dx will also be found when testing for maximum and minimum, it is useful to find any points where the tangent is parallel to the y-axis $(dy/dx=\infty)$, and also the slopes of the curve at the points found in (i) above.

A. 14. Values of y when x is very small or very large

For small values of x, that is, close to the origin, the higher powers of x can be neglected. This often gives a good idea of the shape of the graph near the origin, e.g. if

$$y = 3x^3 - 2x.$$

Near the origin, neglecting the term in x^3,

$$y \simeq - 2x.$$

That is, close to the origin the graph has the approximate shape of the straight line $y = - 2x$.

For large values of x, lower powers of x may be neglected.

In the above example, when x is large $y \simeq 3x^3$; that is, the curve approximates to the shape of $y = 3x^3$ for large x values.

In this connexion it is useful to memorize the shapes of the graphs of $y = \pm ax^2$, $y = \pm ax^3$ ('a' a positive constant). These are sketched in fig. 108.

A. 15. Change of origin

Occasionally the equation of a curve can be much simplified by transferring the origin to another point. An example will serve to show the method.

Consider the equation $3(x-1)^4 = 4(y-2)^3$.

Let O' be the point $(1, 2)$ referrred to the original axes (fig. 109). Take O' as a new origin and $O'X$, $O'Y$ as new axes.

Let P be any point, having co-ordinates (x, y) on the old axes and (X, Y) on the new axes.

Then
$$x = NP + 1 = X + 1$$
and
$$y = MP + 2 = Y + 2.$$

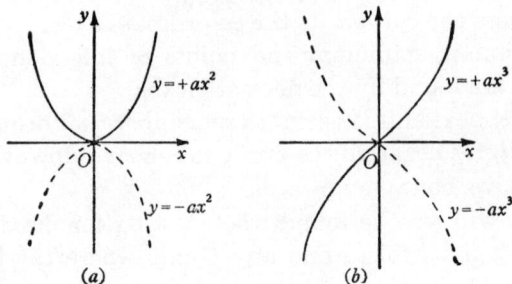

Fig. 108

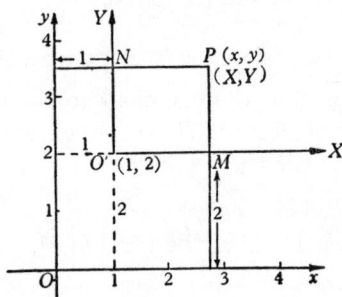

Fig. 109

Thus, with new axes $O'X$, $O'Y$ the original equation becomes
$$3\{X + 1 - 1\}^4 = 4\{Y + 2 - 2\}^3,$$
giving
$$3X^4 = 4Y^3.$$

In general, if the origin is moved to a point (a, b) the new equation is found by substituting $(X + a)$ for the original x and $(Y + b)$ for the original y.

A. 16. Summary

It is obvious that all the points given in A. 1 do not apply to *all* curves. Individual equations are treated on their merits. If a rough sketch only is required the points given in A.11, A. 12, A. 13 (i) and A. 14 are normally the most useful.

EXAMPLES

(1) Sketch roughly the curve $y = 3x^3 - 12x$.

[A. 11] No symmetry about axes; although as only odd powers of x and y occur the curve is symmetrical about the origin.

[A. 12] No asymptotes.

[A. 13] When $x = 0$, $y = 0$.

As $y = 3x(x^2 - 4)$, $y = 0$ when $x = 0$, ± 2.

[A. 14] As $x \to 0$, $\qquad y \to -12x$.

As $x \to \infty$, $\qquad y \to 3x^3$ (see fig. 108).

The points, and parts of the line of the curve which can be deduced from the above, are marked on fig. 110; points by $+$, and line of curve in full line. The curve can be completed, as shown in dotted line. If confirmation is desired, and more accuracy, the maximum and mini-mum points could be found.

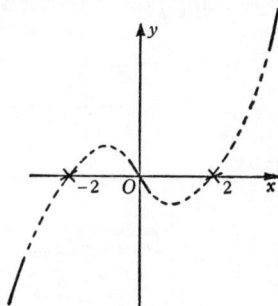

(2) Sketch the curve of

$$x^2 y^2 = 4(x^2 - y^2).$$

[A. 11] The curve is symmetrical about both axes.

Fig. 110

Thus only that part of the curve in the first quadrant (x and y both positive) need be dealt with. The other parts will be sketched from symmetry.

[A. 12] The equation can be written

$$x^2(4 - y^2) - 4y^2 = 0$$

or $\qquad y^2(x^2 + 4) - 4x^2 = 0.$

The coefficient of x^2 is $(4 - y^2)$. Thus $y = \pm 2$ are asymptotes parallel to the x-axis.

The coefficient of y^2 is $(x^2 + 4)$. Now $x^2 + 4 \neq 0$ for real x values; thus there are no asymptotes parallel to the y-axis.

[A. 13] When $x = 0$, $y = 0$.

Solving for y^2

$$y^2 = \frac{4x^2}{x^2 + 4}.$$

Thus real values of y exist for all real x values. Solving for x^2

$$x^2 = \frac{4y^2}{(4 - y^2)}.$$

Thus real x values only exist if $y^2 < 4$.

The curve is therefore completely contained between the two lines $y = \pm 2$ (asymptotes, see [A. 12] above).

The curve is shown in fig. 111. If more accuracy were required the co-ordinates of a few detailed points could be calculated.

A. 2. Curves given by parametric co-ordinates

These are of the form $x = f(t)$, $y = g(t)$, where $f(t)$, $g(t)$ are functions of a variable t.

No hard and fast rules can be laid down here. Tables of values of x and y for various values of t can be drawn up and the corresponding points plotted. This is not always necessary. A detailed

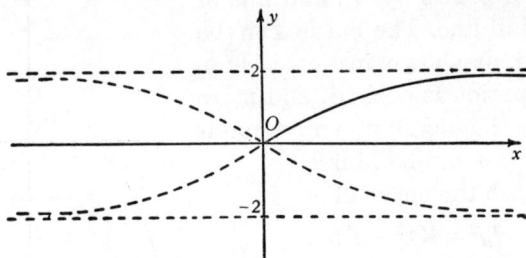

Fig. 111

example of such a curve has already been dealt with in the text (see § 6·13, Example 5).

In simple cases the Cartesian equation can be easily obtained. In such cases it may be easier to work from the Cartesian equation. For example (i) If

$$x = a \cos t, \quad y = a \sin t,$$

then
$$x^2 + y^2 = a^2(\cos^2 t + \sin^2 t)$$

$$= \underline{a^2}.$$

(ii) If
$$x = 4t^2, \quad y = 2t - 3,$$

then
$$2t = y + 3.$$

$$\therefore \ 4t^2 = x = (y + 3)^2,$$

giving the Cartesian equation

$$\underline{(y + 3)^2 = x.}$$

EXERCISE 59

Sketch the following curves:

1. $y^2 = x^2(9 - x^2)$.   **2.** $y = x^2 + \dfrac{1}{x^2}$.

3. $y = \dfrac{x(x - 2)}{(x - 1)}$.   **4.** $3(x - 1)^4 = 4(y + 1)^2$.

5. $y(4 + x^2) = 2x^3$.   **6.** $16x^2 + 9y^2 = 144$.

7. $x = 4 - 2t^2$, $y = 2t$.   **8.** $x = 2\cos t$, $y = 3\sin t$.

9. $y^2 = (x - 2)(4 - x)$.   **10.** $y = x\sin x$.

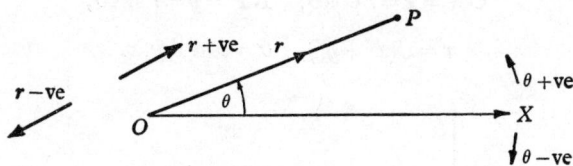

Fig. 112

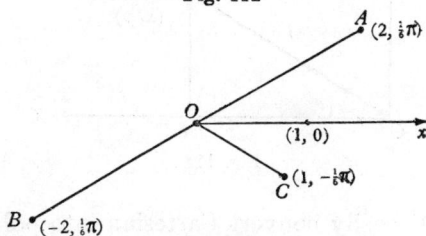

Fig. 113

A. 3. Polar equations

A. 31. Definitions. Sign conventions

An alternative to fixing the position of a point in a plane by means of its distances from two fixed axes (Cartesian co-ordinates method) is to fix it by its distance $OP(r)$ from a fixed point O and the angle θ which $\overrightarrow{OP}$ makes with a fixed direction $\overrightarrow{OX}$. OX is called the initial line and the point O the origin or pole. $\overrightarrow{OP}$ is often called the radius vector. (r, θ) are called the polar co-ordinates of the point (fig. 112).

θ is positive when measured anti-clockwise and negative when measured clockwise.

With θ fixed, r is positive when measured outwards, from O along the arm of the angle, and negative when measured in the opposite sense.

Fig. 113 shows examples of these sign conventions.

As alternatives A could be denoted by $(-2, -\frac{5}{6}\pi)$, C by $(-1, \frac{5}{6}\pi)$, etc.

It is customary to measure r positive and fix the angle either from 0 to π or from 0 to $-\pi$.

A. 32. Relation between Cartesian and polar co-ordinates

If OX, the initial line, is taken as x-axis and O as the origin for a set of rectangular Cartesian co-ordinates, then referring to fig. 114

$$ON = x = r\cos\theta, \quad NP = y = r\sin\theta, \tag{1}$$

$$r = \sqrt{(x^2 + y^2)}, \quad \theta = \tan^{-1} y/x. \tag{2}$$

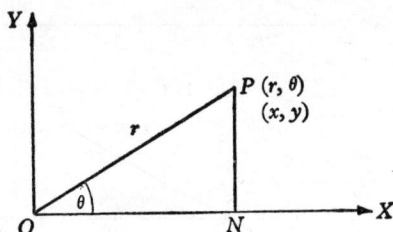

Fig. 114

Equations (1) easily convert Cartesian into polar equations, and equations (2) the reverse.

For example, the circle $x^2 + y^2 = a^2$ becomes $r = a$, $y^2 = 4ax$ becomes

$$r^2 \sin^2\theta = 4ar\cos\theta$$

or

$$r = 4a\cot\theta \, \mathrm{cosec}\,\theta.$$

A. 33. Practical applications

Many curves are more simply expressed in polar rather than Cartesian co-ordinates. For example, spirals of the form $r = a\theta$ and $r = a\,e^{c\theta}$ are important in connexion with shapes of cams and gear-wheels. Polar curves are often used in illustrating the candle-powers of a lamp at varying distances and directions. The theory of the Amsler planimeter, an instrument used in the measurement of area, demands a knowledge of polar co-ordinates. Again, r.m.s. values, and centroids of irregular areas, may be determined using polar diagrams.

A. 34. Tracing curves in polars

Below are listed points to keep in mind:

(i) If the only trigonometric ratio of θ occurring is a cosine, changes in the sign of θ produce no change in r; thus the curve will be symmetrical about the initial line, e.g. $r = 2\cos 3\theta$.

(ii) If the curve is of the form $r^2 = f(\theta)$, then $r = \pm \sqrt{\{f(\theta)\}}$, and it is symmetrical about the origin. In addition, any values of θ which make $f(\theta)$ negative do not give real values for r.

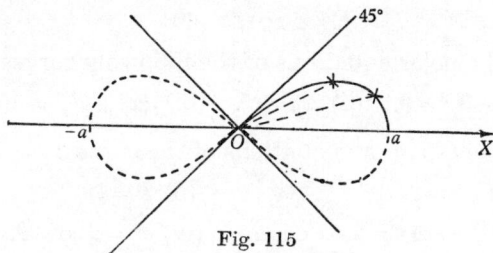

Fig. 115

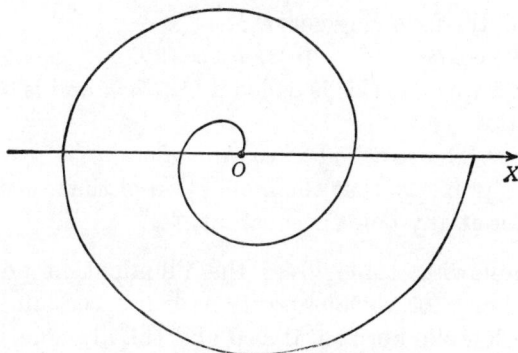

Fig. 116

As an illustration of sketching curves in polar co-ordinates two well-known curves will now be drawn.

The lemniscate. The name comes from a Latin word meaning 'ribbon'. Its equation is of the form $r^2 = a^2 \cos 2\theta$.

(a) From (i) and (ii) above it is seen that the curve is symmetrical about the origin and the initial line. It is therefore only necessary to plot it from $\theta = 0$ to $\frac{1}{2}\pi$.

(b) Within the range given above, $\cos 2\theta$ is negative from $\theta = \frac{1}{4}\pi$ to $\frac{1}{2}\pi$. Within this range, then, real values of r do not exist.

θ	0	15°	30°	45°
r	a	$0.93a$	$0.71a$	0

In fig. 115 the part of the curve derived from the table is shown in full line. The rest of the curve, shown in dotted line, is completed by symmetry.

Archimedes's spiral. This has an equation of the form $r = a\theta$. It is the path of a point moving along a straight line with constant speed, while the line itself rotates about a fixed point in itself with constant angular velocity. The curve is roughly sketched from $\theta = 0$ to 4π in fig. 116.

EXERCISE 60

1. Find the polar equations of the following curves:

(i) $x^2 + (y-3)^2 = 9$.   (ii) $xy = c^2$.   (iii) $(x^2 + y^2)^2 = 4(x^2 - y^2)$.

2. Find the Cartesian equations of the curves:

(i) $r = 2$.   (ii) $\theta = \tfrac{1}{6}\pi$.

(iii) $r\sin\theta = 2 - r\cos\theta$.   (iv) $r^2 = 9\cos 2\theta$.

3. Sketch the following curves:

(i) $r = 3 + \cos\theta$.   (ii) $r = 4\cos 2\theta$.

(iii) $r = a(1 + \cos\theta)$; this is called a *Cardioid* and is of importance in Optics.

(iv) $r = a\,e^\theta$. This is called an *equiangular spiral*.

It has the property that the inclination of the tangent to the radius vector at any point is constant.

4. The following table gives the illumination power of a certain arc lamp for various positions below the lamp. Taking the angle below the horizontal as θ and the illumination power as r, plot the polar diagram.

Take the horizontal as the initial line.

Angle below horizontal	0°	10°	20°	30°	40°
Lux	900	1250	1500	1900	1800
Angle below horizontal	50°	60°	70°	80°	90°
Lux	1700	1400	1250	1000	700

ANSWERS

1. (i) $r = 6\sin\theta$.   (ii) $r^2 = 2c^2\operatorname{cosec} 2\theta$.   (iii) $r^2 = 4\cos 2\theta$.

2. (i) $x^2 + y^2 = 4$.   (ii) $y = \dfrac{1}{\sqrt{3}}x$.   (iii) $x + y = 2$.

(iv) $(x^2 + y^2)^2 = 9(x^2 - y^2)$.

Appendix B

THE STRAIGHT LINE AND THE CIRCLE

B. 1. The straight line

Most of this paragraph should be known already by the student. It may prove useful as revision.

B. 11. Simple equations of a straight line

(a) *Slope and intercept form.* In fig. 117, let slope $= \tan \theta = m$ say. Then

$$m = \frac{LP}{NL} = \frac{MP - ML}{OM}$$

$$= \frac{y - c}{x}$$

Thus

$$\frac{y - c}{x} = m,$$

giving

$$y = mx + c. \tag{1}$$

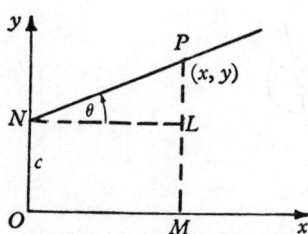

Fig. 117

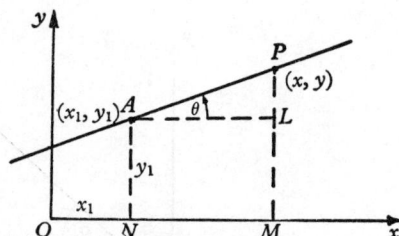

Fig. 118

(b) *Given slope and through a given point.* Given slope m, and fixed point $A(x_1, y_1)$ (fig. 118).

$$m = \tan \theta = \frac{LP}{AL} = \frac{MP - ML}{NM}$$

$$= \frac{MP - NA}{OM - ON}$$

$$= \frac{y - y_1}{x - x_1}.$$

$$\therefore \quad y - y_1 = m(x - x_1). \tag{2}$$

(c) *Line joining two given points.* Given points, $A(x_1, y_1)$ and $B(x_2, y_2)$ (fig. 119).

$$\text{Slope} = \frac{LP}{AL} = \frac{DA}{BD},$$

i.e.

$$\frac{MP - ML}{FM} = \frac{FA - FD}{GF}$$

or

$$\frac{MP - FA}{OM - OF} = \frac{FA - GB}{OF - OG}$$

giving

$$\frac{y - y_1}{x - x_1} = \frac{y_1 - y_2}{x_1 - x_2}. \tag{3}$$

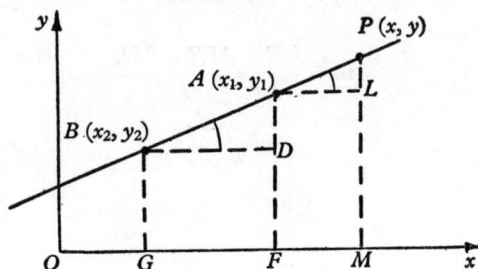

Fig. 119

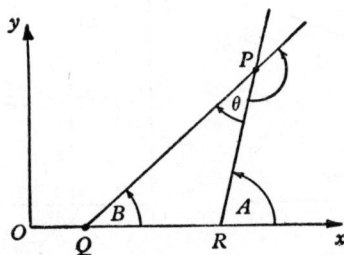

Fig. 120

Note that $\dfrac{y_1 - y_2}{x_1 - x_2}$ is the slope of the line joining the two given points.

B. 12. Angle between two lines

Let the slopes of the given lines (fig. 120) be

$$m_1 = \tan A,$$

$$m_2 = \tan B.$$

Let θ be the angle between the two lines.

$\hat{A} = \hat{B} + \hat{\theta}$ (exterior angle = sum of two interior opposite angles). $\therefore\ \theta = A - B.$

$\therefore$ If $m = \tan\theta$

$$m = \tan(A - B) = \frac{\tan A - \tan B}{1 + \tan A \tan B},$$

i.e.

$$m = \frac{m_1 - m_2}{1 + m_1 m_2}. \tag{1}$$

In general there are two values of the angle between two lines, θ and $(180° - \theta)$.

If the use of the formula in equation (1) gives a negative value for m, it means that the obtuse angle $(180° - \theta)$ has been found.

Parallel lines. In this case $\theta = 0°$ or $180°$.

In each case $\tan\theta = m = 0$.

$$\therefore\ m_1 = m_2. \tag{2}$$

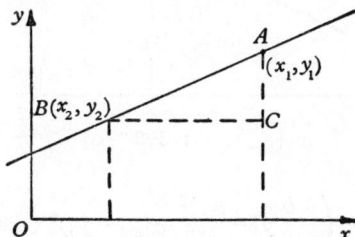

Fig. 121

Perpendicular lines. In this case $\theta = 90°$.

$$\therefore\ m = \tan\theta = \infty.$$

Thus $1 + m_1 m_2 = 0.$ $m_1 m_2 = -1.$ $\tag{3}$

For example, a line perpendicular to a line with slope 3 would have a slope $-\frac{1}{3}$.

B. 13. Distance between two given points

In fig. 121, $AB^2 = AC^2 + BC^2$

$$= (y_1 - y_2)^2 + (x_1 - x_2)^2.$$

Thus $AB = \sqrt{\{(x_1 - x_2)^2 + (y_1 - y_2)^2\}}.$

B. 14. Perpendicular distance from a point on to a line

Let QR, $ax + by + c = 0$, be the given line and $P(x_1, y_1)$ the given point. Draw PR parallel to Ox and PQ perpendicular to QR (fig. 122).

18

Now QR is
$$ax + by + c = 0 \quad \text{or} \quad y = -\frac{a}{b}x - \frac{c}{b}.$$

$\therefore$ Slope of $QR = \tan\theta = -a/b$.

$$\therefore \quad \sin^2\theta = \frac{a^2}{a^2+b^2}. \tag{1}$$

The equation of PR is $y = y_1$.
Where PR cuts QR (at R):
$$ax + by_1 + c = 0.$$

$\therefore$ At R, $\qquad y = y_1 \quad \text{and} \quad x = -\frac{b}{a}y_1 - \frac{c}{a}. \tag{2}$

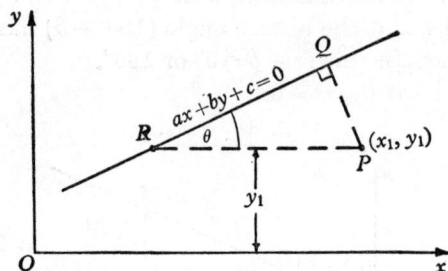

Fig. 122

Thus
$$PR^2 = \left\{x_1 - \left(-\frac{b}{a}y_1 - \frac{c}{a}\right)\right\}^2 + (y_1 - y_1)^2 \quad \text{(see B.13)}.$$

$$= \frac{(ax_1 + by_1 + c)^2}{a^2}. \tag{3}$$

But $\qquad\qquad PQ^2 = PR^2 \sin^2\theta.$

$\therefore$ From (1) and (3):
$$PQ^2 = \frac{(ax_1 + by_1 + c)^2}{a^2} \times \frac{a^2}{(a^2+b^2)},$$

i.e. $\qquad\qquad PQ = \pm\frac{(ax_1 + by_1 + c)}{\sqrt{(a^2+b^2)}}. \tag{4}$

As the *length* of PQ only is required the $+$ or $-$ sign is taken to give a positive result for PQ.

Note that to get the length, the rule is: put the co-ordinates of the given point into the equation of the line and divide by the square root of the sum of the squares of the coefficients of x and y.

In particular, the distance from the origin on to the line is
$$\frac{c}{\sqrt{(a^2+b^2)}}.$$

B. 15. Areas of triangles

These can always be calculated by use of the areas of trape-
ziums and right-angled triangles. A polygon is divided into
triangles, if its area is required. The method is best demonstrated
by a numerical example:

To find the area of triangle ABC where A, B, C are the points
$(1, 2)$, $(3, -1)$ and $(5, 4)$ (fig. 123):

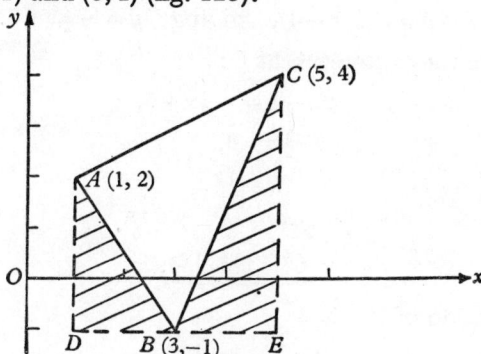

Fig. 123

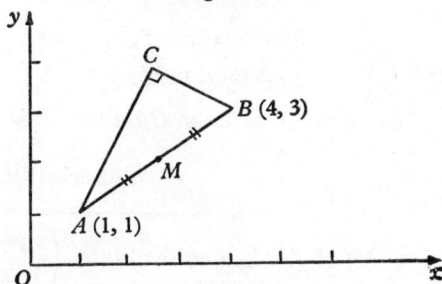

Fig. 124

Draw lines through A and C parallel to Oy, and a line through
B parallel to Ox, to form a trapezium $ADEC$.

$$\triangle ABC = \text{trapezium } ADEC - \triangle ADB - \triangle CBE$$
$$= \tfrac{1}{2}(AD + CE)\,DE - \tfrac{1}{2}.DB.AD - \tfrac{1}{2}BE.CE$$
$$= \tfrac{1}{2}(3 + 5)\,4 - \tfrac{1}{2}.2.3 - \tfrac{1}{2}.2.5$$
$$= 16 - 3 - 5 = \underline{8 \text{ square units.}}$$

EXAMPLES

(1) ABC is a triangle (fig. 124). The co-ordinates of A and B
are $(1, 1)$ and $(4, 3)$. $A\hat{C}B = 90°$ and the gradient of AC is 2.
Find analytically (i) the co-ordinates of C, (ii) the length of the

perpendicular from C to AB, (iii) the distance of C from the mid-point of AB. [Sec. A]

(i) Gradient of AC is 2; but BC is perpendicular to AC, $\therefore$ gradient of BC is $-\frac{1}{2}$.

Equation of AC is

$$y - 1 = 2(x - 1), \quad \text{giving} \quad y = 2x - 1.$$

Equation of BC is

$$y - 3 = -\tfrac{1}{2}(x - 4), \quad \text{giving} \quad y = -\tfrac{1}{2}x + 5.$$

$\therefore$ Where they intersect, at C:

$$2x - 1 = -\tfrac{1}{2}x + 5,$$

$$\tfrac{5}{2}x = 6,$$

$$x = \tfrac{12}{5}.$$

$$\therefore \; y = 2 \cdot \tfrac{12}{5} - 1 = \tfrac{19}{5}.$$

$$C \text{ is } (\tfrac{12}{5}, \tfrac{19}{5}). \tag{1}$$

(ii) Equation of AB is

$$\frac{y-1}{x-1} = \frac{3-1}{4-1} = \frac{2}{3},$$

$$3y - 3 = 2x - 2,$$

$$2x - 3y + 1 = 0.$$

$\therefore$ Length of perpendicular from C on to AB is

$$\frac{2 \cdot \tfrac{12}{5} - 3 \cdot \tfrac{19}{5} + 1}{\sqrt{(2^2 + 3^2)}} = \frac{28}{5\sqrt{(13)}} \quad \text{numerically.} \tag{2}$$

(iii) Mid-point, M, of AB has co-ordinates $\left(\dfrac{x_A + x_B}{2}, \dfrac{y_A + y_B}{2}\right)$,

i.e. M is $\left(\dfrac{1+4}{2}, \dfrac{1+3}{2}\right)$ which gives $(\tfrac{5}{2}, 2)$.

Thus $\qquad CM^2 = (\tfrac{12}{5} - \tfrac{5}{2})^2 + (\tfrac{19}{5} - 2)^2$

$$= \tfrac{1}{100} + \tfrac{81}{25} = \tfrac{13}{4}.$$

$$\therefore \; CM = \tfrac{1}{2}\sqrt{(13)}. \tag{3}$$

(2) Show that the triangle formed by the lines $x + y = 8$, $7x + y = 8$ and $x - 2y + 1 = 0$ is isosceles, and find its area (fig. 125). [Sec. A]

$$x + y = 8 \quad L_1,$$

$$7x + y = 8 \quad L_2,$$

$$x - 2y + 1 = 0 \quad L_3.$$

Let L_1, L_2 meet at A. Then at A, $y = 8 - 7x = 8 - x$.

$$\therefore \quad 6x = 0, \quad x = 0. \quad \text{Thus } y = 8.$$

$$A \text{ is } (0, 8). \tag{1}$$

Let L_2, L_3 meet at B. Then at B,

$$y = \frac{x+1}{2} = 8 - 7x.$$

$$\therefore \quad 15x = 15, \quad x = 1 \quad \text{and thus} \quad y = 1.$$

$$B \text{ is } (1, 1). \tag{2}$$

Let L_1, L_3 meet at C. Then at C,

$$x = 8 - y = 2y - 1.$$

$$\therefore \quad 3y = 9, \quad y = 3 \quad \text{and thus} \quad x = 5.$$

$$C \text{ is } (5, 3). \tag{3}$$

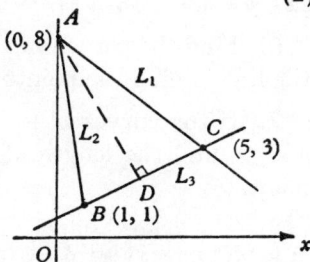

Fig. 125

From (1) and (2):

$$AB^2 = (1-0)^2 + (1-8)^2 = 50.$$

From (1) and (3):

$$AC^2 = (5-0)^2 + (3-8)^2 = 50.$$

$$\therefore \quad \underline{AB = AC} \text{ and the triangle is isosceles.} \tag{4}$$

Now $\qquad BC^2 = (5-1)^2 + (3-1)^2 = 20.$

$$\therefore \quad BC = 2\sqrt{5}. \tag{5}$$

As $AB = AC$, the perpendicular from A bisects the side BC.

$$\therefore \quad \text{Area of } \triangle ABC = \tfrac{1}{2} . AD . BC. \tag{6}$$

Now $\qquad AD$ is $\dfrac{0.1 - 2.8 + 1}{\sqrt{(1^2 + 2^2)}},$ numerically,

i.e. $\qquad AD$ is $\dfrac{15}{\sqrt{5}}$ numerically. $\tag{7}$

From (5), (6) and (7), area of

$$\triangle ABC = \tfrac{1}{2} . 2\sqrt{5} . \frac{15}{\sqrt{5}}$$

$$= \underline{15 \text{ square units.}}$$

Exercise 61

1. Find the gradient of the line joining the two points $(2, 1)$ and $(4, 3)$. Hence find the gradient of a line perpendicular to the given one.

2. Find the equation of the sides of the triangle whose vertices are the points $(2, 1)$, $(-3, -2)$, $(3, -2)$.

3. Find the equation of the line through the point $(2, 1)$, parallel to the line $2x + y + 4 = 0$.

4. Find the equation of the line making equal positive intercepts on the axes and passing through the point $(2, 3)$.

5. Find the distance from the point $(5, 5)$ measured along the line $3x - 4y + 5 = 0$ to the point where this line cuts the line $2x + y = 6$.

6. Find the perpendicular distance from the point $(4, 2)$ on to the line joining the points $(2, 1)$ and $(5, 4)$.

7. If the curves $x^2 + y^2 = 4$ and $y^2 = 3x$ intersect at points P and Q, find the length of PQ and the co-ordinates of its mid-point.

8. The co-ordinates of the vertices of a triangle ABC are $A(2, 5)$, $B(6, 1)$, $C(8, 7)$. Find (i) the equations of AB and AC. (ii) Angle BAC. (iii) The length of the perpendicular from C to AB.   [Sec. A]

9. The co-ordinates of the vertices of a triangle are $(-1, 2)$, $(5, 0)$ and $(3, 6)$. Find the equations of the perpendicular bisectors of two of the sides of the triangle, and show that they intersect at the point $(2\frac{1}{2}, 2\frac{1}{2})$.

10. Show that $ax + by + c + \lambda(a_1x + b_1y + c_1) = 0$, where λ is an arbitrary constant, represents a line through the point of intersection of the two lines $ax + by + c = 0$ and $a_1x + b_1y + c_1 = 0$. Find the equation of the line through the point of intersection of $3x - 2y + 5 = 0$ and $x - 4y - 2 = 0$, which passes through the point $(1, 2)$.

ANSWERS

1. $1; -1$.

2. $3x - 5y - 1 = 0$; $3x + y - 7 = 0$; $y + 2 = 0$.

3. $2x + y - 5 = 0$.   **4.** $x + y - 5 = 0$.   **5.** $\frac{45}{11}$.

6. $\dfrac{1}{\sqrt{2}}$.   **7.** $2\sqrt{3}$; $(1, 0)$.

8. $x + y - 7 = 0$; $x - 3y + 13 = 0$; $\tan^{-1}2 = 63° 28'$; $4\sqrt{2}$.

10. $31x - 34y + 37 = 0$.

B. 2. The Circle

A circle is the curve traced out by a point which moves so that its distance from a fixed point is constant. The fixed point is the centre of the circle, and the fixed distance the radius.

B. 21. Standard equations of a circle

(a) *Circle, centre* $(0, 0)$ *and radius* r.

Let P be any point (x, y) on the circle (fig. 126). Then

$$OM^2 + MP^2 = r^2,$$

i.e.

$$x^2 + y^2 = r^2. \tag{1}$$

(b) *Circle, centre* (u, v), *radius* r.

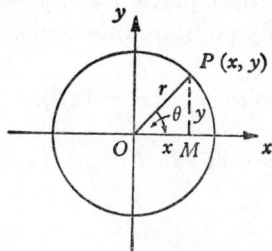

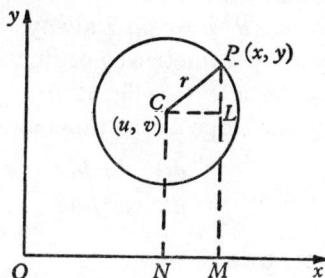

Fig. 126   Fig. 127

In fig. 127, $CL^2 + LP^2 = CP^2$.

$$(OM - ON)^2 + (MP - NC)^2 = CP^2.$$

$$(x - u)^2 + (y - v)^2 = r^2. \tag{2}$$

(c) $x^2 + y^2 + 2gx + 2fy + c = 0$ *as the general equation of a circle.*

Consider the equation $x^2 + y^2 + 2gx + 2fy + c = 0$.

Completing the squares in x and y:

$$(x + g)^2 + (y + f)^2 - g^2 - f^2 + c = 0.$$

i.e.

$$(x + g)^2 + (y + f)^2 = (g^2 + f^2 - c). \tag{3}$$

Comparing with equation (2), this represents a circle, centre $(-g, -f)$ and radius $\sqrt{(g^2 + f^2 - c)}$.

If $g^2 + f^2 - c = 0$, the circle reduces to a point $(-g, -f)$.

If $g^2 + f^2 - c < 0$, the circle has an imaginary radius and is not the locus of points with real numbers for co-ordinates.

Note the conditions in the general equation:

(i) Coefficients of x^2 and y^2 the same.

(ii) No term in xy.

For example:

If
$$4x^2 + 4y^2 + 8x + 20y + 8 = 0,$$

then
$$x^2 + y^2 + 2x + 5y + 2 = 0.$$

This is a circle, centre $(-1, -\frac{5}{2})$, radius

$$\sqrt{(1^2 + (\tfrac{5}{2})^2 - 2)} = \tfrac{1}{2}\sqrt{(21)}.$$

B. 22. Tangents and normals

The slope of the tangent may be found by differentiating the equation and finding dy/dx at any given point. Thus the equations of tangents and normals are easily found.

Alternatively, the normal is along a radius and the tangent perpendicular to it.

$x = r\cos\theta$, $y = r\sin\theta$ always satisfy the equation $x^2 + y^2 = r^2$. They are parametric co-ordinates for any point on this circle.

θ is the $x\hat{O}P$ (see fig. 126).

For the slope of the tangent at the point $(r\cos\theta, r\sin\theta)$:

$$\frac{dy}{dx} = \frac{dy}{d\theta} \bigg/ \frac{dx}{d\theta} = \frac{r\cos\theta}{-r\sin\theta} = -\cot\theta.$$

EXAMPLES

(1) Given $x = 3 + 2\sin\theta$, $y = 2(1 + \cos\theta)$, find an equation connecting x and y. Draw the graph of y against x.   [Sec. A]

$$x - 3 = 2\sin\theta,$$

$$y - 2 = 2\cos\theta.$$

$$\therefore \ (x-3)^2 + (y-2)^2 = 4(\sin^2\theta + \cos^2\theta) = 4.$$

The curve is therefore a circle, centre $(3, 2)$ and radius 2.

Fig. 128 shows the graph of y against x.

(2) $A(1, 1)$ and $B(5, 7)$ are two given points. Find (i) the equation of the circle on AB as diameter. (ii) The co-ordinates of the extremities of the diameter of this circle which is at right angles to AB.   [Sec. A]

(i) The centre C of the circle is the mid-point of AB.

$$\therefore \ C \text{ is } \left(\frac{1+5}{2}, \frac{1+7}{2}\right) \text{ which is } (3, 4). \tag{1}$$

The radius of the circle is the length of CA (or CB).

$$\therefore \ \text{radius is } \sqrt{\{(3-1)^2 + (4-1)^2\}} = \sqrt{(13)}. \tag{2}$$

From (1) and (2), the equation of the circle is

$$(x-3)^2 + (y-4)^2 = 13.$$

i.e. $$x^2 + y^2 - 6x - 8y + 12 = 0. \qquad (3)$$

(ii) Slope of line AB is $\dfrac{7-1}{5-1} = \dfrac{3}{2}$.

$\therefore$ Slope of a line at right angles to AB is $-\frac{2}{3}$.

For this to be a diameter it must pass through centre $C\,(3, 4)$.

$\therefore$ Equation of required diameter is:

$$y - 4 = -\tfrac{2}{3}(x - 3),$$

giving $$2x + 3y - 18 = 0. \qquad (4)$$

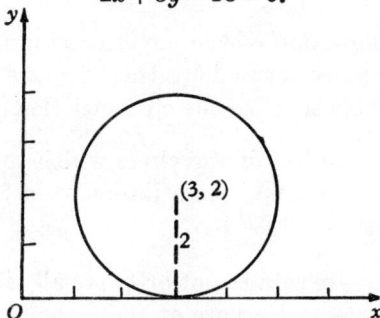

Fig. 128

Where this diameter meets the circle, from (4) and (3), substituting $y = \dfrac{18 - 2x}{3}$ into the equation of the circle:

$$x^2 + \frac{(18 - 2x)^2}{9} - 6x - \tfrac{8}{3}(18 - 2x) + 12 = 0,$$

giving $$13x^2 - 78x = 0,$$

$$13x(x - 6) = 0.$$

$$\therefore \quad x = 0 \text{ or } 6.$$

$\therefore$ From (4): $\qquad y = 6 \text{ or } 2.$

The extremities of the diameter are the points $(0, 6)$; $(6, 2)$.

Exercise 62

1. Find the equations of the tangent and normal to the circle $x^2 + y^2 = 25$ at the point $(3, 4)$.

2. Find the value of c if $y = mx + c$ is a tangent to the circle $x^2 + y^2 = r^2$.

Also show that $xx_1 + yy_1 = r^2$ is the tangent at the point (x_1, y_1) on the circle $x^2 + y^2 = r^2$.

3. Write down the centres and radii of the circles:

$$\text{(i)} \qquad x^2 + y^2 - 4x - 6y + 4 = 0,$$

$$\text{(ii)} \quad 3x^2 + 3y^2 + 9x + 12y + 6 = 0.$$

4. Sketch the two possible ways in which two circles can touch each other. Deduce that two circles touch if the distance between their centres is either the sum or the difference of their radii. Prove that the two circles

$$x^2 + y^2 - 4x - 4y + 7 = 0, \quad x^2 + y^2 - 10x - 12y + 45 = 0$$

touch each other externally.

5. Find the equation of the circle passing through the vertices of the triangles formed by the lines $x + y = 8$, $7x + y = 8$, $x - 2y + 1 = 0$. (This is called the circumcircle of the triangle.)

6. Find the equation of the circle which touches the x-axis at the point $(3, 0)$ and makes an intercept of 8 on the positive side of the y-axis.

7. Two circles are said to cut orthogonally if the tangents at their points of intersection are at right angles. The tangent to one circle thus passes through the centre of the other. Deduce that the condition for two circles to cut orthogonally is that the sum of the squares on their radii is equal to the square of the distance between their centres. Show that the circles

$$x^2 + y^2 - 2x - 4y + 4 = 0, \quad x^2 + y^2 - 6x - 8y + 18 = 0$$

cut orthogonally.

8. Find the equation of the circle whose centre is the point $(4, 5)$ and whose circumference passes through the centre of the circle $x^2 + y^2 + 4x - 6y = 12$.

9. Find the Cartesian equation of a circle whose parametric co-ordinates are given by $x = 4 - 3\cos\theta$, $y = 2 + 3\sin\theta$. Find also the equation of the diameter of this circle which passes through the point on the circumference given by $\theta = \frac{1}{3}\pi$.

10. Sketch the circle whose equation is

$$x^2 + y^2 + 4x - 6y + 9 = 0.$$

Making use of Pythagoras's theorem, find the length of the tangent from the point $(3, 4)$ to this circle.

11. The distance of a moving point P from a fixed point $(3, -2)$ is equal to half its distance from a fixed point $(0, 4)$. Show that the locus of P is a circle and find its centre and radius.

[Sec. A]

12. A and B are the points $(0, 1)$ and $(2, 0)$ and a point P moves in the plane of the axes so that $PA^2 + PB^2 = k^2$. Show that the path of P is a circle for all values of k^2 greater than 2·5. Find the centre and radius of the circle. If $k = 5$ find the co-ordinates of the extremities of the diameter which is perpendicular to AB.

[Sec. A]

Answers

1. $3x + 4y = 25$; $3y = 4x$.   **2.** $c = \pm r \sqrt{(1 + m^2)}$.

3. (i) $(2, 3)$; 3.   (ii) $(-\frac{3}{2}, -2)$; $\frac{1}{2} \sqrt{(17)}$.

5. $9x^2 + 9y^2 - 30x - 84y + 96 = 0$. Note that the centre is the point of intersection of the perpendicular bisectors of any two sides.

6. $x^2 + y^2 - 6x - 10y + 9 = 0$.   **8.** $x^2 + y^2 - 8x - 10y + 1 = 0$.

9. $x^2 + y^2 - 8x - 4y + 11 = 0$; $2 - y = \sqrt{3}(x - 4)$.

10. $\sqrt{(22)}$.   **11.** $(4, -4)$; $\sqrt{(20)}$.

12. Centre $(1, \frac{1}{2})$; radius $\frac{1}{2} \sqrt{(2k^2 - 5)}$; $(\frac{5}{2}, \frac{7}{2})$ and $(-\frac{1}{2}, -\frac{5}{2})$.

SPECIMEN A1 PAPERS

PAPER I (MECHANICAL)

Answer 6 *Questions only.* *Time:* 3 *hours*

1. Differentiate:

(a) $\dfrac{2\sin^{-1}x}{\sqrt{(1-x^2)}}$;

(b) $\tanh^{-1}2x$;

(c) $\log_e \cos 3x$;

(d) $e^{-2x}(1-x^2)^3$.

2. Evaluate:

(a) $\displaystyle\int_1^2 (3x^2-x+1)^2(6x-1)\,dx$;

(b) $\displaystyle\int_0^{\frac{1}{4}\pi} x^2\cos 2x\,dx$;

(c) $\displaystyle\int \dfrac{(2x+3)}{(x-2)(x+3)}\,dx$;

(d) $\displaystyle\int \dfrac{1}{4x^2+4x+17}\,dx$.

3. (a) Using Maclaurin's theorem, or otherwise, expand the following functions as far as the term in x^4:

(i) e^{-2x};　　(ii) $\log_e(1-x)$;　　(iii) $\cos 3x$.

(b) Use your expansion of $\log_e(1-x)$ to find an approximate value for $\log_e 0{\cdot}95$.

4. Sketch the curves of $y=\cosh x$ and $y=x+2$ for values of x from -1 to $2{\cdot}5$. Use Newton's approximation to find the positive and negative roots of the equation $\cosh x = x+2$, correct to 3 decimal places.

5. (a) Prove the formula

(i) $\cosh^2 u - \sinh^2 u = 1$;

(ii) $\cosh(u+v) = \cosh u \cosh v + \sinh u \sinh v$.

(b) Evaluate:

(i) $\displaystyle\int_0^1 \dfrac{1}{\sqrt{(x^2+4x+5)}}\,dx$;　　(ii) $\displaystyle\int \dfrac{1}{\sqrt{(x^2-4x)}}\,dx$.

6. Find the second moment of volume of a right circular cone, base radius r, height h, about an axis through the vertex and parallel to the base.

If the cone is of uniform density and is suspended from the vertex, find the period of small oscillations in a vertical plane through the vertex.

7. Solve the following differential equations:

(a) $2x^2\dfrac{dy}{dx} - y^2 = 4$;   (b) $x\dfrac{dy}{dx} + 4y = x$,

given that when $x = 1$, $y = 1$;

(c) $\dfrac{d^2y}{dx^2} + 2\dfrac{dy}{dx} + 5y = 0$,

given that when $x = 0$, $y = 0$ and $dy/dx = 1$.

8. Prove that the radius of curvature R at any point along the curve $y^2 = 16x^5$ is given by $R^2 = \dfrac{(1 + 100x^3)^3}{225x}$. Find the value of x for which the radius of curvature is a minimum.   [Sec. A]

9. (a) Simplify $(2+j3)(3-j2)$, expressing the result in polar form. Find the three cube roots of $(2+j3)(3-j2)$ in their polar form. Exhibit the number and its cube roots on an Argand diagram.

(b) Evaluate $\displaystyle\int e^{ax}\cos bx\,dx$.

ANSWERS

1. (a) $\dfrac{2}{(1-x^2)}\left(1 + \dfrac{x\sin^{-1}x}{\sqrt{(1-x^2)}}\right)$;   (b) $\dfrac{2}{1-4x^2}$;

(c) $-3\tan 3x$;   (d) $-2e^{-2x}(1-x^2)^2(1+3x-x^2)$.

2. (a) $\dfrac{1304}{3}$;   (b) $-\tfrac{1}{4}\pi$;

(c) $\tfrac{1}{5}\log_e\{(x-2)^7(x+3)^3\} + C$;   (d) $\tfrac{1}{8}\tan^{-1}\left(\dfrac{2x+1}{4}\right) + C$.

3. (a) (i) $1 - 2x + 2x^2 - \tfrac{4}{3}x^3 + \tfrac{2}{3}x^4$;   (ii) $-x - \tfrac{1}{2}x^2 - \tfrac{1}{3}x^3 - \tfrac{1}{4}x^4$;

(iii) $1 - \tfrac{9}{2}x^2 + \tfrac{27}{8}x^4$.

(b) -0.05129 to 5 decimal places.

4. 2.085; -0.726.

5. (b) (i) 0.375;   (ii) $\cosh^{-1}\left(\dfrac{x-2}{2}\right) + C$.

6. $\dfrac{\pi r^2 h^3}{5}\left(\dfrac{r^2}{4h^2} + 1\right)$; $2\pi\sqrt{\left(\dfrac{(r^2+4h^2)}{5gh}\right)}$.

7. (a) $y = 2\tan\left(\dfrac{Cx-1}{x}\right)$;   (b) $y = \dfrac{1}{5}\left(x + \dfrac{4}{x^4}\right)$;

(c) $y = \tfrac{1}{2}e^{-x}\sin 2x$.

8. $x \simeq 0 \cdot 1077$.

9. (a) $13 \angle 22° 37'$; $2 \cdot 3513 \angle 7° 32'$; $2 \cdot 3513 \angle 127° 32'$;
$2 \cdot 3513 \angle 247° 32'$.

(b) $\dfrac{e^{ax}}{(a^2+b^2)} (a \cos bx + b \sin bx)$.

PAPER II (MECHANICAL)

1. Differentiate:

(a) $\log_e \left(\dfrac{1-2x}{4+3x} \right)$;

(b) $e^{-x^2} \sin 3x$;

(c) $\cosh^2 x + \sinh^2 2x$;

(d) $\dfrac{\tan^{-1} x}{x^3}$.

2. Evaluate:

(a) $\displaystyle\int x^2 e^{2x} \, dx$;

(b) $\displaystyle\int \dfrac{1}{(3x^2+4x-5)} \, dx$;

(c) $\displaystyle\int_0^5 \dfrac{1}{(x^2+5x+6)} \, dx$.

(d) By substituting $x = \sin \theta$, evaluate $\displaystyle\int_0^1 x^3 (1-x^2)^{\frac{3}{2}} \, dx$.

3. Solve the differential equations:

(a) $\dfrac{1}{2x} \dfrac{dy}{dx} = \sqrt{(1+y^2)}$;

(b) $x \dfrac{dy}{dx} + 2y = \sin x$;

(c) $\dfrac{d^2y}{dx^2} + 4 \dfrac{dy}{dx} + 4y = 0$, given that when $x=0$, $y=0$ and
$dy/dx = 2$.

4. Expand by any means (a) $\sec x$, (b) $\cosh x$, (c) $\tan^{-1} 2x$, as far as the third term. Use the expansion of $\cosh x$ to show that when x is small the equation $y = c \cosh x/c$ becomes approximately the equation of a parabola.

5. (a) If $z = (x+y) \log (x/y)$, prove that $x \dfrac{\partial z}{\partial x} + y \dfrac{\partial z}{\partial y} = z$.

(b) If the power required to propel a steamer varies as the cube of the velocity and the square of the length, prove that a 2% increase in velocity and a 3% increase in length will require approximately a 12% increase in power.

6. A hollow cylinder, open at both ends, is to be made from a given quantity of metal. Its external diameter is to be twice the internal diameter. If the given volume is 6π m^3, find the dimensions for minimum surface area.

7. (a) Find, by integration, the distance from the plane face of the centroid of a solid hemisphere.

(b) Show that the second moment of area of a triangle, base b, height h, about an axis through the vertex parallel to the base is $\frac{1}{4}bh^3$. Deduce its second moment about the base as axis.

8. State the formula giving the bending moment at any distance along a thin uniform horizontal beam slightly bent under gravity.

A uniform beam, length l, weight per unit length w, and flexural rigidity EI, is clamped horizontally at one end and rests on a prop at the same level at the other end. Find the deflexion at any distance x along the beam.

9. (a) Show that $e^{jx} = \cos x + j \sin x$. Hence show that $3\,e^{(-2+j5)x} + 4\,e^{(-2-j5)x}$ may be written in the form

$$e^{-2x}(A \cos 5x + B \sin 5x)$$

and give the values of A and B.

(b) Express the roots of the equation $3x^2 + 6x + 7 = 0$ in the form $r \angle \theta$.

ANSWERS

1. (a) $\dfrac{-11}{(1-2x)(4+3x)}$; (b) $e^{-x^2}(3 \cos 3x - 2x \sin 3x)$;

(c) $\sinh 2x + 2 \sinh 4x$; (d) $\dfrac{x - 3(1+x^2) \tan^{-1} x}{x^4(1+x^2)}$.

2. (a) $\dfrac{e^{2x}}{4}(2x^2 - 2x + 1) + C$;

(b) $\dfrac{1}{2\sqrt{(19)}} \log_e C\left(\dfrac{3x+2-\sqrt{(19)}}{3x+2+\sqrt{(19)}}\right)$;

(c) $0 \cdot 272$; (d) $\frac{2}{35}$.

3. (a) $y = \sinh(x^2 + C)$; (b) $y = \dfrac{1}{x^2}(\sin x - x \cos x + C)$;

(c) $y = 2x\,e^{-2x}$.

4. (a) $1 + \frac{1}{2}x^2 + \frac{5}{24}x^4$;   (b) $1 + \frac{1}{2}x^2 + \frac{1}{24}x^4$;

(c) $2(x - \frac{4}{3}x^3 + \frac{16}{5}x^5)$. Parabola is $y = c + \dfrac{x^2}{2c}$.

6. External diameter 4 m; internal diameter 2 m; height 2 m.

7. (a) $\frac{3}{8}$ of the radius;   (b) $\frac{1}{12}bh^3$.

8. $\dfrac{wx^2(l-x)(3l-2x)}{48}$.

9. (a) $A = 7$, $B = -j$;   (b) $1 \cdot 53 \angle 130° 54'$ and $1 \cdot 53 \angle 229° 6'$.

PAPER III (ELECTRICAL)

Answer 6 Questions only.   Time: 3 hours

1. (a) Differentiate:

(i) $x^2 \sinh^{-1} x$;   (ii) $e^{2x} \log_e \sqrt{\dfrac{1+x}{1-x}}$;   (iii) $\dfrac{\cosh 3x}{x^2}$.

(b) If $y = \dfrac{x}{\sqrt{(1-x^2)}}$, show that $x^4 \dfrac{d^2 y}{dx^2} = 3y^5$.   [Sec. A]

2. (a) Evaluate:

(i) $\displaystyle\int x \log_e 2x \, dx$;   (ii) $\displaystyle\int \dfrac{(2x+1)}{\sqrt{(x^2+x-3)}} \, dx$;   (iii) $\displaystyle\int \dfrac{1}{\sqrt{(2+x-x^2)}} \, dx$.

(b) If the power in a certain circuit is given by

$$p = EI \sin \omega t \sin (\omega t + \phi),$$

find the mean value of p from $t = 0$ to $t = 2\pi/\omega$.

3. Two opaque spheres, radii 20 mm and 80 mm, are fixed with their centres A and B 9 m apart. A point source of light P is placed between them on the line AB. Find the distance AP when the total surface area illuminated is a maximum.

4. (a) Write $\dfrac{(1+j2)(2-j)}{(3+j4)}$ in the form $r\,e^{j\theta}$.

(b) If P represents the complex number $z = x + jy$, find the locus in the x, y plane of the point P if

(i) $|z-1| = 1$;   (ii) $\arg(z-1) = \frac{1}{4}\pi$;   (iii) $\left|\dfrac{z-1}{z+1}\right| = 1$.

Write down the x, y equations of these three loci.   [Sec. A]

5. An inductive resistance has inductance L and resistance R. It is connected across an alternating voltage $E_0 \sin pt$. Write down the differential equation satisfied by the current and solve it. Take the current zero at time zero.

6. (a) Solve the following differential equation:

$$(1+x^2)\frac{dy}{dx} = \sqrt{(1-y^2)}.$$

(b) The current i in a certain circuit satisfies the equation

$$\frac{d^2i}{dt^2} + \frac{di}{dt} + \tfrac{17}{4}i = 0.$$

Solve the equation.

As t becomes large, show that the current tends to zero.

7. State the expansion of $\log_e(1+x)$ as a power series in x. For what range of values of x is the expansion valid?

Obtain an expansion for $x^2 \log_e(1+x)$.

Use this to find a value for $\displaystyle\int_0^{\frac{1}{2}} x^2 \log_e(1+x)\,dx$ correct to 3 decimal places.

Verify the result by evaluating the integral, using integration by parts.

8. (a) Differentiate $\sinh^{-1} x/a$, using the definition of an inverse function.

(b) Sketch the curve $y^2 = \dfrac{16}{4+x^2}$ for values of x from 0 to 4. Find the area between the curve and the lines $x=0$, $x=4$.

9. (a) If $u = \sin^{-1}(x/y)$, prove that $x\dfrac{\partial u}{\partial x} + y\dfrac{\partial u}{\partial y} = 0$.

(b) The area of a triangle is calculated from the formula $\Delta = \frac{1}{2}ab\sin C$. Percentage errors of 1% in a, 2% in b and 1% in the angle C are made. If C is measured as $30°$, find the maximum percentage error in the calculated area of the triangle.

Answers

1. (a) (i) $2x\sinh^{-1}x + \dfrac{x^2}{\sqrt{(1+x^2)}}$;

(ii) $e^{2x}\left[\log_e\left(\dfrac{1+x}{1-x}\right) + \dfrac{1}{(1-x^2)}\right]$;

(iii) $\dfrac{3x\sinh 3x - 2\cosh 3x}{x^3}.$

2. (a) (i) $\frac{1}{2}x^2(\log_e 2x - \frac{1}{2}) + C$; (ii) $2\sqrt{(x^2 + x - 3)} + C$;

(iii) $\sin^{-1}\frac{1}{3}(2x - 1) + C$.

(b) $\dfrac{EI \cos \phi}{2}$.

3. 1 m.

4. (a) $1\,e^{-j0\cdot2839}$.

(b) (i) Circle $(x - 1)^2 + y^2 = 1$;

(ii) the line $y = x - 1$; (iii) the line $x = 0$ (y-axis).

5. $i = \dfrac{E_0}{R^2 + p^2 L^2}[pL\,e^{-Rt/L} + R \sin pt - pL \cos pt]$.

6. (a) $\tan^{-1}x = \sin^{-1}y + C$ or $y = \sin(\tan^{-1}x + A)$.

(b) $i = e^{-\frac{1}{2}t}(A \cos 2t + B \sin 2t)$; as $t \to \infty$, $e^{-\frac{1}{2}t} \to 0$.

7. $x - \frac{1}{2}x^2 + \frac{1}{3}x^3 - \ldots$; $-1 < x \leqslant 1$; $0\cdot013$.

8. (a) $\dfrac{1}{\sqrt{(x^2 + a^2)}}$; $11\cdot55$. **9.** (b) $\left(3 + \dfrac{\pi\sqrt{3}}{6}\right)\% \simeq 3\cdot91\%$.

MISCELLANEOUS EXERCISES

Note. The order of the questions approximately follows the same order as the chapters in the text.

1. If $x = \cos 2\theta$ and $y = \theta + \sin 2\theta$, find dy/dx and also d^2y/dx^2.

2. (i) Differentiate $\sec^m x \tan^n x$ with respect to x, where m and n are constants.

(ii) If $x = \log_e(1+t)$ and $y = at^2 + bt + c$, where a, b, c are constants, show that $dy/dx - 2y = (2a-b)e^x - 2(a-b+c)$.

[L.U.]

3. (*a*) Differentiate with respect to x:

(i) $\tan 3x$;   (ii) $\log_e(1 + \sin^2 x)$;

(iii) $\dfrac{x}{\sqrt{(4-x^2)}}$.

(*b*) The velocity of a piston is approximately given by: $v = \omega R\{\sin\theta + (R/2L)\sin 2\theta\}$, in which θ is the crank angle from the inner dead centre, ω is the constant angular velocity of the crank, and R and L denote the lengths of the crank and connecting rod, respectively. Assuming $L = 4R$, find the first positive value of the crank angle θ for which the piston velocity has a maximum value. [Sec. A]

4. The graph of $y = \dfrac{(a+bx)}{(x-1)(x-4)}$ has a turning point at the point $P(2, -1)$. Find the values of a and b, and show that y is a maximum at P. [L.U.]

5. By means of a sketch of the graph of $y = \tan x$, or otherwise, show that the equation $\tan x = kx$ (where k is a constant exceeding unity) has a root between 0 and $\frac{1}{2}\pi$. If $k = 1\cdot3$, find the value of this root correct to three significant figures, expressing the result in degrees. [L.U.]

6. A motor M exerts a constant torque T, and is geared to a shaft S. The speed of the motor is n times the speed of the shaft. If I_M, I_S are the effective moments of inertia of the moving parts of the motor and of the shaft, respectively, obtain an expression for the angular acceleration of the shaft in terms of I_M, I_S, n and T. Hence, show that the angular acceleration of the shaft is a maximum when $n = \sqrt{(I_S/I_M)}$. [Neglect any frictional torques.]

7. In the manufacture of a certain article the number produced in one week is proportional to $x/(a+x)$, where a is a constant and x is the amount paid in wages per article per week. The selling price of the article (less costs other than wages) is b, in the same units as x and a. Find the value of x for which the net profit on one week's production is a maximum. Show that, if b is small compared with a, this value of x is approximately $\frac{1}{2}b$.

[L.U.]

8. Verify that $y = (A + Bx)\, e^{2x}$, A and B constants, is a solution of the equation $\dfrac{d^2y}{dx^2} - 4\,\dfrac{dy}{dx} + 4y = 0$.

9. Evaluate:

(i) $\displaystyle\int_1^4 \frac{(1 + 2\sqrt{x})^2}{\sqrt{x}}\, dx$;

(ii) $\displaystyle\int_0^{\frac{1}{4}\pi} (1 + \cos x)^2\, dx$.

[Sec. A]

10. Solve the equation $\displaystyle\int_1^x (13 - 4x)\, dx = 9$.

[Part I]

11. Evaluate:

(i) $\displaystyle\int_0^{\frac{1}{4}\pi} (\sin 2x - \sin^2 x)\, dx$;

(ii) $\displaystyle\int_0^{\frac{1}{4}\pi} (\cos^2 2x - x)\, dx$;

(iii) $\displaystyle\int (\tan^2 2x - \sec^2 3x)\, dx$;

(iv) $\displaystyle\int_0^{\pi/\omega} 4 \sin^2 (\omega t + \tfrac{1}{3}\pi)\, dt$.

12. Evaluate:

(i) $\displaystyle\int (e^x + e^{-x})^2\, dx$;

(ii) $\displaystyle\int \left\{ \frac{1}{\sqrt{(2x+1)}} + \frac{1}{2x+1} \right\} dx$;

(iii) $\displaystyle\int_0^{\frac{1}{12}\pi} \frac{1 - \cos^3 3x}{\cos^2 3x}\, dx$.

13. Find the root-mean-square value of a current given by $i = I_1 \sin \omega t + I_2 \sin 2\omega t$.

14. Prove that $\sinh 3u = 4 \sinh^3 u + 3 \sinh u$.

15. Differentiate with respect to x:

(i) $e^x \sinh 2x$; (ii) $\cosh 3x \tanh 2x$;

(iii) $\sinh^{-1}(e^x)$; (iv) $\cosh^{-1}\left(\dfrac{1}{\sqrt{x}}\right)$.

16. Evaluate:

(i) $\displaystyle\int (\sinh^2 2x + \operatorname{sech}^2 x)\, dx$;

(ii) $\displaystyle\int \tanh^2 x\, dx$; (iii) $\displaystyle\int_0^1 (\cosh 3x + \cos \tfrac{1}{6}\pi x)\, dx$.

17. Find the value of $\dfrac{d^2}{dx^2}(x^2 \tan^{-1}x)$ when $x = \dfrac{1}{\sqrt{3}}$.

18. Differentiate with respect to x:

(i) $\sin^{-1}\left(\dfrac{1-x^2}{1+x^2}\right)$; (ii) $\tan^{-1}\left(\dfrac{2}{\sqrt{x}}\right)$;

(iii) $\cosh^{-1}x \sinh^{-1}x^2$.

19. Show that $\tanh^{-1}(\sin \theta) = \cosh^{-1}(\sec \theta)$, $0 < \theta < \tfrac{1}{2}\pi$.

20. Using the logarithmic form of $\cosh^{-1}(\sec \theta)$, show that the equation $\cosh^{-1}(\sec \theta) + \log_e \sin 2\theta = 0$ can be reduced to the equation $2 \sin^2 \theta + 2 \sin \theta - 1 = 0$. Hence, show that the solution of the given equation is given by $\theta = \sin^{-1}\{\tfrac{1}{2}(\sqrt{3} - 1)\}$.

21. If $y = (1 + \sinh^{-1}x)^2$, show that $(1+x^2)\dfrac{d^2y}{dx^2} + x\dfrac{dy}{dx} = 2$.

22. Evaluate:

(i) $\displaystyle\int_0^2 \dfrac{1}{\sqrt{(16-x^2)}}\, dx$; (ii) $\displaystyle\int_0^7 \dfrac{1}{49+x^2}\, dx$;

(iii) $\displaystyle\int_0^3 \dfrac{1}{\sqrt{(x^2+9)}}\, dx$; (iv) $\displaystyle\int_5^6 \dfrac{1}{\sqrt{(x^2-25)}}\, dx$.

23. Evaluate:

(i) $\int \dfrac{1}{\sqrt{(4-2x^2)}}\, dx$;

(ii) $\int_0^5 \dfrac{1}{25+4x^2}\, dx$;

(iii) $\int \dfrac{5}{\sqrt{(3x^2+12)}}\, dx$;

(iv) $\int \dfrac{7}{\sqrt{(3x^2-7)}}\, dx$.

24. Evaluate:

(i) $\int_{0\cdot18}^{0\cdot40} \dfrac{dx}{\sqrt{(1-4x^2)}}$;

(ii) $\int_0^1 x\log_e x\, dx$;

(iii) $\int_0^2 \dfrac{(2x-1)^2}{(2x+1)}\, dx$;

(iv) $\int_0^{\frac{1}{4}\pi} x^2 \sin x\, dx$.   [Part I]

25. Evaluate:

(i) $\int_0^{\frac{1}{4}} t \sin 2\pi t\, dt$;

(ii) $\int_5^{11} \dfrac{4}{x^2-2x-3}\, dx$

and give each of these answers correct to two decimal places.

[Part I]

26. Evaluate:

(i) $\int_0^1 \dfrac{x^2}{x^3+1}\, dx$;

(ii) $\int \dfrac{4x-1}{3x+2}\, dx$;

(iii) $\int_3^4 \dfrac{1}{(x-2)(x+3)}\, dx$;

(iv) $\int_2^3 \dfrac{x+3}{x^2+3x-4}\, dx$.

27. Evaluate:

(i) $\int \dfrac{1}{5+4x-x^2}\, dx$;

(ii) $\int_2^4 \dfrac{1}{\sqrt{(12+4x-x^2)}}\, dx$;

(iii) $\int_2^7 \dfrac{1}{x^2-4x+29}\, dx$;

(iv) $\int \dfrac{1}{\sqrt{(x^2+2x+10)}}\, dx$.

28. Evaluate:

(i) $\int \dfrac{x+1}{\sqrt{(x^2+1)}}\, dx$;

(ii) $\int \dfrac{2x-1}{\sqrt{(x^2-4)}}\, dx$;

(iii) $\int \dfrac{2x+3}{\sqrt{(6x-x^2)}}\, dx$;

(iv) $\int \dfrac{x+3}{\sqrt{(4x^2+4x+5)}}\, dx$.

29. Evaluate:

(i) $\displaystyle\int 5\cos 2x\cos 8x\,dx;$ (ii) $\displaystyle\int_0^{\frac{1}{2}\pi}\sin^5 x\cos^3 x\,dx;$

(iii) $\displaystyle\int_2^4 \sqrt{(16-x^2)}\,dx;$ (iv) $\displaystyle\int_0^{\frac{1}{2}\pi} x^2\cos 3x\,dx.$

30. Evaluate:

(i) $\displaystyle\int_0^1 x^3 e^{2x}\,dx;$ (ii) $\displaystyle\int \sinh^2 x\cosh^3 x\,dx;$

(iii) $\displaystyle\int \frac{1}{\sqrt{(x-4)}}\,dx;$ (iv) $\displaystyle\int_0^{\frac{1}{2}\pi}\sin x\log(\cos x)\,dx.$

[L.U.]

31. Evaluate:

(i) $\displaystyle\int_0^1 \frac{1}{\sqrt{(2x-x^2)}}\,dx;$ (ii) $\displaystyle\int_1^2 \frac{(x+2)}{x(x+1)^2}\,dx;$

(iii) $\displaystyle\int_0^1 x\tan^{-1}x\,dx;$ (iv) $\displaystyle\int_3^4 \frac{x^2}{(x-2)}\,dx.$ [L.U.]

32. Sketch that part of the curve $9y^2 = 16x^3$ between $x = 0$ and $x = 2$ for which y is positive.

Calculate:

(i) The length of this arc of the curve.

(ii) The area of the figure enclosed by this arc, the line $x = 2$ and the x-axis.   [Part I]

33. O is the origin of co-ordinates, A is the point at which the line $y = x + 1$ meets the y-axis, and B is the point at which the line meets the curve $y = 2x^2$ and for which x is positive.

Calculate:

(i) The area of the figure OAB.

(ii) The distance of its centroid from OA.

(iii) The volume generated by this figure when it is rotated through four right angles about the y-axis.   [Part I]

34. The equation of a curve is $y^2 = (x-1)/(x-2)$. Show that no part of the curve lies between $x = 1$ and $x = 2$, and sketch the curve for the remaining values of x. The portion of the curve between $x = 0$ and $x = 1$ is rotated through 360° about the axis of x. Find the volume generated and the distance of its centroid from the origin. [L.U.]

35. The magnetic field strength δH produced at a point P distant x from a current element $I\delta l$ may be expressed, in SI units, as: $\delta H = (I\delta l/4\pi x^2)\sin\theta$, where θ is the angle between the current element and the line joining P to it. Deduce the field strength at a distance a from the centre of a straight conductor of length b due to a current I flowing through it. Hence, calculate the field strength at the centre of a square loop of side 0·3 m when it carries a current of 5 A, and estimate the percentage error that would be introduced in this value if the loop were assumed instead to be a circle of diameter 0·3 m.

36. A current I is flowing along a wire of length l and circular cross-section. The radius of the wire increases uniformly with distance, from a at one end to $2a$ at the other end. If the resistivity is ρ, find the potential difference between the ends of the wire. How far along the wire from the narrower end will be the mid-potential?

37. A solid hemisphere of radius a is just immersed in water so that the surface of the liquid is a tangent plane to the hemisphere. The plane base of the hemisphere makes an angle of 60° with the horizontal. Find in magnitude, direction and line of action, the thrust on the plane base and the resultant thrust on the curved surface. [L.U.]

38. A solid of revolution is generated by rotation about the axis of x of the area bounded by the curve $y = \sec x$ and the lines $x = 0$, $x = \frac{1}{4}\pi$ and $y = 0$. Find the volume of the solid, the position of its centroid and its radius of gyration about its axis. [L.U.]

39. A shell is formed by rotating the portion of the parabola $y^2 = 4x$ for which $0 \leqslant x \leqslant 1$ through two right angles about its axis. Prove that the area of its curved surface is $\frac{8}{3}\pi(2\sqrt{2} - 1)$ and that its radius of gyration k about the axis of rotation is given by $35k^2 = 8(5 + 3\sqrt{2})$. [L.U.]

40. Find the radius of curvature at the point $x = 1$, $y = 3$ on the curve $y^2 = x^3 + 8$.

41. Find the radius of curvature and the co-ordinates of the centre of curvature at the point $(4, 2)$ on the curve whose equation is $x^2 = 8y$.   [Sec. A]

42. The transition curve on a railway track has the shape of an arc of the curve $y = \frac{1}{4}x^3$. At what rate is a truck on this track changing its direction when it is passing through (a) the point $(4, 16)$; (b) the point $(2, 2)$, and (c) the point $(1, \frac{1}{4})$? Take the unit of length as one kilometre and give the answers in degrees per kilometre.

43. A circle of radius a rolls inside a fixed circle of radius $3a$. Show that, referred to suitably chosen perpendicular axes through the centre of the fixed circle, the parametric equations to the curve C described by a point P on the circumference of the rolling circle can be expressed in the form

$$x = 2a \cos \theta + a \cos 2\theta, \quad y = 2a \sin \theta - a \sin 2\theta.$$

Find the angle ψ which the tangent to C at P makes with the axis of x and, by obtaining $ds/d\psi$ or otherwise, show that the magnitude of the radius of curvature at P is the numerical value of $8a \sin \frac{3}{2}\theta$.   [L.U.]

44. Show that if x is small enough for powers of x beyond the third to be negligible, $e^x - \log_e \sqrt{\left(\dfrac{1+x}{1-x}\right)}$ may be replaced by

$$1 + \frac{x^2}{2} - \frac{x^3}{6}.$$ [Part I]

45. Using standard series, find the first three non-vanishing terms when $e^x \sin 3x$ is expanded as a series of ascending powers of x.

46. Use Maclaurin's theorem to obtain the first three terms in the expansion of $\log_e (1 + e^{-x})$.

47. By expanding the integrand in powers of x, evaluate $\displaystyle\int_0^{\frac{1}{4}} \frac{1}{x} \log_e (1 + x)\, dx$ correct to three decimal places.   [L.U.]

48. By expanding the integrands in powers of x, evaluate correct to three decimal places:

(i) $\displaystyle\int_0^{\frac{1}{2}} \frac{\sin x}{x}\, dx$; (ii) $\displaystyle\int_0^1 \sqrt{x}\, \cosh x\, dx$.

49. Show that $\displaystyle\lim_{x \to 0} \frac{2x - 3\sin x + x\cos x}{x^3(1 - \cos 2x)} = \frac{1}{120}$.

50. In a circle, centre C, two radii CP and CQ contain a small angle x radians. PR is perpendicular to CQ, and the tangent to the circle at P meets CQ produced in T. By using the expansions of $\sin x$ and $\tan x$ in ascending powers of x, prove that, if powers of x higher than the fourth are neglected, the length of the arc PQ is equal to $\frac{1}{3}(2PR + PT)$.

51. If a chain has a span of 10 m between two points at the same level and its ends slope at $45°$, prove that its length is about $11\cdot34$ m.

52. A uniform cable of length $2l$ is suspended between two points A and B in the same horizontal line at a distance $2a$ apart. If the sag is small, show that the parameter c of the catenary is approximately given by $c^2 = a^3/6(l - a)$.

A uniform cable of length 50 m is stretched between two points at the same level distant 49 m apart. Find an approximate value for the sag of the cable.

53. On an Argand diagram mark the points P and Q to represent $1 + j$ and $3 + j2$, respectively. Complete the square $PQRS$ so that RS lies above PQ. Express the vectors OR, OS and OC, where C is the centre of the square, in the form $a + jb$.

[Part I]

54. Express $z = 2 - j2$ in the form $r \angle \theta$ and find z^4, $(1/z^2)$ and the three cube roots of z, all in the form $r \angle \theta$.

55. Express $(4 + j3)$ and $(4 + j3)^2$ in the form $r\,e^{j\theta}$. Find the three values of $(4 + j3)^{\frac{2}{3}}$ in the form $r\,e^{j\theta}$.

56. Given that $u + jv = (2 + j2)/(z + 3)$, where $z = x + jy$, express u and v as functions of x and y. [Sec. A]

57. If $(z-1)/(z-j)$ is real, show that the locus of z $(=x+jy)$ in the Argand diagram is a straight line, and find its x, y equation.

58. Given that $V = (1/r)\sin(r+ct)$, show that

$$\frac{\partial^2 V}{\partial r^2} + \frac{2}{r}\frac{\partial V}{\partial r} = \frac{1}{c^2}\frac{\partial^2 V}{\partial t^2}.$$

59. If $z = e^x \cos y$ and $u = e^x \sin y$, show that $\dfrac{\partial z}{\partial x} = \dfrac{\partial u}{\partial y}$ and $\dfrac{\partial z}{\partial y} = -\dfrac{\partial u}{\partial x}$. Also show that $\dfrac{\partial^2 z}{\partial x^2} + \dfrac{\partial^2 z}{\partial y^2} = 0 = \dfrac{\partial^2 u}{\partial x^2} + \dfrac{\partial^2 u}{\partial y^2}$.

60. (i) If $z = t^{-\frac{1}{2}} e^{-x^2/t}$, prove that $\dfrac{\partial^2 z}{\partial x^2} = 4\dfrac{\partial z}{\partial t}$.

(ii) If $V = f(x-y) + g(x+2y)$, where f and g are arbitrary functions, prove that V satisfies a partial differential equation of the form $\dfrac{\partial^2 V}{\partial y^2} = A\dfrac{\partial^2 V}{\partial x^2} + B\dfrac{\partial^2 V}{\partial x\,\partial y}$, where A and B are constants, and find the values of A and B.   [L.U.]

61. (i) If $u = e^{x+y}\cos(x-y)$, prove that:

(a) $\dfrac{\partial u}{\partial x} + \dfrac{\partial u}{\partial y} = 2u,$ (b) $\dfrac{\partial^2 u}{\partial x^2} + \dfrac{\partial^2 u}{\partial y^2} = 0.$

(ii) If $z = (xy)^{\frac{1}{2}} f\left(\dfrac{x}{y}\right)$, prove that $x\dfrac{\partial z}{\partial x} + y\dfrac{\partial z}{\partial y} = z$.   [L.U.]

62. If $z = x^2 \sin(x+y)$, find $\dfrac{dz}{dt}$ when $x = \dfrac{\pi}{3}$, $y = \dfrac{\pi}{2}$ and

$$\frac{dx}{dt} = \frac{1}{2}, \quad \frac{dy}{dt} = \frac{1}{3}.$$

63. Given that $y = \dfrac{\omega^2 C^2 R}{R^2 + \omega^2 L^2}$, find the approximate percentage change in y if ω increases by 5%, L decreases by 5% C decreases by 3% and R remains constant.

64. The voltage amplification in the case of a tuned anode coupling of two triode valves is given by $A = \dfrac{\mu}{1 + (CRr/L)}$.

If C, L and R are each increased by 1%, find the approximate percentage change in A, taking μ and r as constants. Give the answer in terms of A and μ only.

65. The acceleration due to gravity is calculated by timing the oscillations of a smooth sphere on a large horizontal concave mirror. The formula used is: $g = \dfrac{28\pi^2(R - r)}{5T^2}$, where

T = periodic time of the oscillations (in sec.); R = radius of the concave mirror (in metres); r = radius of the sphere (in metres); g is in m/s^2.

The readings taken, with the estimated accuracy in brackets, were:

$T = 1 \cdot 76$ sec. ($\pm 0 \cdot 005$); $R = 0 \cdot 562$ m ($\pm 0 \cdot 0001$); $r = 0 \cdot 0124$ m ($\pm 0 \cdot 0001$).

Find the calculated value of g in metres/sec.^{2}, and estimate the maximum possible error.

66. Solve the equation $t(dv/dt) = v(1 - t)$.

67. If $b(d\theta/dt) + h(\theta - \theta_0) = a$, express θ in terms of t, given that $\theta = \theta_0$ at $t = 0$ (a, b, h and θ_0 are constants).

68. Solve the differential equation

$$(1 + x)\,(dy/dx) - xy = x. \qquad \text{[L.U.]}$$

69. The potential V in a certain region $0 \leqslant x \leqslant a$ satisfies the equation $d^2V/dx^2 = \lambda V^{-\frac{1}{2}}$, where λ is a constant. By multiplying both sides of the equation by dV/dx and integrating, prove that if V and dV/dx both vanish when $x = 0$, and $V = E$ when $x = a$, then $16E^3 = 81\lambda^2 a^4$. [Part I]

70. Solve the equation $x(dy/dx) + 2y = x^3$, given that $y = 0$ when $x = 1$. [L.U.]

71. Solve the differential equations:

(i) $x\dfrac{dy}{dx} + 2y = \cos x$;

(ii) $\dfrac{d^2y}{dt^2} + 10\dfrac{dy}{dt} + 29y = 0$, given that $y = 0$ and $\dfrac{dy}{dt} = 4$ when $t = 0$. [L.U.]

72. Solve the equation $\dfrac{d^2y}{dt^2} + y = 2\dfrac{dy}{dt}$, given that $y = 2$ and $\dfrac{dy}{dt} = 5$ when $t = 0$. [Part I]

73. Find the general solutions of the equations:

(i) $\dfrac{d^2y}{dx^2} - 3\dfrac{dy}{dx} - 10y = 0$;   (ii) $9\dfrac{d^2y}{dx^2} + 6\dfrac{dy}{dx} + y = 0$;

(iii) $\dfrac{d^2y}{dx^2} + 6\dfrac{dy}{dx} + 25y = 0$.

74. The differential equation $dv/dr + v/2r = 0$ represents a particular form of circular vortex motion, v being the velocity at radius r. Find the relation between v and r, and given that $v = 3$ m/s, when $r = 0.3$ m, find the velocity when $r = 0.6$ m.

75. A constant electromotive force V is applied to a circuit consisting of an inductance L in series with a resistance R. At $t = 0$ the current $i = 0$. Show that, at time $t = 5L/R$, the current is within 1 % of its steady value.

76. The relation between the radial compressive stress p and the circumferential or hoop stress f at any radius x in a thick cylinder is given by the differential equation: $f + p = -x(dp/dx)$, and it is further known that $f - p = 2A$, the term A being a constant.

If $p = 0$ when $x = R$, obtain expressions giving the values of f and p in terms of x, A and R.   [Sec. A]

77. A beam of flexural rigidity EI, length l, and weight per unit length w, is clamped horizontally at one end and is freely supported at the same horizontal level at the other end. Starting from the equation $EI\dfrac{d^4y}{dx^4} = w$, prove that

$$y = \frac{wx^2}{48EI}(2x^2 - 5lx + 3l^2).$$

78. In a closed $L - C - R$ series circuit, $R = 2$ ohm, $L = 1$ milli-henry and $C = 100\ \mu$F. If q is the charge on the capacitor at any time t, show that $\dfrac{d^2q}{dt^2} + 2 \times 10^3 \dfrac{dq}{dt} + 10^7 q = 0$. If the charge is initially Q and the initial current zero, find q in terms of t.

79. A spring of stiffness 40 N/m lies on a smooth horizontal table with one end fixed to the table. A mass of 250 g is attached to the other end of the spring. When in motion, there is a resistance proportional to the velocity of the body which is 2 newtons when the speed is 1 m/s. Show that the displacement x metres, measured from the unstretched position, at time t sec. satisfies the equation:

$$\frac{d^2x}{dt^2} + 8\frac{dx}{dt} + 160x = 0.$$

If $x = 0.09$ and $dx/dt = 0$ when $t = 0$, find x in terms of t.

80. Given that

$$f(x) = x \qquad \text{for} \quad 0 < x < \pi,$$

$$f(x) = x + \pi \quad \text{for} \quad -\pi < x < 0,$$

and that $f(x)$ is periodic of period 2π, sketch the graph of $f(x)$ against x and deduce that $f(x) - \frac{1}{2}\pi$ is an odd function.

Express $f(x)$ as a Fourier series.

81. Given that

$$f(x) = 0 \quad \text{for} \quad -\pi \leqslant x \leqslant 0,$$

$$f(x) = x \quad \text{for} \quad 0 \leqslant x \leqslant \pi,$$

express $f(x)$ as a Fourier series within the given range.

82. Find a Fourier series, as far as the second harmonic, to represent the periodic function i, of period 0.02 sec., given by the values in the following table:

t	0	0.002	0.004	0.006	0.008	0.01	0.012	0.014	0.016	0.018	0.02
i	0	1	1.6	2.2	1.9	0	-2	-2.3	-1.5	-1.1	0

83. ABC is a triangle and G is its centroid. AG meets BC in D and BG meets AC in F. A, B and G are the points $(-2, -3)$, $(8, 11)$ and $(4, 5)$, respectively.

(i) Calculate the co-ordinates of the points D, C and F.

(ii) Prove that A is the acute angle whose tangent is $3/55$.

[Part I]

84. Sketch the curves whose polar equations are $r = 2\sin\theta$, $r = 1.5\cos\theta$, showing that they are circles. Find the centres and radii of these two circles and deduce that they cut at right angles.

85. Find the equation of the circle which touches the x-axis at $x = 2$ and cuts the y-axis at $y = 1$. State the co-ordinates of the centre of this circle and the length of its radius, and also prove that the circle again cuts the y-axis at $y = 4$.   [Sec. A]

86. Find the length of the line $12x + 5y = 39$ which is intercepted by the circle $x^2 + y^2 = 25$.

87. A ray of light passes through the point $(2, 3)$ and is reflected at a point A on the line $y = 1$; it then passes through the point $(6, 4)$. Find the co-ordinates of the point A.

Answers

1. $-\dfrac{(1 + 2\cos 2\theta)}{2\sin 2\theta}$; $-\dfrac{(2 + \cos 2\theta)}{2\sin^3 2\theta}$.

2. (i) $\sec^m x \tan^{n-1} x \, (m\tan^2 x + n\sec^2 x)$.

3. (a) (i) $3\sec^2 3x$;   (ii) $\dfrac{\sin 2x}{1 + \sin^2 x}$;   (iii) $\dfrac{4}{(4 - x^2)^{\frac{3}{2}}}$.

   (b) $\theta \backsimeq 77°$.

4. $a = 0$; $b = 1$.   **5.** $x \backsimeq 0.813$ rad. $\backsimeq 46° \, 35'$.

6. $\dfrac{nT}{n^2 I_M + I_S}$.   **7.** $x = \sqrt{(a^2 + ab)} - a$.

9. (i) $32\frac{2}{3}$;   (ii) $\frac{3}{8}\pi + \sqrt{2} + \frac{1}{4} \backsimeq 2.84$.

10. $x = 2\frac{1}{2}$ or 4.

11. (i) $1 - \frac{1}{4}\pi \backsimeq 0.215$;   (ii) $\dfrac{\pi}{12} + \dfrac{\sqrt{3}}{16} - \dfrac{\pi^2}{72} \backsimeq 0.233$;

   (iii) $\frac{1}{2}\tan 2x - x - \frac{1}{3}\tan 3x + C$;   (iv) $\dfrac{2\pi}{\omega}$.

12. (i) $\dfrac{e^{2x}}{2} + 2x - \dfrac{e^{-2x}}{2} + C$;

   (ii) $\sqrt{(2x + 1)} + \frac{1}{2}\log_e(2x + 1) + C$;

   (iii) $\dfrac{1}{3}\left(1 - \dfrac{1}{\sqrt{2}}\right) \backsimeq 0.098$.

13. $\sqrt{\left(\dfrac{I_1^2+I_2^2}{2}\right)}$.

15. (i) $e^x(\sinh 2x + 2\cosh 2x)$;

(ii) $3\sinh 3x\tanh 2x + 2\cosh 3x\,\mathrm{sech}^2\,2x$;

(iii) $\dfrac{e^x}{\sqrt{(e^{2x}+1)}}$; (iv) $\dfrac{-1}{2x\sqrt{(1-x)}}$.

16. (i) $\tfrac{1}{8}\sinh 4x - \tfrac{1}{2}x + \tanh x + C$;

(ii) $x - \tanh x + C$; (iii) $\tfrac{1}{3}\sinh 3 + 3/\pi \simeq 4\cdot294$.

17. $\dfrac{\pi}{3} + \dfrac{7\sqrt{3}}{8} \simeq 2\cdot563$.

18. (i) $\dfrac{-2}{1+x^2}$; (ii) $\dfrac{-1}{\sqrt{x(x+4)}}$;

(iii) $\dfrac{\sinh^{-1}x^2}{\sqrt{(x^2-1)}} + \dfrac{2x\cosh^{-1}x}{\sqrt{(x^4+1)}}$.

22. (i) $\dfrac{\pi}{6}$; (ii) $\dfrac{\pi}{28}$; (iii) $0\cdot8812$; (iv) $0\cdot6222$.

23. (i) $\dfrac{1}{\sqrt{2}}\sin^{-1}\dfrac{x}{\sqrt{2}} + C$; (ii) $\tfrac{1}{10}\tan^{-1}2 \simeq 0\cdot1107$;

(iii) $\dfrac{5}{\sqrt{3}}\sinh^{-1}\dfrac{x}{2} + C$; (iv) $\dfrac{7}{\sqrt{3}}\cosh^{-1}\sqrt{\dfrac{3}{7}}x + C$.

24. (i) $0\cdot28$; (ii) $-0\cdot25$; (iii) $1\cdot22$; (iv) $1\cdot14$;
all correct to 2 decimal places.

25. (i) $\dfrac{1}{4\pi} \simeq 0\cdot08$; (ii) $\log_e 2 \simeq 0\cdot69$.

26. (i) $\tfrac{1}{3}\log_e 2 \simeq 0\cdot231$; (ii) $\tfrac{4}{3}x - \tfrac{11}{9}\log_e(3x+2) + C$;

(iii) $\tfrac{1}{5}\log_e\tfrac{12}{7} \simeq 0\cdot108$; (iv) $\tfrac{1}{5}\log_e\tfrac{56}{3} \simeq 0\cdot585$.

27. (i) $\tfrac{1}{6}\log_e C\left(\dfrac{x+1}{5-x}\right)$; (ii) $\dfrac{\pi}{6} \simeq 0\cdot524$;

(iii) $\dfrac{\pi}{20} \simeq 0\cdot157$; (iv) $\sinh^{-1}\left(\dfrac{x+1}{3}\right) + C$.

28. (i) $\sqrt{(x^2+1)} + \sinh^{-1}x + C$;

(ii) $2\sqrt{(x^2-4)} - \cosh^{-1}\dfrac{x}{2} + C$;

(iii) $-2\sqrt{(6x-x^2)}+9\sin^{-1}\left(\dfrac{x-3}{3}\right)+C$;

(iv) $\frac{1}{4}\sqrt{(4x^2+4x+5)}+\frac{5}{4}\sinh^{-1}\left(\dfrac{2x+1}{2}\right)+C$.

29. (i) $\frac{1}{12}(3\sin 10x+5\sin 6x)+C$;

    (ii) $\frac{13}{6144}$;   (iii) $8\left(\dfrac{\pi}{3}-\dfrac{\sqrt{3}}{4}\right)\simeq 4\cdot913$;

    (iv) $-0\cdot748$.

30. (i) $\frac{1}{8}(e^2+3)\simeq 1\cdot299$;   (ii) $\frac{1}{3}\sinh^3 x+\frac{1}{5}\sinh^5 x+C$;

    (iii) $2\sqrt{(x-4)}+C$;   (iv) $\dfrac{1}{\sqrt{2}}-1+\dfrac{1}{2\sqrt{2}}\log_e 2\simeq -0\cdot048$.

31. (i) $\frac{1}{2}\pi$; (ii) $2\log_e\frac{4}{3}-\frac{1}{6}\simeq 0\cdot408$; (iii) $\frac{1}{4}\pi-\frac{1}{2}\simeq 0\cdot285$;

    (iv) $5\cdot5+4\log_e 2\simeq 8\cdot273$.

32. (i) $4\frac{1}{3}$;   (ii) $\dfrac{32\sqrt{2}}{15}\simeq 3\cdot016$.

33. (i) $\frac{5}{6}$;   (ii) $\frac{2}{5}$;   (iii) $\frac{2}{3}\pi$.

34. $0\cdot307\pi\simeq 0\cdot964$; $0\cdot370$.

35. $\dfrac{bI}{2\pi a\sqrt{(4a^2+b^2)}}$; 15·01 ampere turns/metre; 11·06 %.

36. $\dfrac{i\rho l}{2\pi a^2}$; $\dfrac{l}{3}$.

37. Referred to axes in vertical plane of symmetry, with origin at centre of base and x-axis along diameter:

Thrust on plane base is $\pi a^3 w$ at point $\left(\dfrac{\sqrt{3}}{8}a,\,0\right)$, perpendicular to base.

Thrust on curved surface is $\dfrac{\pi a^3 w\sqrt{7}}{3}$ at point $\left(\dfrac{\sqrt{3}}{8}a,\,\dfrac{a}{4}\right)$, at angle $\tan^{-1}\dfrac{1}{3\sqrt{3}}$ below the horizontal. [w=weight of unit volume of water.]

38. π; $\frac{1}{4}\pi-\frac{1}{2}\log_e 2\simeq 0\cdot439$; $\sqrt{\frac{2}{3}}\simeq 0\cdot817$.

40. $\dfrac{15\sqrt{5}}{22} \simeq 1\cdot525.$ **41.** $8\sqrt{2} \simeq 11\cdot314;\quad (-4, 10).$

42. (a) $0\cdot197°;$ (b) $5\cdot44°;$ (c) $44\cdot0°.$

43. $\psi = \pi - \tfrac{1}{2}\theta.$ **45.** $3x + 3x^2 - 3x^3.$

46. $\log_e 2 - \tfrac{1}{2}x + \tfrac{1}{4}x^2.$ **47.** $0\cdot448.$

48. (i) $0\cdot493;$ (ii) $0\cdot817.$

52. $4\cdot3$ m.   **53.** $2+j4; j3; 1\cdot5+j2\cdot5.$

54. $2\cdot828 \angle -45°; 64 \angle 180°; \tfrac{1}{8} \angle 90°; 1\cdot414 \angle -15° + k120°,$
   $k = 0, 1, 2.$

55. $5e^{j0\cdot6435};\quad 25e^{j1\cdot287};\quad 2\cdot924e^{j(0\cdot429+\frac{2}{3}k\pi)},\ k = 0, 1, 2.$

56. $u = \dfrac{2(x+y+3)}{(x+3)^2+y^2};\qquad v = \dfrac{2(x-y+3)}{(x+3)^2+y^2}.$

57. $x+y-1=0.$ **60.** (ii) $A=2; B=1.$

62. $\dfrac{\pi}{6} - \dfrac{5\pi^2}{108}\sqrt{3} \simeq -0\cdot268.$ **63.** $+4\ \%.$

64. $\left(\dfrac{A}{\mu} - 1\right).$ **65.** $9\cdot81 \pm 0\cdot11.$

66. $Ate^{-t}.$ **67.** $\theta = \theta_0 + \dfrac{a}{h}(1 - e^{-ht/b}).$

68. $y = \dfrac{Ae^x}{(x+1)} - 1.$

70. $y = \dfrac{1}{5}\left(x^3 - \dfrac{1}{x^2}\right).$

71. (i) $y = \dfrac{1}{x^2}(x\sin x + \cos x + C);$

   (ii) $y = 2e^{-5t}\sin 2t.$

72. $y = (2+3t)\,e^t.$

73. (i) $y = Ae^{-2x} + Be^{5x};$ (ii) $y = (Ax+B)\,e^{-\frac{1}{3}x};$
   (iii) $y = e^{-3x}(A\cos 4x + B\sin 4x).$

74. $v^2r = C;\quad 3\sqrt{5}/10 \simeq 0\cdot67$ m/s.

76. $p = A\left(\dfrac{R^2}{x^2} - 1\right); \quad f = A\left(\dfrac{R^2}{x^2} + 1\right).$

78. $q = Q\,e^{-kt}\,(\cos nt + \tfrac{1}{3}\sin nt)$, where $k = 1000$, $n = 3000$.

79. $x = \tfrac{3}{100}\,e^{-4t}\,(3\cos 12t + \sin 12t).$

80. $f(x) = \tfrac{1}{2}\pi - (\sin 2x + \tfrac{1}{2}\sin 4x + \tfrac{1}{3}\sin 6x + \ldots).$

81. $f(x) = \dfrac{\pi}{4} - \dfrac{2}{\pi}\left(\cos x + \dfrac{1}{3^2}\cos 3x + \dfrac{1}{5^2}\cos 5x + \ldots\right)$

$$+ (\sin x - \tfrac{1}{2}\sin 2x + \tfrac{1}{3}\sin 3x - \ldots).$$

82. $i \simeq -0{\cdot}02 + 0{\cdot}012\cos 100\pi t + 2{\cdot}15\sin 100\pi t$
$$-\,0{\cdot}012\cos 100\pi t - 0{\cdot}51\sin 100\pi t.$$

83. $(7, 9); \qquad (6, 7); \qquad (2, 2).$

84. Centre $(1, \tfrac{1}{2}\pi)$, radius 1; centre $(\tfrac{3}{4}, 0)$, radius $\tfrac{3}{4}$.

85. $(x-2)^2 + (y-2{\cdot}5)^2 = 6{\cdot}25; \quad$ centre $(2, 2{\cdot}5); \quad$ radius $2{\cdot}5$.

86. 8 units. **87.** $(\tfrac{18}{5}, 1).$

INDEX

(Numbers refer to the pages)

Angle between two lines, 272
Approximate changes in a function, 12, 25, 189
Approximate integration, 151
Approximation to roots of an equation, 25
Arbitrary constants, 10, 59, 196
Arc, length of, 90
Archimedes' spiral, 270
Areas by integration, 32, 33, 83, 85
Argand diagram, 167, 170
Asymptotes, 262, 266
Auxiliary equation, 209

Beams: bending moments in, 227; deflexion of, 229 et seq.; problems, 231 et seq.
Boundary conditions, 230
Bridge, suspension, 161

Capacitor, discharge of, 242
Cardioid, 270
Cartesian co-ordinates, 268
Catenary, 157 et seq.
Centres of gravity, 95 et seq.
Centres of pressure, 111 et seq.
Centroid, 95
Change of origin, 263
Circle, equations of, 279 et seq.
Complete primitive, 196
Complex numbers, 163 et seq.
Compound pendulum, 108
Concavity and convexity, 20, 263
Conjugate complex numbers, 165
Constants, arbitrary, 10, 59, 196
Continuous functions, 15
Contraflexure, points of, 20
Convergence of series, 135 et seq.
Convexity and concavity, 20, 263
Corrections, small, 12, 190
Critical damping, 222, 240, 241
Curvature, 122 et seq.; radius of, 122 et seq.; circle of, 123
Curved surface, resultant thrust on, 114
Cycloid, 123
Curves: length of, 90; tracing of, 262 et seq.

Damped oscillations, 221, 239
Damping, critical, 222, 240, 241
Decrement, logarithmic, 223
Definite integral, 84
Deflexion of beams, 229 et seq.
Derivative, total, 192
Differentiability, 17
Differential coefficient, 13
Differential equations: exact, 200; first order, 197; formation of, 10, 56, 195; linear, of first order, 203; linear, of second order, 207 et seq.; second order, 205; variables separable, 198
Differentials, 12, 192; total, 192
Differentiation: of function of function, 1; of hyperbolic functions, 40; of inverse functions, 45, 46, 47, 50, 51, 53; logarithmic, 8; of logarithmic functions, 8; partial, 183 et seq.; of products, 1; of quotients, 1; standard forms, 1, 2
Discontinuities, 15, 248
Disk, moment of inertia of, 101
Distance between two points, 273
Divergence of series, 135 et seq.
Dynamical formulae, 100

Electrical circuits, simple, 238, 239
Energy, kinetic, 100
Envelopes, 129
Equations: approximation to roots of, 25; auxiliary, 209; of circles, 279 et seq.; differential, 195 et seq.; intrinsic, 123, 158; polar, 267 et seq.; of straight lines, 271 et seq.
Equiangular spiral, 270
Even functions, 148, 245
Evolutes, 128 et seq.
Exact differential equations, 200
Expansion in power series, 145 et seq.
Exponential form of a complex number, 175 et seq.
Exponential series, 147, 176

Flexural rigidity, 229
Fluid thrusts, 109 et seq.
Force, shearing, 226
Forced oscillations, 241

Fourier series, 244 *et seq.*

Fractions, partial, 63

Functions: approximate changes in, 189; continuous, 15; differentiable, 15; discontinuous, 15, 248; expansion of, 145 *et seq.*; explicit, 5; hyperbolic, 37 *et seq.*; implicit, 5; inverse, 44 *et seq.*; limiting values of, 152; odd and even, 148, 245; periodic, 244; rates of change of, 28, 192

Gravity, centres of, 95 *et seq.*

Gyration, radius of, 100

Harmonic analysis, 244 *et seq.*

Hyperbolic functions, 37 *et seq.*

Increments, small, 12, 25, 189; total, 189

Inertia, moments of, 100 *et seq.*

Inflexion, points of, 20

Integrals: definite, 84; particular, 196; standard, 32, 48, 58

Integrating factor, 200, 203

Integration: approximate, 151; areas, 32, 33, 83, 85; of irrational functions, 71; by parts, 78; of rational functions, 62 *et seq.*; standard forms, 32, 48, 58; by substitution, 60, 77; of trigonometrical functions, 73; use of partial fractions, 63

Intrinsic equations, 123, 158

Inverse functions, 44 *et seq.*; differentiation of, *see* Differentiation; logarithmic form of, 52

Involutes, 130 *et seq.*

Kinetic energy, 100

L-C-R circuit, 239

Lemniscate, 269

Length of arc, 90 *et seq.*

Limiting values of functions, 152

Linear differential equations, *see* Differential equations

Logarithmic decrement, 221

Logarithmic differentiation, 8

Logarithmic series, 148

Maclaurin's expansions, 146

Maxima and minima, 18

Mean value, 32

Mean value theorem, 24

Moments, first, 95; second, 100; of inertia, *see* Inertia

Motion: simple harmonic, 206, 220, 221; resisted, 216 *et seq.*

Newton's method: for root of an equation, 25; for radius of curvature, 127

Normals, 129, 280

Numbers: complex, 163 *et seq.*; rational, 163; surds, 163; transcendental, 163

Odd functions, 148, 245

Origin, change of, 263

Oscillations: damped, 221, 239; forced, 241; simple harmonic, 206, 220

Pappus's theorems, 84

Parabola, 263, fig. 108, 131

Parallel axes theorem, 103

Parameters, 7

Partial differentiation, 183 *et seq.*

Partial fractions, 63

Particular integral, 196

Pendulum, compound, 108

Periodic functions, 244

Perpendicular axes theorem, 102

Perpendicular from point to line, 273

Points of inflexion, 20

Polar equations, 267 *et seq.*

Polar form of complex number, 168

Polynomials, 62

Power series, 145 *et seq.*

Pressure, centre of, 111 *et seq.*

Primitive, complete, 196

Principal values, 45, 46, 47, 51

Radius of curvature, 125

Radius of gyration, 100

Rates of change of function, 28, 192

Rational numbers, 163

Rectangle, second moment of, 101

Resisted motion, 216 *et seq.*

Revolution, volumes of, 32, 83; surface area of volumes of, 93

Rod, second moment of, 101

Root mean square, 32

Roots, nth, of complex numbers, 177 *et seq.*

Routh's Rule, 101

Series: binomial, 144; convergence and divergence of, 135 *et seq.*; ex-

Series: *cont.*
ponential, 147, 176; Fourier, 244;
logarithmic, 148; power, 145 *et
seq.*; sine and cosine, 147; sinh and
cosh, 147; tangent, 150
Simple harmonic motion, 206, 220;
damped, 221
Small changes, 12, 25, 189
Solids, surface area of, 93
Spirals: Archimedes, 270; equiangular,
270
Standard integrals, 32, 48, 58
Straight line, equations of, 271 *et seq.*
Substitutions, *see* Integration
Surds, 163
Surface area of solids, 93
Suspension bridge, 161

Tabulatory method for Fourier series,
257
Tests: for convergency, 136 *et seq.*;
for maxima and minima, 19
Thrust on curved surface, resultant,
114
Torque, 100
Total derivative, 192
Total differential, 192
Total increment, 189
Transcendental numbers, 163
Transition curve, 297
Triangle, area of, 275

Values: mean, 32; principal, 45, 46,
47, 51; root mean square, 32
Variables, functions of several, 183